T0133069

Show Me the Bone

Show Me the Bone

Reconstructing Prehistoric Monsters in Nineteenth-Century Britain and America

GOWAN DAWSON

University of Chicago Press
Chicago and London

Gowan Dawson is professor of Victorian literature and culture at the University of Leicester. He is coeditor of *Victorian Scientific Naturalism*, also published by the University of Chicago Press, and is the author of *Darwin, Literature and Victorian Respectability*.

The University of Chicago Press, Chicago 60637
The University of Chicago Press, Ltd., London
© 2016 by The University of Chicago
All rights reserved. Published 2016.
Printed in the United States of America

25 24 23 22 21 20 19 18 17 16   1 2 3 4 5

ISBN-13: 978-0-226-33273-4 (cloth)
ISBN-13: 978-0-226-33287-1 (e-book)
DOI: 10.7208/chicago/9780226332871.001.0001

Library of Congress Cataloging-in-Publication Data

Dawson, Gowan, author.
  Show me the bone : reconstructing prehistoric monsters in nineteenth-century Britain and America / Gowan Dawson.
    pages cm
  Includes bibliographical references and index.
  ISBN 978-0-226-33273-4 (cloth : alkaline paper)—ISBN 978-0-226-33287-1 (e-book)
1. Paleontology–Great Britain–History–19th century.   2. Paleontology–United States–History–19th century.   3. Paleontologists–Biography.   I. Title.
  QE705.A1D39 2016
  560 — dc23

                                        2015025656

In loving memory of my grandparents, Jan and Mary Bełej

Contents

Cuvier's Law of Correlation

In the crowded lectures he gave at New York's College of Physicians and Surgeons each winter during the 1810s, the politician and polymath Samuel Latham Mitchill regaled his audiences with an audacious test of his knowledge of natural history. Already famous for his flamboyance and erudite eccentricities, Mitchill would declare:

> "*Ex pede Herculem*," said the ancient artist, "I can tell Hercules from his toe"—"*Ex dente animal*," says Cuvier, "give me the bone, and I will describe the animal"—"I go further . . . show me a single scale, and I will let you know the fish that owned it."[1]

Having honed his ichthyological expertise in the pungent fish markets of Lower Manhattan, Mitchill avowed that his powers of identification surpassed the classical claim—generally attributed to the Greek mathematician Pythagoras rather than an unnamed ancient artist—that the pagan divinity Hercules could be measured, proportionally, solely by his foot.[2] More significantly, Mitchill even proposed that he could go further than the most revered naturalist in Europe, who since the final years of the eighteenth century had used his unrivaled mastery of comparative anatomy to make sense of an array of strange and often only fragmentary fossil remains. The enigmatic prehistoric creatures Georges Cuvier was able to reconstruct helped establish his preeminence in the academic institutions of Paris, which, despite the tumultuous disruptions of the French Revolution and subsequent decades of war, remained the world's leading center of scientific research.[3] Mitchill's brash assertion of his own superiority was quickly ridiculed by his political enemies in New York, who complained of his "trifling vanity—his eternal egotism," and questioned in verse "How Mitchill knows a fish, from head to tail, / On bare inspection of one single scale."[4] Even his friends acknowledged that Mitchill's "manner . . . as an instructor was calculated to attract the attention of the students" and involved "more freedom of expression than rigid induction might justify."[5] The notoriously hyperbolic declaration that was sometimes misremembered as "Show me a fin, and I will point out the fish" reflected the febrile atmosphere of early nineteenth-century America, encapsulating what D. Graham Burnett has called the "boosterism for natural-historical investigation in the young Republic."[6] This distinctively American boosterism, however, would have important implications for the study of natural history, and especially the fledgling science of paleontology, on both sides of the Atlantic.

It is uncertain whether Cuvier actually made the declaration "Give me the bone, and I will describe the animal" that Mitchill attributed to him. It does not appear in any of his published writings, nor in his extant correspondence, and if he uttered it at all, it could only have been as a conversational remark or an oral aside during a lecture.[7] While training in medicine at Edinburgh during the 1780s, Mitchill had often "resided in Paris," although this, as he conceded, was at a time when the "present constellation of science, had not then risen," and long before Cuvier began giving lectures on comparative anatomy and his researches on fossil bones.[8] A later Edinburgh student who did attend Cuvier's lectures in the early 1820s, the anatomist Robert Knox, once asserted, "'Give me the teeth', said Cuvier, 'and I will describe the stomach, the feet, the skeleton,'" but Knox, an acerbic opponent of Cuvierian methods, is hardly a dependable witness (as will be seen in chapter 8) and, in any case, did not make this claim until more than thirty years after he left Paris.[9] It seems far more likely that Mitchill, in his lectures in New York, simply coined a brasher, more pithy paraphrase of Cuvier's avowal that from only "isolated bones . . . he who possesses rationally the laws of organic economy would be able to reconstruct the whole animal."[10] This claim was articulated in the "Discours préliminaire" from *Recherches sur les ossemens fossiles* (1812), an English translation of which Mitchill edited for American readers in 1818. Despite Mitchill's evident embellishment, it was not long before the apocryphal story of Cuvier's daring boast spread across the Atlantic.

It was even accepted in France, where, in the early 1850s, the politician and erstwhile prime minister Adolphe Thiers translated Mitchill's Americanization of Cuvier back into French ("Donnez-moi un os . . ."), perceiving such hauteur as a source of Gallic national pride.[11] After all, an earlier icon of French science, René Descartes, had been still more hubristic in proclaiming "Give me matter and motion and I will construct the universe," and the tradition of making audacious scientific claims beginning "Give me . . ." went all the way back to the Greek mathematician Archimedes.[12] As part of this tradition, and notwithstanding its origins amid the self-aggrandizing boosterism of the New World, Cuvier's putative proposition "Give me the bone . . . ," along with Mitchill's own "Show me a single scale . . . ," came to symbolize the remarkable new insights into anatomy that, during the nineteenth century, enabled naturalists to infer the size, appearance, and even life habits of animals from just a single part of their anatomy. Such astonishing inferences testified to the seemingly unlimited powers of scientific reasoning, especially in predicting things that were beyond direct experience, as well as affording a performative dimension to natural history similar to that which enabled exponents of the physical sciences to astound audiences with their carefully choreographed mastery over nature.[13]

In particular, practitioners of the nascent science of paleontology, a term coined in 1822 to distinguish the study of fossil organisms from geology's traditional emphasis on rock strata, were heralded as scientific virtuosos, sometimes even as veritable wizards, who could resurrect the hitherto unknown denizens of the ancient past from merely a glance at a fragmentary bone (fig. I.1).[14] Such extraordinary displays of predictive reasoning were accomplished, it was claimed, through the law of correlation, which Cuvier formulated following his arrival in Paris in 1795 only months after the end of the revolutionary Reign of Terror. This law was an essential component of the new approach to comparative anatomy that Cuvier first adumbrated in *Leçons d'anatomie comparée* (1800–1805), but it subsequently became still more integral to the paleontological practices presented in his *Recherches sur les ossemens fossiles*. It proposed that each element of an animal corresponds mutually with all the others, so that a carnivorous tooth must be accompanied by a particular kind of jawbone, neck, stomach, and so on, that facilitates the consumption of flesh, and equally cannot be matched with bones and organs adapted to a herbivorous diet. As such, a single part, even the merest fragment of fossilized bone, necessarily indicates the configuration of the whole. It was on this basis that Cuvier could allegedly declaim: "Give me the bone, and I will describe the animal."

Cuvierian Mania

The law of correlation, which will be discussed in more detail in chapter 1, was one of the fundamental axioms of nineteenth-century science, and arguably plays a role in the development of paleontology equivalent to that of evolution in biology or the nebular hypothesis in astronomy. Like them, it was both a theoretical breakthrough that transformed specialist research in

FIGURE 1.1. Théobald Chartran, *Cuvier réunit les documents devant servir à son ouvrage sur les ossements fossiles (1823)*. Oil on canvas, circa 1888. Sorbonne, Paris. This late nineteenth-century painting, part of a sequence of nine portraits of French scientific heroes, shows Cuvier focusing intently on a single bone, presumably mentally reconstructing the creature from which it came, while preparing the second edition of his monumental work on fossil quadrupeds *Recherches sur les ossemens fossiles* (1821–24). © 2014 White Images / Scala, Florence.

the area and an iconic doctrine that epitomized the discipline for the wider public.[15] At a time when almost all fossil quadrupeds were found only as partial fragments or disarticulated bones and teeth, the law of correlation allowed inferences about the remainder of the skeleton to be made with a high degree of certainty.[16] Reconstructions of an animal's skeletal configuration could now be made by piecing together osseous fragments that, when in situ, were often mixed up with other remains, without the fear of producing merely a monstrous amalgam of incongruous parts; "out of collected fossil remains creating to ourselves a monster," as one critic put it.[17] The overdetermined term *monster* nonetheless continued to be widely employed as a convenient synonym for almost all prehistoric creatures, which is how it is used in the subtitle of this book.[18] Cuvier, as Martin Rudwick has proposed, brought to the late eighteenth-century "debate on fossil bones a potentially decisive technique, applying comparative anatomy with unprecedented precision to settle issues that had long been contentious and unresolved."[19] Most notably, Cuvier deployed his law of correlation to glean details of the structure of a range of new fossil mammals, revealing, for the first time, the vast differences between present and prehistoric faunas, and thereby providing irrefutable evidence for the previously disputed concept of extinction.[20] For Cuvier himself, all of his revolutionary scientific findings—the vast extent of the earth's geohistory and the fixity of species, as much as extinction—were dependent on his ability to identify previously unknown creatures from just isolated fragments.[21]

The depredations of the Reign of Terror had left a power vacuum in postrevolutionary intellectual life, and Cuvier, having rapidly established his reputation in Paris, compared his predictive powers to those of the mathematicians who dominated the French capital's scientific institutions. He insisted that the organic law of correlation was a rational axiom no less unerring than the fundamental principles of physical sciences such as astronomy and chemistry. In order to sustain such claims, natural history had to exhibit the same precision and methodological rigor as other mathematically based sciences, and Cuvier was disdainful of the imprecise, descriptive accounts of nature with which his predecessors at the Muséum d'histoire naturelle had engaged mass audiences, considering it demeaning to cultivate such a public following.[22] Even while restricting himself to addressing colleagues and peers, however, Cuvier remained alert to the theatrical potential of his paleontological predictions, in which the conventionally distinct acts of discovery and demonstration were inextricably intertwined. When, having begun lecturing to influential members of Parisian high society, Cuvier adopted a new, more demonstrative style of writing in the "Discours préliminaire," his ebullient claims to be able to reconstruct an entire animal from only a single bone—even if never quite as dramatic as Mitchill's New York lectures implied—captured the attention of audiences far beyond scientific specialists, transforming the law of correlation from a recondite concern of anatomical experts to one of the most celebrated, or at least widely invoked, scientific axioms of the nineteenth century.

Although it was formulated at a time when, following the Revolution and then the rise to power of Napoleon, France and Britain were almost continuously at war, it was across the Channel that Cuvier's law of correlation took particular hold. Indeed, Knox exasperatedly diagnosed a veritable "Cuvierian mania" among his compatriots in the late 1830s.[23] As a Protestant born outside of France's prerevolutionary borders, it was possible to domesticate Cuvier as a kind of British—or at least non-French—hero, and, by the time of Knox's barbed observation, his law of correlation had more supporters in Britain than on the Continent. Part of the reason for this conspicuous enthusiasm for Cuvier's understanding of animal structure was that the harmonious organic correspondences on which it was predicated seemed to demonstrate that only providential design could have produced such perfectly integrated mechanisms. Regardless of Cuvier's own ambivalent silence with regard to religion, the law of correlation became increasingly central to the Anglican tradition of natural theology maintained, albeit with different inflections, at the ancient universities of Oxford and Cambridge.[24] One of the principal upholders of this tradition, the Oxford geologist William Buckland, was himself hailed as "the English Cuvier," a cherished sobriquet that was subsequently, and more enduringly, assumed by his London-based protégé Richard Owen.[25]

Translations were integral in instigating this anglophone "mania" for all things Cuvierian, and the equipping of the so-called "anglicized Cuvier" with the theological accoutrements to render the French savant's work appropriate for the pious sensibilities of British audiences is well known to historians.[26] What has never previously been recognized is that such orthodox appropriations of the law of correlation were far from uncontested. As several philosophers of science have noted (although without historians taking heed), Cuvier's formulation of the laws of organic form, and especially in his partial and inconsistent use of Kantian principles, "resides precisely on the common ground shared by the teleologist and the materialist."[27] His organic laws were more amenable to divergent interpretations than any other contemporary scientific principles, and could therefore be repackaged, with no less validity, to endorse radical and heretical purposes, as much as conservative and religious ones. Indeed, with the first translations of Cuvier's anatomical lectures marketed to medical students and Whig radicals, he was initially perceived in Britain as espousing a dangerous, atheistic materialism. Cuvier's scientific authority was eventually appropriated for the conservative establishment by gentlemen of science such as Buckland and Owen, but this was not as inevitable as historians have hitherto assumed, and instead was only the final outcome of a bitter turf war over the proprietorship of the law of correlation between militant materialists and Anglican natural theologians. While this materialist incarnation of the great "conservative creationist . . . villain" of conventional historiographies may seem surprising, the complexity and volatility of both British science and society in the early nineteenth century make it perilous to attach particular scientific approaches, especially those imported from France, to fixed ideological positions.[28] After all, biblical literalists, as is shown in chapter 2, condemned the law of correlation as innately irreligious because it was incompatible with scriptural accounts of animal forms and habits, and actually preferred the evolutionary—and implicitly atheistic—views of Cuvier's Parisian rival Jean-Baptiste Lamarck, which, notably, they considered more congruent with Scripture.

The eventual consolidation of the position of conservative gentlemen of science as the exclusive custodians of Cuvier's legacy is attributable, in large part, to Owen's combination of anatomical prowess and sheer strategic guile. Even in his own lifetime, Owen was notorious for being unscrupulous or even disingenuous, especially in his bitter opposition to Darwinism, and for a long time he was either written out of triumphalist narratives of scientific progress predicated on evolution or else cast as the malevolent enemy of everything that is enlightened and secular.[29] More recently, historians have endeavored to rehabilitate aspects of Owen's scientific reputation, including his own saltational model of evolutionary change, although the perception remains that his willing acceptance of the patronage of the conservative Anglican establishment rendered him little more than an obscurantist preserver of antiquated nepotism and privilege.[30] Far from fearing modernity, however, Owen, when he first began endorsing the law of correlation in the 1830s, was a self-possessed tyro eager for publicity, and in the metropolis in the early years of Queen Victoria's reign he was at the epicenter of a publishing industry undergoing profound transformations following the advent of the steam-powered printing press, reductions in taxes on printed materials, and the consolidation of enormous new reading audiences.[31] It has previously been assumed that this industrialized print culture helped usher in the new developmental theories presented in *Vestiges of the Natural History of Creation* (1844), and later Charles Darwin's *On the Origin of Species* (1859), by "forging 'a reading public' for liberal scientific views of progress."[32] But the skillful manipulation of innovative publishing formats by Owen and his acolytes shows that large reading audiences could still also be recruited for overtly nonprogressive concepts such as Cuvier's law of correlation. In fact, Owen's feats of paleontological reconstruction, including the spectacular discovery of a giant struthian bird from New Zealand from just a broken piece of femur, became some of the most sensational and widely reported events of the entire nineteenth century. Owen's mastery of the law of correlation was also celebrated in the new forms of visual spectacle that attracted huge crowds in the mid-nineteenth century, both at the Great Exhibition of 1851 and at the reconstructed Crystal Palace at Sydenham.[33] From stage-managed news reports and fashionable serial novels to the life-size brick-and-mortar models of enormous prehistoric monsters that lurked in the gardens at Sydenham, Owen's deployment of correlation was a highly conspicuous part of the emergent consumer culture of Victorian Britain.[34]

As with Mitchill's flamboyant boosterism in New York, the law of correlation was no less acclaimed on the other side of the Atlantic, where, in the absence of international copyright restrictions, British books and periodicals were rapidly reprinted for American audiences.[35] An indigenous tradition of Cuvierian paleontology also developed with Edward Hitchcock's interpretation, beginning in the 1830s, of fossil footprints found in New England as the only surviving remnants of unprecedentedly large birds similar to the gigantic struthian Owen had inferred back in London. This transatlantic Cuvierianism was consolidated by the arrival of Cuvier's Swiss protégé Louis Agassiz, who emigrated to America in the 1840s and extolled, with only a little less hubris than Mitchill, his own capacity to infer the features of fossil fish from isolated scales. Looking back across the Atlantic, Agassiz was profoundly unimpressed by Cuvier's successors in Paris, and by the 1850s he was confident that American natural history—albeit principally represented by his own researches—was now superior to that practiced in France.[36] Agassiz became as revered in his adoptive country as Cuvier in France or Owen in Britain, and by the outbreak of the American Civil War in 1861, his endorsement of the law of correlation was even invoked on behalf of the Confederacy and the cause of slavery.

Undead Science

It is this distinctive anglophone engagement with Cuvier's method of paleontological reconstruction that is the principal focus of *Show Me the Bone*, the title of which amalgamates Mitchill's actual avowal with the putative one he attributed to Cuvier. The book begins, though, by examining the Franco-German contexts in which the law of correlation originated, and it also considers subsequent events in France and elsewhere in Europe. The purpose of *Show Me the Bone* is not to debunk nor to affirm Cuvier's much-contested ability to reconstruct extinct creatures from just single bones. Rather, as with James Secord's approach to the controversial transmutationism of *Vestiges* in *Victorian Sensation* (2000), it investigates how the law of correlation was successively repackaged by different anglophone audiences, including those in the farthest outposts of the British Empire.[37] Such appropriations were diverse and often contradictory, and the same Cuvierian law that was espoused by Anglican Oxford dons upholding the natural theological argument from design could, as noted above, be embraced with equal enthusiasm by plebeian radicals touting unauthorized reprints of blasphemously materialistic lectures. Similarly, Cuvier's understanding of the harmonious correlation of the animal frame, in which no part could change without altering all the others, was frequently employed as a bulwark against the evolutionary speculations of both Lamarck and *Vestiges*, but it also piqued the interest of the young Darwin and was later incorporated into his own law of correlated variability. At the same time, Darwin's presence does not overshadow the story of correlation (if anything, he remains merely a supporting actor in the larger drama), and Owen too could accommodate the Cuvierian law with his own orthogenetic evolutionism.[38] Many other lesser-known historical actors also made hugely significant interventions, and the celebrated claim about reconstructing from a single bone, which was debated by evangelicals in squalid Edinburgh courtyards, puffed anonymously in the London press, discussed among strangers in railway carriages, conflated with occult intuitions by spiritualists in America, and tested with painful hands-on experiments by colonial settlers in Australia, affords a highly effective "cultural tracer" that can be followed through a wide range of geographical settings and divergent social circumstances.[39]

Indeed, Cuvier's distinctive understanding of anatomical structure was readily adapted to particular local circumstances, and its reception in the English-speaking world was frequently determined by geographical, political, and religious factors.[40] The emphasis on revealed theology in Presbyterian Edinburgh, for instance, meant that the demonstration of the perfect and harmonious correlation of the animal frame was less meaningful than south of the border in Oxford, where it was seen to afford irrefutable proof of divine design. Significantly, rather than simply considering these varying local contexts in isolation, *Show Me the Bone* also examines their interaction, with developments in the Scottish capital, as well as in the radical purlieus of London, often shaping responses to Cuvier within Oxford's ancient seminaries. Larger international crosscurrents, whether scientific disputes or actual wars, similarly had an important impact on how the law of correlation was received in different locations and by particular groups. For example, Owen's adherence to the Cuvierian law even after the deaths of his erstwhile Anglican patrons—something that has long perplexed historians—was determined, in part, by its enhancement of the authority of museum-based experts such as himself and marginalization of the local knowledge garnered by colonial naturalists, who, unsurprisingly, grew suspicious of such metropolitan methodologies.[41] Secord has contended that "every local situation has within it connections with and possibilities for interaction with other settings," and he proposes replacing the customary emphasis on "science in context" with a new attention to what he terms "knowledge in transit." In order to fully understand the range of interconnections possible within a particular context, Secord also urges the benefits of "creating a history that keeps the virtues of the local but operates at a unit of analysis larger than a single country."[42] Ranging from the smuggling of French science books during the cross-Channel blockades of the opening years of the nineteenth century, to the pervasive reprinting of British works in America, and the growing interpretative autonomy of naturalists in both Australia and New Zealand, *Show Me the Bone* follows the transit of scientific knowledge far beyond conventional national boundaries, examining how such knowledge is not merely passively communicated but is actually made in the process of global circulation.

In four parts, *Show Me the Bone* examines the (I) arrival, (II) triumph, (III) overthrow, and then (IV) afterlife of Cuvier's famous reconstructive method. The book charts the law of correlation's changing fortunes across the nineteenth century, from its troubled importation into Britain during the Napoleonic Wars, through the ferocious siege mounted by its critics at a time when similar military tactics were being deployed for real in the Crimea, to the final years of the First World War, when the Cuvierian law was still being invoked in interpretations of huge dinosaurs from the American West that were compared to the tanks lumbering across the muddy quagmire of the western front. Such an extended time frame—if not quite the *longue durée*, then certainly the whole of the long nineteenth century— is unusual in a book that, as with *Show Me the Bone*, offers finely textured accounts of the broader social and cultural contexts of nineteenth-century science, and documents the interactions between the various sites and settings in which it was conducted.[43] By focusing on a particular concept such as the law of correlation, and tracking its various incarnations and transformations across a prolonged period of time, it is nonetheless possible to present a picture of the world of nineteenth-century science that, as *Show Me the Bone* endeavors to do, combines detailed analysis with a sweeping historical narrative.[44]

Such an approach can also transform conventional historiographic assumptions, revealing, in the case of the Cuvierian method, the remarkable afterlife of the law of correlation in the latter part of the nineteenth century. The concept of afterlife, or "undead science," is familiar in social studies of contemporary science, where it signifies the strange persistence of theories such as cold fusion that have been rejected at the formal closure of controversies, but it has never previously been utilized in the history of science.[45] When combined with recent developments in the study of science popularization, however, it can yield important new insights. Claims about Cuvier's unerring and almost prophetic powers continued to circulate in the nonspecialist formats in which science was brought to new audiences after midcentury, even long after the law of correlation had been decisively refuted by a new generation of expert practitioners.[46] From the mid-1850s, Thomas Henry Huxley contended that anatomical correlations were simply customary correspondences deduced by empirical observation rather than infallible a priori laws. Within a few years almost the entire scientific community, especially Huxley's fellow scientific naturalists, had shifted to this position, and in many respects Huxley's onslaught against Cuvierian correlation can be considered as the opening salvo in his wider campaign to reform and modernize Victorian science.[47]

But just as Owen, the ostensible obscurantist, in fact embraced the innovations of the mid-Victorian publishing and media industries, so Huxley's apparently modernizing zeal, when viewed through the prism of the law of correlation, can actually appear curiously old-fashioned, with one defender of the Cuvierian law in the 1850s accusing him of wanting to "put back the hand of the rational dial" and reinstate "anachronisms" that were no less absurd than denying the "circulation of the blood."[48] Certainly, Huxley's initial attacks on Cuvier were not connected, as even recent historians have tended to assume, to his support for evolution, as they predated his conversion to Darwin's approach to species.[49] Augmenting this paradoxical perception of his outmodedness, Huxley, despite his subsequent reputation as an exceptional communicator of science, was much less adept at manipulating the press than Owen (at least in this instance), and failed to articulate his intense opposition to the law of correlation in ways that were appropriate for the new broader audiences for science. Instead, Huxley left the field open to popularizers, journalists, and literary writers, who, regardless of the caveats raised by experts, continued to eulogize Cuvier's vaunted abilities.[50]

That Feat of Palæontology

This disparity between the scientific community's relinquishment of the Cuvierian method and its continued currency among several other groups had
hugely significant consequences, and it certainly challenges long-standing assumptions about the power of the scientific naturalists to simply take the Victorian public with them in their endeavor to subjugate older scientific views.[51]
Most notably, *Show Me the Bone* reveals how the practices of elite men of science were actually shaped by their involvement with the nonscientific public.
Huxley, faced with the prevalence and persistence of popular claims about
Cuvier's abilities, had no choice but to curtail his research and engage in his
own popularizing activities. He was obliged, however, to tailor his approach
to suit the expectations of audiences more usually addressed by popularizers
than practitioners.[52] Remarkably, Huxley, as *Show Me the Bone* reveals, reversed his vaunted condemnation of the law of correlation when addressing
plebeian audiences, and, notwithstanding his usual insistence on prioritizing
intellectual truth over all other considerations, continued to treat the topic
in dramatically different ways according to whom he was speaking to. Significantly, the demarcations between popularization and practice that Huxley
strenuously sought to maintain were never quite as clear as he supposed, and
his strategic volte-face on correlation could not be prevented from leaching
across into publications regarded as part of his more specialist oeuvre, or
even impacting on his own paleontological practices. In this way, *Show Me
the Bone* offers a radically new perspective on several aspects of Huxley's career, especially his famous rivalry with Owen, which now no longer appears
such a one-sided mismatch. It also affords an especially conspicuous example
of what Roger Cooter and Stephen Pumphrey have termed the "power of
non-learned culture to intrude its practices into the fabric of science."[53]

Developing Cooter and Pumphrey's influential argument, Jonathan
Topham has recently proposed that breaking down the conventional binary
opposition between popular and so-called real science requires envisaging
a "feedback mechanism, by which what is conceived of as 'popular science'
crucially shapes the thought style of research scientists."[54] Ralph O'Connor
likewise contends that historians must assume "influence in both directions
between elite and nonelite practices, including the often underappreciated
processes by which popular science shapes the practice of elite science."[55] As
O'Connor's acknowledgment that such processes are "often underappreciated" implies, this is not something that historians of science have been conspicuously successful in doing, notwithstanding the rich and nuanced body
of work on popular science produced since the mid-1990s.[56] Even when a

simple top-down process of disseminating authorized enlightenment to a passive and grateful public is discarded, too often the focus has been on either the ways that elite science was appropriated and transformed by its audiences, or the development of an indigenous "ethnoscience" or "low scientific culture" that remains defiantly independent of elite knowledge.[57] The popular is certainly given agency in such models, but generally only within its own autonomous realm, and it has little or no effect on the equally separate elite scientific culture. While designations such as "popular science" and "popularization" increasingly took on divisive social implications in the second half of the nineteenth century, the awkward persistence of what Huxley ruefully called "that feat of palæontology which has so powerfully impressed the popular imagination, the reconstruction of an extinct animal from a tooth or a bone," shows clearly that avowedly popular conceptions continued to play a highly significant role in shaping science, and that even authoritative scientific naturalists were implicated in this dialectical process.[58]

The interdisciplinary field of "literature and science" has perhaps been more successful in supplanting its own earlier unidirectional conceptions of scientific "influence" on the literary, and instead demonstrating that the interaction between literature and science, even when they can be considered distinct entities, is very much a reciprocal process.[59] The intellectual "traffic," as Gillian Beer put it in *Darwin's Plots* (1983), is "two-way."[60] Topham's "feedback mechanism" and O'Connor's "influence in both directions" evidently have much in common with Beer's "two-way traffic," not least in O'Connor's necessary caveat that such models "should not be taken too literally as a map of the terrain" because the reciprocal traffic is never uniform and rarely without conflict.[61] *Show Me the Bone* brings these parallel approaches into still closer conjunction by showing how literary factors helped shape perceptions of the law of correlation, among practitioners as much as the public, and often in ways that diverged from the opinions of the elite scientific community. As noted earlier, Cuvier's shift to a new, more rhetorical style in the "Discours préliminaire" in 1812 had a decisive impact on the future fate of his celebrated law, and issues of style, genre, and form remained integral to correlation's continuing success (or otherwise) for the rest of the nineteenth century. This is especially evident in the intimate connection between the practices of Cuvierian paleontology and the new methods of serialization, in which books were published in sequential parts, that revolutionized British publishing from the 1830s. When announced and then corroborated across a number of installments, the daring inferences made by paleontologists employing Cuvier's law were rendered far more striking, as well as demonstrably verifiable, than if they had been expounded only in a single monograph. At the same time, Owen's own enthusiastic reading of the monthly numbers of Charles Dickens's serial novels, in which he endeavored to anticipate details of the plot by predicting the relation of the part to a larger narrative whole, closely paralleled his inferences from only single bones, and sheds important new light on his paleontological procedures. Indeed, Owen, who also wrote his own serialized ghost stories, spent much of his career waiting, with apparently equal anticipation and excitement, for fossilized remains coming bit by bit and novels arriving part by part, in the expectation that both would verify earlier predictions.

Disputes over the putative powers of paleontologists were conducted in novels and poems—including works by Honoré de Balzac, William Makepeace Thackeray, Henry James, Arthur Conan Doyle, and Emily Dickinson, as well as lesser-known writers such as Samuel Warren—as much as scientific publications, and, crucially, literary works impinged on the activities of practitioners in the same way as more overt forms of popularization.[62] Huxley, for instance, had continually to negotiate what he called "'Science as she is misunderstood' in the . . . novel," although it was his own muddled obfuscation of his actual opinions on correlation that helped perpetuate literary endorsements of Cuvierian methods.[63] This was especially apparent in the new and hugely popular genre of detective fiction, where, into the twentieth century, Sherlock Holmes and other heroic sleuths frequently compared their forensic abilities to those of paleontologists. Almost a century after Samuel Latham Mitchill's audacious assertion, in a lecture hall in Lower Manhattan, of what could be deduced from just one bone or scale, the legendary detective created by Doyle was no less certain that "Cuvier could correctly describe a whole animal by the contemplation of a single bone" and that his own powers of detection were equally prodigious and unerring.[64]

PART I

Arrival, 1795–1839:
Translations and Appropriations

Correlation Crosses the Channel

On 5 April 1800 Napoleon Bonaparte attended the annual meeting of the Institut national, the centralized replacement for the five separate Académie royales that were suppressed after the French Revolution. Only two months earlier he had been confirmed as first consul, giving him absolute control of all aspects of the French state, and this was his first visit to the *institut*, of which he had been elected a member in 1797, since assuming these new authoritarian powers. Having himself recently been appointed secretary to the *institut*'s scientific First Class, Georges Cuvier was required to deliver ornate eulogies to its deceased members. His first of these formal *éloges*, read at the April 1800 meeting in Napoleon's presence, was for Louis-Jean-Marie Daubenton, Cuvier's predecessor as professor of natural history at the Collège de France. Among Daubenton's many achievements, Cuvier proclaimed:

> He was the first who applied the knowledge of comparative anatomy to the determination of species of quadrupeds whose remains are met with in a fossil state. . . . His most remarkable achievement of this kind, was the determination of a bone . . . as the leg-bone of a giant. He discovered, by means of comparative anatomy, that it could only be the radius of a giraffe, although he had never seen that animal, and although no figure of its skeleton existed.[1]

Daubenton had died on the final day of the eighteenth century, but Cuvier, who astutely manipulated the rhetoric of his *éloges* to promote his own scientific interests, considered his deceased predecessor's remarkable identification of the unknown giraffe from merely a single leg bone, which had occurred back in 1762, as opening an important field of scientific investigation for the new century.[2] Significantly, it was a field that Cuvier, having arrived in Paris only five years before, intended to claim as his own.

While Napoleon's response to this particular *éloge* is not known, Cuvier's proposal that Daubenton's remarkable anatomical feat could become the basis for a new scientific program deploying the methods of comparative anatomy to determine the identity of even the most fragmentary fossils accorded with the first consul's own expansive ambitions. As David Brewster later remarked, "Work like Cuvier's Comparative Anatomy . . . could not but attract the attention of Napoleon."[3] In fact, the two men soon became closely acquainted, with Cuvier being appointed to several administrative posts within the Napoleonic regime. Science, after all, was perceived as a heroic embodiment of the French state, and even during the disruptions of the postrevolutionary period continued to receive munificent governmental funding.[4] Napoleon and Cuvier had actually been born within a week of each other in August 1769 (although neither on the French mainland), and the close parallels between them were apparent to contemporaries. The Scottish anatomist Robert Edmond Grant, who visited Paris regularly and knew Cuvier well, reflected in 1830:

> M. Cuvier was born in the same year, in the same month, and nearly on the same day with Napoleon, whose intimacy he enjoyed during a great part of his perilous career; and thus . . . the same period gave birth to the two illustrious foreigners whose lives must be so much identified with the history of France, and whose thoughts and actions have had so great an influence on the scientific and political world.[5]

The nascent scientific field outlined at the *institut* in April 1800 would be ruled by Cuvier with the same authoritarian power that Napoleon, rapidly expanding France's borders by military force, exercised over much of Europe. Indeed, by 1802 Cuvier's critics were depicting him as "a pedagogue who would turn his rod into a scepter of domination."[6] There was, however, one important difference between the listening first consul and the *Napoléon de l'intelligence*—as an early twentieth-century biographer dubbed Cuvier— who delivered the *éloge* for Daubenton.[7]

When, at the end of 1804, Napoleon augmented his powers as first consul by restoring France's hereditary monarchy and crowning himself emperor, he was amassing a huge flotilla in the Channel to once more attempt an invasion of Britain. Following previous abortive invasion plans in 1798 and 1802, the emperor took personal control of this new Armée de l'Angleterre, prompting widespread panic and vitriolic loathing of the French leader across the Channel.[8] By 1805 the Royal Navy's continuing blockade of the French fleet, bolstered by its decisive victory at the Battle of Trafalgar, had forestalled this third and final attempt, and in the following years both France and Britain imposed rigorously enforced commercial embargoes of each other's cross-Channel trade. Despite these considerable naval impediments, which prohibited official importations of French books and periodicals, and notwithstanding the virulent Francophobia that they elicited, Cuvier was considerably more successful than his bellicose sovereign in breaching British defenses and establishing a scientific bridgehead across the Channel in the initial decades of the nineteenth century.

Even his *éloge* for Daubenton was published in an English translation, albeit almost three decades after it was first delivered in Napoleon's presence and long after the emperor's final defeat by the British and their allies at the Battle of Waterloo. It was in the April 1828 number of the *Edinburgh New Philosophical Journal* that the passage relating to Daubenton's identification of the giraffe appeared, for the first time, in English. The same section was soon picked out as warranting particular attention by periodicals in Britain's other center of publishing, with the *London Medical Gazette* recognizing the original strategic intentions of the "*éloge* pronounced by Cuvier" and observing that "Daubenton may be said to have been the founder of that beautiful superstructure which his pupil afterwards raised, and which enabled him, from the inspection of a single phalanx, to determine the species and characters of the animal to which it had belonged."[9] It was this vaunted ability to surpass even Daubenton by identifying creatures from the most diminutive remnants of their remains that was integral to the British acknowledgment of Cuvier's scientific preeminence.

The translation in the *Edinburgh New Philosophical Journal* was penned by the journal's editor Robert Jameson, who, fifteen years earlier, had edited an English rendition of the "Discours préliminaire" from Cuvier's *Recherches sur les ossemens fossiles* (1812), which he titled *Essay on the Theory of the Earth* (1813). This had gone through five editions by the time of Jameson's translation of the *éloge* for Daubenton, and, along with similarly influential English-language versions of *Leçons d'anatomie comparée* (1800–1805) and *Le règne animal* (1817), incited what the Scottish anatomist Robert Knox dismissively termed a "Cuvierian mania." Knox was an admirer of Cuvier's French rivals and critics, and in 1839 he reflected bitterly that it was only "in Britain, where almost universally Cuvier's writings have been substituted for the book of Nature."[10] Five years earlier, in the wake of Cuvier's death in May 1832, the naturalist William Swainson similarly complained that "amateurs of zoology in this country, ever prone to judge in extremes . . . have invested his memory with a universality of talents almost superhuman; and are now ready to bow to his authority with . . . blind and implicit homage."[11]

The subject of such fulsome and unconditional remembrance was considered part of a scholarly republic of letters that apparently transcended petty national rivalries. Even during the Napoleonic Wars, Cuvier maintained friendly relations with men of science on the other side of the Channel and was elected a fellow of the Royal Society in 1806.[12] Despite his involvement with the Napoleonic regime, Cuvier's own private reservations about its schemes of conquest and the egoistic manner of the empire helped him to cultivate such cosmopolitan intellectual allegiances.[13] In public, however, Cuvier sycophantically praised Napoleon's wartime leadership, and at times he could be no less dismissive of the British than his famously haughty emperor.[14] In *Leçons* he insinuated that the enemy "nation" had fallen into the parochial "kind of pride, useful perhaps in a political view" of "despising foreigners too much; and in valuing and consulting only its own countrymen." This had resulted, Cuvier alleged, in the "dryness which forms the character of some of its natural historians and comparative anatomists."[15] In 1804, as the invasion forces were massing in the Channel, the *Monthly Review* patriotically protested at this "very ill-founded aspersion against English philosophers," which it considered "not an extraordinary one, as coming from a Frenchman, though . . . far from being congenial to M. Cuvier's usual candour, liberality and correctness."[16] The distinctly anglophone "Cuvierian mania" diagnosed by Knox three decades later nevertheless makes it clear that, unlike Napoleon, Cuvier had gone on to triumphantly vanquish such British resistance.

As Knox and Swainson's disaffection with their glibly Cuvierian compatriots suggests, though, the response to the celebrated French savant in Britain was far from uniform, particularly in relation to the purportedly a priori principle that facilitated his capacity to identify creatures from only isolated parts of their anatomy. Cuvier's so-called law of correlation was readily adapted to particular local circumstances, and its reception in Britain was frequently determined by geographical, political, and religious factors. These factors, moreover, were not as exclusively conservative or orthodox as previous historians of British responses to Cuvier have tended to assume.[17] Such assumptions are, in part, the result of historians overlooking the medical context of the initial English translations of Cuvier's anatomical writings, and instead focusing exclusively on the second wave of translations of his geological works.[18] What Martin Rudwick has termed the "anglicized Cuvier" was actually a myriad of different, and often competing, interpretations of the same savant, whose works could be repackaged to endorse both conservative and radical, as well as religious and heretical purposes.[19] In fact, in the opening decades of the nineteenth century, the law of correlation was often more amenable to the interests of radical materialists than to conservative theologians, who instead preferred other aspects of Cuvier's work that could bolster revealed theology.

Although specialist booksellers continued to supply wealthy francophone men of science with the original editions of Cuvier's works even during the cross-Channel blockades (French books were regularly smuggled via the Netherlands), most British readers relied on translations that, in the absence of international copyright restrictions, were attuned to domestic concerns and appropriated Cuvier's scientific authority for their own ends.[20] Many of these translations, both in periodicals and in books, were published in Edinburgh, a manifestation of the close connections between Scottish and French intellectual culture that had first developed in the eighteenth century.[21] The local context of the Scottish capital's institutional politics and theological sensibilities was, as this chapter will show, no less significant in shaping British responses to Cuvierian correlation in the opening decades of the nineteenth century than the more dramatic events of the Napoleonic Wars.

The Germ of the Principle of Coexistence

Daubenton's celebrated identification of the giraffe in the 1760s was not the only precursor of the principle of correlation that Cuvier began to develop thirty years later. He himself also acknowledged, in an unpublished lecture given at the Institut national in the mid-1790s, that Antoine-Laurent de Jussieu's *Genera plantarum* (1789) showed how an apparently slight peculiarity in the major organs of a plant enabled botanists to anticipate much of the structure of the rest of it.[22] Cuvier enthusiastically embraced Jussieu's contentious replacement of Linnaeus's emphasis on external characteristics with a new focus on internal structures. What he gleaned from *Genera plantarum* was that each family of plants is marked by a correlation of perceptible characters in which the other characters are subordinated to, and can therefore be inferred from, those of greatest generality.[23] Cuvier later adopted this as his "principle of the *subordination of characters*," which facilitated both a natural classification of animals according to the relative importance of the organs that performed their most characteristic functions and an assumption that the more functionally significant organs would necessarily entail the form of the subordinate ones.[24]

As well as Jussieu, the embryologist Antoine Étienne Serres claimed that "there was . . . in the great work of Vicq-d'Azyr the germ of the principle of coexistence and harmony of parts whose demonstration constitutes one of the finest glories of Cuvier."[25] Félix Vicq-d'Azyr had been Daubenton's protégé before his untimely death in 1794, and from him, as Serres intimated,

Cuvier learned both the importance of comparison in anatomical study, an approach that Vicq-d'Azyr had himself developed from the practices of his mentor, and the necessity of examining vital functions ahead of the organs that performed them.[26] Vicq-d'Azyr had also recognized the implications of Jussieu's botanical taxonomy for understanding anatomical structures. In fact, Pietro Corsi has contended that it was Vicq-d'Azyr who first "established the principles of the subordination of organs and correlation of parts," and that Cuvier, in subsequently asserting his own priority, deliberately tried to "minimize the importance of his predecessor's work."[27]

But, notwithstanding Cuvier's reluctance to acknowledge the conveniently deceased Vicq-d'Azyr, the antecedents of his particular understanding of animal structure were not exclusively French. Having been born in a region of the German duchy of Württemberg that was only annexed by France during the Revolutionary Wars of the early 1790s, Cuvier had greater access to German thought than most of his French contemporaries. It was Teutonic philosophy that proved most advantageous for the nascent principle of correlation. In particular, Immanuel Kant's *Kritik der Urtheilskraft* (1790), which Cuvier likely read soon after it was first published, afforded an understanding of the intricately interlocking functions by which living beings operated.[28] Acknowledging that mechanism alone could not adequately account for the unity of organic structures, which instead also required a teleological explanation, Kant enjoined that it was through these self-perpetuating internal economies, rather than the externally imposed purposiveness discerned in traditional natural theology, that providential intention was expressed. Cuvier's conception of the carefully balanced inner economy of each organism, according to Dorinda Outram, was heavily indebted to this aspect of Kant's philosophical critique of early modern teleology.[29] However, while Cuvier notionally accepted Kant's insistence that providential intention of whatever sort could be used only as a "regulative" principle that provided a heuristic rather than a real understanding of animal structure (or, in other words, that teleological final causes offered an analogy of how organic forms are constructed but remained only an unprovable hypothesis), he frequently, as Michael Letteney has pointed out, reverted to conceiving of teleology as a "constitutive" category that, by necessity, entailed the reality of a creative agency.[30] This partial and inconsistent Kantianism meant that the concept of correlation that Cuvier developed from, among other things, his reading of *Kritik der Urtheilskraft* could be embraced, with equal enthusiasm, both by those who disavowed any externally imposed providential intention and, as will be seen later, by proponents of precisely the natural theological tenets that Kant vigorously refuted.

Cuvier, who both regarded the adult organism as his principal focus and rigorously prioritized function over form, had little interest, despite his bilingualism, in either the innovative embryology or the idealist *Naturphilosophie* that dominated German science in the early nineteenth century.[31] His particular approach to Kant was nevertheless shared with a tendency of the *Naturphilosophie* developed by Carl Friedrich Kielmeyer, whom Cuvier had studied with in Württemberg and, in the mid-1790s, still regarded as his master.[32] Kielmeyer employed Kant's regulative combination of mechanical and teleological principles to invoke a unifying vital force according to which matter became organized in particular forms, but that itself had no existence independent of matter. In practice, this "teleomechanism" or "vital materialism," as Timothy Lenoir has termed it, relied on the same conflation of Kant's "regulative" and "constitutive" principles that Cuvier frequently committed. Indeed, Lenoir has proposed that "Cuvier himself came very close to adopting the principle of teleomechanism" in positing a nonmechanical formative agency immanent in matter and assuming that life presupposes organization.[33] In endorsing this view, John Reiss has avowed that Cuvier's formulation of the laws of organic form "resides precisely on the common ground shared by the teleologist and the materialist."[34] At the same time, Kant's "regulative" insistence that our understanding is not real or objective allowed Cuvier, according to Letteney, to similarly "have his cake and eat it too," by permitting him to examine organisms as if they were determined only by laws of matter while circumventing charges of materialism by maintaining that there is no direct proof of the actual existence of matter.[35] Significantly, as with the inconsistencies in his own reading of *Kritik der Urtheilskraft*, this teleo-materialist common ground and materio-idealist cake consumption entailed that—more than other contemporary scientific approaches—Cuvierian anatomy could be perceived, without inconsistency, as supporting diametrically opposed viewpoints. This ambiguity, as will be seen in this and the next chapter, would have hugely important consequences for the reaction to Cuvier's understanding of animal structure in Britain in the first three decades of the nineteenth century.

From this conjunction of various French and German sources that reflected his own personal and intellectual hybridity, Cuvier fashioned a principle in natural history that, notwithstanding its many antecedents, was entirely new and distinctive. Starting from Xavier Bichat's premise that life was characterized by the repulsion of the chemical laws that incessantly break down organized bodies, Cuvier posited that only the harmonious cooperation of all the body's constituent parts could ward off, albeit temporarily, the inevitable processes of decomposition.[36] These corresponding parts, which constituted life itself, were adapted to, and their shapes determined by, the functions that they performed. Functions such as respiration and digestion that were essential for a particular organism to actually exist—what Cuvier termed the "conditions necessary for existence"—dictated its organization in accordance with the most efficient means of accomplishing those necessary functions, as well as the mutual compatibility of the different functions with one another.[37] This harmony of functions demanded a corresponding mutual dependence of parts, ensuring there was an invariable correlation between particular organs in the animal economy as well as an equally consistent exclusion of any other potential combinations. Thus (in Cuvier's own favorite example), a sharp tooth, adapted for the function of cutting flesh, must be accompanied by a particular kind of foot, as well as jawbone, neck, stomach, and so on, that facilitates the consumption of meat. At the same time, the sharp tooth cannot be matched with a foot, such as a hoof, that does not have the capacity to grasp prey, and itself corresponds only to teeth adapted to grinding a herbivorous diet. With animal structures shaped so strictly by their functional needs, certain parts such as teeth and feet, as Cuvier quickly recognized, would necessarily indicate the configuration of the integrated whole.

Cuvier first began to articulate his new understanding of organic structure when, toward the end of 1795, the failing health of the elderly professor Antoine-Louis Mertrud required him to take over the lectures on animal anatomy at the Muséum d'histoire naturelle. In these lectures, Cuvier avowed that, when acquainted with the "laws of coexistence," an experienced observer "from the appearance of a single bone, will be often able to conclude, to a certain extent, with respect to the form of the whole skeleton to which it belonged."[38] Significantly, by claiming the status of scientific laws for his physiological axioms, Cuvier implied that they were permanent and absolute, and therefore universally applicable. His conception of what constituted a natural law, however, was much more precise and particular than the overarching abstract laws governing all of aspects of nature invoked by Enlightenment deists such as the Comte de Buffon, and, more recently, the advocates of German *Naturphilosophie*. In 1804 Cuvier observed pointedly that "superior men of genius . . . who tire of the futility of abstract speculation have abandoned the heights of a too general philosophy and instead sought nature's true laws through the scrutiny of her works."[39]

Cuvier's more specific and limited "true laws" still entailed that just a single bone could reveal the remainder of the skeleton from which it came, an arresting and bold assertion that helped him rapidly establish a reputation as a lecturer of note amid the power vacuum of postrevolutionary intellectual life.[40] As one British admirer later observed of Cuvier's lectures on the "anatomy of animals," the "number of his audience (often exceeding a thousand) sufficiently proves the ability of the teacher, and the interest inspired by the subject."[41] But for several years Cuvier's audacious claim was confined to "manuscript copies" of his lectures "taken by pupils from his oral demonstrations," which, Robert Edmond Grant claimed, were "circulated in Paris, and even cited in works, before he undertook to publish them."[42] Even when the statement did, finally, appear in print, in March 1800, in the opening volume of *Leçons d'anatomie comparée*, it was reconstituted from the notes taken by Cuvier's dutiful student Constant Duméril, to whom he had delegated the cumbersome arrangements for publication.[43]

One Bone Fragment Is Often Sufficient

What had "distracted his attention from the compilation of the Leçons," as Grant later reflected, was Cuvier's "frequent excursions . . . to the excavations in the gypsum strata of Montmartre."[44] Here, in the quarries north of Paris that supplied the city's building materials, laborers had begun to encounter strange animal remains embedded in the gypsum matrix. Having previously examined the colossal bones—or at least their pictorial proxies—of mammoths and megatheriums discovered on different continents, Cuvier now oversaw the careful extraction of smaller but no less peculiar fossil fauna just a short journey from his home in the Jardin des plantes. However, whereas the skeletal remains of the mammoth and megatherium were both relatively complete (having been found in loose deposits from more recent strata), the hard plaster stone quarried at Montmartre yielded only partial fragments and disarticulated bones and teeth that gave little sense of how the animals from which they came had once looked. Ironically, it was the laws of animal economy and organization that Cuvier adumbrated in the lectures on comparative anatomy whose publication had been delayed by his involvement with the Montmartre excavations that afforded the interpretative solution to these enigmatic fossil remains.

In October 1798 Cuvier gave the first public account of the method for examining fossil bones that he would employ for the next three decades. Addressing a meeting of the Institut national, he began by proudly proclaiming:

> Today comparative anatomy has reached such a point of perfection that, after inspecting a single bone, one can often determine the class, and sometimes

even the genus of the animal to which it belonged. . . . The bones that compose
each part of an animal's body are always in a necessary relation to all the other
parts, in such a way that—up to a point—one can infer the whole from any
one of them.

In this oral lecture, which was never published in his lifetime, Cuvier made
it clear that his principal concern was simply with the identification and classification of fossil remains. He also employed precisely the same language—
especially the circumspect disclaimer "jusqu'à un certain point" (variously
translated as "to a certain extent" and "up to a point")—he was simultaneously using to describe the organization of living animals in the lectures on
comparative anatomy that, at this stage, were themselves still unpublished.
Cuvier assured his auditors that this extrapolation from the laws of coexistence was the "method that is employed in the research I am going to tell
you about" on the strange animal remains of Montmartre. Conscious of the
audacity of the claims he was making for this new method of interpreting fossils, Cuvier carefully moderated them by acknowledging that, when inferring
from a single bone, he would at times be "reduced to more subtle conjectures
and to less firm conclusions" that would have only an "appropriate degree of
probability."[45]

In a concluding section of the same lecture, though, he became less circumspect, speculating that, in cases where a large proportion of an animal's
bones had been discovered and were "well-known," it might be possible to
determine, or even actually reconstruct, the forms of its muscles and skin. In
such circumstances, he declared,

one would thus have an image not only of the skeleton that still exists but of
the entire animal as it existed in the past. One could even, with a little more
boldness, guess some of its habits; for the habits of any kind of animal depend
on its organization, and if one knows the former one can deduce the latter.

There was no suggestion that there was any connection between this possibility that an animal's form and habits might be "restored" from an understanding of its internal organization, and the more limited capacity to identify its
species and genus "after inspecting a single bone."[46] Cuvier, after all, considered them at opposite ends of his address to the *institut*. Their inclusion in
the same lecture, especially when Cuvier's audacious rhetoric insisted that
with only one part the anatomist could "infer the whole," nevertheless made
it possible to conceive of not merely identifying but actually reconstructing
all facets of an animal from just a single bone.

In the following year, Cuvier was more specific about how his new method actually worked in practice when writing to Johann Heinrich Autenrieth, a physician and fossil collector in his native Württemberg. But, tellingly, in this private letter, written in September 1799, he gave an explanation of his ability to identify animals from only a single fragmentary bone that was very different from the bold claims he had made in his public address to the *institut* less than a year before. Asking Autenrieth for information on the fossil bones in Stuttgart's Königlichen Naturalien-Cabinet, and assuring his correspondent that he would be grateful for details of even the most fragmentary remains, Cuvier observed:

> My collection of skeletons is now so complete that one bone fragment is often sufficient, provided that there are facet joints, to determine what genus and even what species it belonged to, especially when it is one of the bones of the feet, or jaw.[47]

It was the extensive collection of animal skeletons that Cuvier was amassing in one location, at the Muséum d'histoire naturelle, that enabled him to compare the morphological characteristics of a single fossil bone with those of similarly shaped components of complete skeletons of extant species. If there were constantly recurring relations between the bones in these particular skeletons, then it could be assumed that the isolated fragments were once part of a cognate integrated framework that had otherwise since disappeared. Crucially, this pragmatic method of identification involved merely the observation of recurrent morphological features rather than the mastery of the invariable laws of coexistence that was prioritized in both Cuvier's lectures on comparative anatomy and his public announcements on the Montmartre fossils.

None of the claims Cuvier made for his new method during the late 1790s were published in that decade, and instead they remained restricted to oral lectures, student notes in manuscript, and private letters. In addition to the distractions of the Montmartre fossils, Cuvier's still precarious status among the Parisian intellectual elite, as Rudwick has argued, obliged him to be careful of publishing anything that might prompt accusations of mere speculation.[48] Publication was, in any case, far from being a necessity in many areas of science in the late eighteenth century, with more significance given to the delivery of spoken papers or informal conversation among peers.[49] Indeed, the *institut* did not issue any printed transactions, and Cuvier was adamant that journalists not be admitted to its meetings, fearful of the unauthorized reports of its oral deliberations they might publish.[50] In the absence of any published statements, it is hardly surprising that the ambivalence regarding whether the identification of an animal from just a single bone was predicated on simple empirical observation or rational physiological laws appears not to have been recognized at this stage, even, perhaps, by Cuvier himself. Meanwhile, the elision of identification and reconstruction encouraged by Cuvier's rhetorical flourishes, which David Brewster described as "sparkling with playfulness and fancy," also took root.[51] It was precisely these issues, both of which were already apparent by the late 1790s, that would become pivotal to the fierce controversies over Cuvierian correlation that, as will be seen in later chapters, persisted throughout the entire nineteenth century.

The Power of the Principle

Even after the eventual appearance of the initial two volumes of *Leçons d'anatomie comparée* in the opening months of the new century, it would be a further four years before Cuvier published a detailed description of how he deployed his distinctive method when examining the fossil remains extracted from the Montmartre quarries. When he finally did, his scientific paper narrated a dramatic test of his anatomical principles that left little doubt that they were underwritten by immutable natural laws rather than mere observation. In a paper given at the *institut* on the final day of 1804 and subsequently published in the *Annales du Muséum national d'histoire naturelle*, Cuvier announced that the teeth of a diminutive recent specimen from the Paris gypsum indicated that it was a marsupial, even though, in its extant forms, this order was almost entirely confined to Australia and America. While this supposition was geologically momentous, indicating the vast differences between present and fossil faunas and therefore providing compelling evidence for the reality of extinction, the particular circumstances in which this tiny opossum's remains were discovered also afforded an especially potent means of verifying Cuvier's interpretative method.

The remainder of the putatively marsupial skeleton was still encased in a second section of gypsum matrix, so Cuvier, having already predicted that it would reveal the distinctive epipubic pelvic bones that support the pouches of living marsupials, invited his colleagues to witness his careful extraction of the rest of the stone. Cuvier, according to Outram, modeled his lecturing style on the Comédie-Française's principal tragic actor François-Joseph Talma, and he evidently realized the theatrical potential of this dramatic demonstration of his predictive powers.[52] As he recounted:

> This operation was done in the presence of some persons to whom I had announced the result in advance, with the intention of proving to them—by the act—the justice of our zoological theories, since the true hallmark of a theory is unquestionably the power it provides to predict phenomena.[53]

As an exponent of the new emphasis on specialist expertise that came to dominate French natural history from the 1790s, Cuvier was contemptuous of the mass audience to whom Buffon and his successors addressed their more descriptive accounts of nature's wonders, and did not consider it necessary to cultivate a public following.[54] At the same time, he recognized the crucial role played by audiences in validating the success of scientific prognostications (which, like an actor's performance, are otherwise rendered void), recruiting specialist peers already primed about the anticipated result—who, notably, were never identified—to not just witness but actively collaborate in his carefully choreographed performance.[55] In this way, the conventionally distinct acts of discovery and demonstration occurred simultaneously, becoming inextricably intertwined.[56] As Cuvier already appreciated, the predicted marsupial pelvis that his fine steel needle duly made visible to his audience represented a spectacular vindication of the "zoological theories" he had begun to formulate a decade earlier. In the *Annales du Muséum* an accompanying engraving (fig. 1.1) included two separate figures that presented a "before and after" sequence of the pelvis and epipubic bones, shown first in isolation and then reincorporated with the remainder of the now identifiably marsupial skeleton. This skillful piece of "visual rhetoric," as Rudwick has proposed, was "designed to convert . . . readers into 'virtual witnesses' of that fulfilled prediction."[57] As such, the readers of Cuvier's paper assumed the same corroborating role as his original audience of colleagues and peers.

Cuvier's report to the *institut* concluded with a comparison of his own predictive powers with the "degree of probability" permitted by the "exact sciences," prophetically avowing:

> There is no science that cannot be made almost geometrical: the chemists have proved as much in recent times for theirs, and I hope that the time is not far off when one will be able to say as much of the anatomists.[58]

The perceived precision and rigor of mathematically based sciences such as chemistry afforded a model for savants who, as with Cuvier, endeavored to reform natural history and rid it of the hypothetical haziness of Buffon.[59] Prediction, which evinced the absolute regularity of the phenomena and methods invoked, was the exemplar of the intellectual authority of such geometrical disciplines.[60] In *Leçons* Cuvier had already claimed that the "laws, which regulate the relations of living beings . . . are themselves founded on the same necessary relations as the laws of . . . mathematics," and his successful prediction of the Parisian marsupial now gave credence to the quasi-mathematical precision of his zoological theories, even if, significantly, he refrained from actually calling them laws.[61]

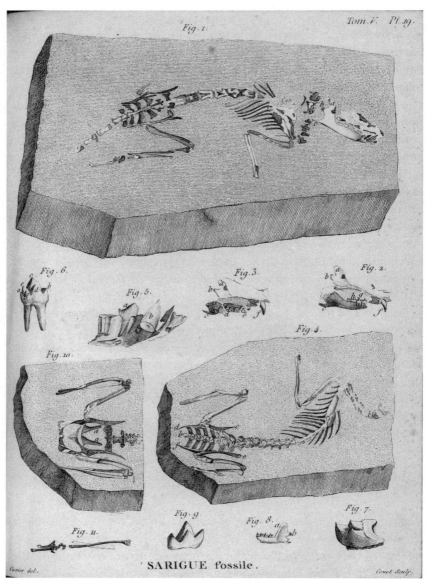

This improbable opossum, as will be seen in subsequent chapters, re-mained the paradigmatic test case for the success of Cuvierian correlation for the next eighty years. In the mid-1840s, Richard Owen, by then Cuvier's principal advocate in Britain, insisted that the "circumstances attending the discovery . . . furnish so striking an exemplification of the power of the prin-ciple which guided that great anatomist in the interpretation of fossil bones" that they could not be considered "entirely foreign" even to a discussion of "British Fossil Mammalia." While the iconic marsupial remains "buried in two portions of a split block of gypsum" were integral to the domestication of Cuvier by British naturalists, they were still more significant in ensuring that, when details of his distinctive method first began to appear in print in the opening years of the nineteenth century, it was seen as an expression of the same unvarying rational axioms studied by mathematicians and chem-ists rather than, as Cuvier had actually acknowledged in his private letter to Autenrieth, the more prosaic powers of empirical observation.[62]

The Valuable Dissertations of Foreigners

Although the Revolutionary Wars of the 1790s drastically reduced the import of books and periodicals from France to Britain, news of Cuvier's innovative fossil research soon began to cross the Channel. As well as the devastating impact of the wars, the initial excavations in the quarries of Montmartre also coincided with several important new developments in British publishing. Most notably, the final years of the eighteenth century witnessed the emer-gence of new commercial science journals that were both more timely and less expensive than the conventional transactions of scientific societies. Like the Continental periodicals on which they were modeled, they aimed to keep their readers abreast of discoveries from around the world by anthologiz-ing, and translating where necessary, articles from a wide range of interna-tional sources.[63] Thomas Thomson, editor of the *Annals of Philosophy*, later reflected that these new "philosophical journals" were chiefly notable for circulating the "valuable dissertations of foreigners through Britain, which might otherwise remain in great measure unknown to us."[64] Cuvier's haughty insinuations about British parochialism in *Leçons* suggested how needful this more cosmopolitan approach was, and the new specialist commercial jour-nals were the principal means by which Continental science reached British readers at the close of the eighteenth century.[65]

In 1797 the chemist William Nicholson began his *Journal of Natural Phi-losophy, Chemistry and the Arts*, which, while inevitably prioritizing Nichol-son's own specialism, also included details of recent developments in other areas.[66] As well as original contributions, much of Nicholson's innovative *Journal* consisted of translations of articles from the French periodicals that did manage to get across the Channel. In February 1799 it carried a translation of one of Cuvier's lectures published in the *Bulletin des sciences*. In it, as the translation informed anglophone readers, Cuvier pledged to "collect as much as it was in his power all the fossil bones appertaining to each species of animal . . . to form or recompose the skeletons of these species." As with the French original, though, there was barely any mention of the particular techniques that could enable disarticulated and dispersed skeletal remains to be recomposed, beyond a passing allusion to how "Citizen Cuvier proves, by osteological comparison," that certain fossil craniums could not have come from buffalo.[67]

While the Revolutionary Wars had not entirely stopped the flow of periodicals across the Channel, they did make it virtually impossible—especially once the Traitorous Correspondence Act was passed in 1793—for British men of science to visit Paris and avail themselves of the latest scientific hearsay. Without access to Cuvier's oral lectures or the handwritten student notes then circulating in the French capital, even the most well-connected British readers had to rely on what appeared in printed sources such as the *Bulletin des sciences*. Significantly, this meant that reports of Cuvier's identifications of particular fossil species appeared in the British press ahead of any explanation of the methods he had used to make them, which, as seen earlier, were confined during the 1790s to unpublished sources. The consequences of this cross-Channel disparity were still apparent into the 1830s, when a writer for the *British Critic* heralded "Cuvier's own great laws . . . of the co-existence and co-ordinate relations of every part of the animal frame," before erroneously claiming:

> It is remarkable that these very important generalizations appear to have been forced on the mind of Cuvier by his efforts to class the fossil mammalia of the basin of Paris while yet he possessed only a few teeth and scattered bone—so often does the very poverty of the materials afforded to scientific research, call forth the most powerful intellectual resources.[68]

As the Parisian auditors of Cuvier's lectures on comparative anatomy at the Muséum d'histoire naturelle could have confirmed, the generalizations had actually preceded the fossil classifications, although the curtailment of travel to France during the 1790s meant it was impossible to have known this in Britain.[69]

As well as the new commercial science periodicals such as Nicholson's *Journal*, reports of Cuvier's identifications of fossil species also appeared in general literary reviews whose encyclopedic ambitions prompted them to devote considerable attention to foreign science.[70] The *Monthly Magazine* was founded in 1796, and its regular retrospect of French literature soon included a translation of Cuvier's lecture on the megatherium carried in its French equivalent, the *Magasin encyclopédique*. Tellingly, the sole occasion in which its translation departed from the original abstract was when the "writer [i.e., Cuvier] proceeds to a detailed comparison" of each skeletal part with those of analogous species. This, as the translator noted impatiently, "for the sake of brevity, we omit."[71] Even when French printed sources did make some reference to the comparative methods Cuvier employed, the British press, during the 1790s, considered them of little interest.

Once again, it was the spectacular discovery of the diminutive Parisian marsupial at the end of 1804 that focused journalistic attention on Cuvier's techniques rather than just the strangeness or size of the ancient creatures he was excavating. Launched in 1798 as a competitor to Nicholson's *Journal*, Alexander Tilloch's *Philosophical Magazine* proved even more successful than its rival (which it eventually subsumed), in part because it utilized new printing technologies to ensure that readers were kept up to date with the very latest news of scientific discoveries.[72] In April 1805 it recalled to readers how "Cuvier was able to discover and restore entirely the fossil skeletons of several animals found in the quarries of Montmartre," before excitedly announcing: "The method by which he effected this restoration has been confirmed in a striking manner, by the discovery he has made of a skeleton of the opossum." Adapting rather than translating Cuvier's recent paper in the *Annales du Muséum*, the *Philosophical Magazine* offered the following précis of his method:

> The relations which M. Cuvier knew to exist between the different organs, and which he calls the *zoological laws*, enabled him to judge from what he saw of what he did not see. Such is the certainty of these relations, that M. Cuvier was able to predict, that in searching further in the quarry the two characteristic bones of the bag in which the opossum carries its young, would be found. Experience confirmed what theory had foreseen.[73]

In his original paper Cuvier had himself referred simply to "zoological theories," and it was only in *Leçons* four years before that he had alluded to the "laws of coexistence" and the "laws, which regulate the relations of living beings." Tilloch's *Philosophical Magazine* nevertheless informed anglophone readers that the remarkable prescience involved in the opossum's discovery exemplified what Cuvier, purportedly, "calls the *zoological laws*," even adding its own italicization to emphasize their exceptional status. These laws, moreover, had apparently equipped Cuvier with the ability to judge of things that were invisible, as well as an almost preternatural power of foresight.

When Cuvier's 1804 paper on the opossum was reprinted in the revised second edition of *Recherches sur les ossemens fossiles* in 1822, he boldly prefaced it: "I leave this article as it originally appeared in the *Annales du Muséum* as a monument . . . to the power of zoological laws."[74] This, though, was a retrospective appraisal, and while the unaltered original concluded by invoking the potentially geometrical precision of the anatomist's predictive powers, it conspicuously eschewed, as was noted earlier, any allusions to more general laws. In neither this nor any of his other papers on fossil species for the *Annales du Muséum*, as William Coleman has noted, did Cuvier "explicitly develop his procedure for reconstruction."[75] In fact, facilitated by the rapidity with which Tilloch's *Philosophical Magazine* brought scientific news across the Channel, it was in Britain where the successful anticipation of the iconic opossum's taxonomic classification was first ascribed exclusively to rational scientific laws.[76]

Only a year after the *Philosophical Magazine*'s account of Cuvier's decisive confirmation of his method, however, the Royal Navy began a blockade of French ports that, especially once Napoleon responded with his own embargo against Britain in November 1806, greatly disrupted cross-Channel trade. The weight of the books officially imported from France in 1806 fell to just a few hundredweight, when it had been more than a thousand in 1801.[77] By drastically reducing the already restricted import of French books and periodicals, this mutual and rigorously enforced commercial blockade curtailed any further discussion of Cuvier's ostensible zoological laws in Britain for the remainder of the decade.

Not for Naturalists

The *Philosophical Magazine* made no further mention of Cuvier's so-called zoological laws, and the only previous occasions on which its anglophone readers might have encountered a more detailed explanation of them was in the translations of *Leçons d'anatomie comparée* that had appeared during the brief interlude of peace in the opening years of the nineteenth century. Copies of the much-delayed initial two volumes published in Paris in March 1800 were soon imported by the London bookseller Joseph De Boffe as part of the increased cross-Channel trade permitted by the diplomatic negotiations that would eventually result in the Treaty of Amiens. Within five months, lengthy translated excerpts were included in a complimentary notice in the *Critical Review*, another of the general literary journals that shared the *Monthly*

Magazine's encyclopedic ambitions. It noted that the account of the "animal œconomy" in the first volume raised the "very important consideration, viz. the relation between the various systems of organs, showing how the existence of one is connected with that of every other." The reviewer, however, could not explore Cuvier's account of these particular mutual relations any further, explaining: "To have followed him more closely would have engaged more of our attention and space than we could well bestow, and would have prevented that variety so essentially necessary for a literary journal, designed for general readers as well as scientific inquirers."[78] Such scientific inquirers had instead to turn to two book-length translations that were very much intended for specialist readers and made few concessions to a more general audience.

These translations of *Leçons* appeared in 1801 and 1802 and were aimed, respectively, at students in the two centers of medical education in Britain: Edinburgh and London. In the Scottish capital, the fierce competition between extramural lecturers excluded from the city's ancient university and medical corporations encouraged constant innovation to maintain the interest of fee-paying students. John Allen delivered private lecture courses on physiology that, according to the *Gentleman's Magazine*, "were of such excellence as to have induced M. Cuvier eagerly to seek his acquaintance."[79] Although the Traitorous Correspondence Act was briefly suspended after the Treaty of Amiens, such a personal meeting remained highly impractical. Allen instead scoured the latest French scientific imports for new understandings he could pass on to his students. In a "work, published about eighteen months ago at Paris," as he reflected in the summer of 1801, he found a "comprehensive view of the relations that subsist between the different functions of the Animal Economy," and "was induced . . . to attempt the translation . . . into English." This translation was in fact merely an abridgment of the initial chapter of the first volume of *Leçons*, which, as Allen acknowledged in the dedication, was principally "UNDERTAKEN FOR" the "USE" of the "GENTLEMEN ATTENDING HIS LECTURES ON THE ANIMAL ECONOMY." Even if Allen's *An Introduction to the Study of Animal Economy* (1801) sold more widely, it was, he warned in the preface, "intended for the use of Students of Physiology, and not for Naturalists." The first extended account of what Allen termed Cuvier's "laws of co-existence" to appear in Britain was published in a format that avowedly excluded nonmedical readers.[80]

In order to make his *Introduction* suitable for the "use of Students," Allen "suppressed the subordinate divisions" and abridged the 521 pages of the initial volume of *Leçons* into just 79 octavo pages. Despite this drastic truncation, Allen's concise translation, which cost a relatively meager two shillings,

retained all the sections in the original pertaining to the harmony of functions and mutual dependence of parts. Edinburgh's anglophone medical
students were therefore apprised of how in a skeleton the "forms of all the
different parts are strictly related. There is hardly a bone that can vary in its
surfaces, its curvatures, in its protuberances, without corresponding variations in other bones." And this, crucially, meant that "from the appearance
of a single bone" one could infer the "form of the whole skeleton to which it
belonged."[81] In the French original, which was available only in an expensive
two-volume set that in Britain cost fourteen shillings, this audacious assertion
was offset by lengthy descriptive passages detailing the organs of movement.
In Allen's brief *Introduction*, however, the claim—appearing in English for
the very first time—assumed a considerably more conspicuous position,
and, in its cheaper format, was more readily accessible.

Scottish undergraduates were generally not as wealthy or genteel as their
southern counterparts in Oxford and Cambridge, and Allen's concern with
the financial constraints of his students reflected his distinctly liberal political
opinions.[82] A decade earlier, he had publicly celebrated the French Revolution, and, during the early 1790s, taken an active role in the demands for a
more democratic politics in Britain. His "opinions were essentially republican" as a contemporary later recalled, and he was "violent often in language . . . uttering the most terrifick expressions" of his radical convictions.[83]
It was this that had precluded Allen from teaching at Edinburgh's ancient
university, and compelled him to instead undertake private extramural lecturing. Even while he was translating Cuvier's *Leçons*, Allen remained under
the suspicion of the authorities in the Scottish capital, and the emphasis of
his own lectures on the physicality of the mind, including bold assertions that
man was nothing but an "Automaton . . . endowed with Sensation," gained
him a reputation as an atheist and materialist.[84]

What he termed Cuvier's "Gallicisms" certainly posed no problems for
Allen, whose long-standing faith in the "honest and philanthropic republican government . . . in France" was only rescinded in 1804 when "Napoleon
assumed the purple" as emperor.[85] In fact, there were, as L. S. Jacyna has
proposed, clear ideological reasons "why Allen responded with such enthusiasm to Cuvier's ideas." The Cuvierian "idea of the body as an interde-
pendent whole in which there was no predominant centre" that was foregrounded in the *Introduction* accorded with Allen's distaste for constitutional
"doctrines which located the source of sovereignty, not in the people, but
in some transcendent authority" in his later political treatise *Inquiry into
the Rise and Growth of the Royal Prerogative in England* (1830).[86] The necessary cooperation between a collective of organs that enabled the integrated

whole to be inferred from a single part was, in Allen's translation of Cuvier's lectures (which, notably, did not address the more hierarchical principle of the subordination of characters), entirely consistent with the democratic political agenda that Scottish liberals were also importing from France during the 1790s.[87]

Implicit in this view of both physiological and political interdependence was the disavowal of a hierarchical sovereign power, whether a temporal monarch or supernatural deity. In an unpublished lecture, Allen made it clear that this precluded the natural theological understanding of divine design that, in the late eighteenth century, was regularly invoked by professors at the neighboring university. He berated "those Pseudo-Philosophers who contemplating the similitude between the works of art, and the productions of Nature, have presumptuously insisted, that, because, in their opinion human Intelligence is the cause of Art that, a similar Intelligence must, of necessity, be the productive Cause of Nature."[88] While this was consistent with Kant's opposition to an externally imposed and constitutively conceived providential intention, it was less compatible with Cuvier's own more inconsistent Kantianism, and, most conspicuously, was entirely at odds with how his work was soon after appropriated by the powerful proponents of another new work published a year after Allen's translation. This was William Paley's *Natural Theology* (1802), whose advocates, as will be discussed in the following chapter, saw in correlation an incontrovertible proof of divine design. Such Cuvierian providentialism, moreover, upheld a conservative vision of the social order that was the complete opposite of Allen's demands for democratic reform.

As well as medical students, the audiences for Allen's extramural lectures also included local Whig politicians, and when he first introduced Cuvier's views of animal structure in his purposefully cheap *Introduction*, they assumed distinct ideological implications. These radical connotations were certainly evident to Allen's more conservative contemporaries in Edinburgh. His fellow extramural anatomy teacher John Barclay, who attended Allen's lectures in the 1790s, condemned his translation as "nothing but a heap of words; or, to borrow a term from the language in which the original is written, mere *verbiage*." Even worse, Barclay then insisted that the translation of *Leçons* showed only that "were any person capable of supporting, with any tolerable degree of success, the doctrines of materialism, it would certainly be Cuvier."[89] Materialism was perhaps the most contentious issue within the medical community in the early nineteenth century, especially as its insistence that life was merely a correlate of physical organization was considered threatening to religion, morality, and even the very foundations of social order.[90]

When, a year after Allen's translation, one of Cuvier's friends published their correspondence on the issue of phrenology as *Lettre de Charles de Villers à Georges Cuvier* (1802), the rabidly conservative *Anti-Jacobin Review*, which had apparently never heard of Cuvier before, warned its readers:

> Who this Charles Villiers [*sic*] or this George Cuvier is, we do not find out in this pamphlet, but that both are of the sect of materialists is pretty apparent. . . . The correspondence is between two young surgeons . . . both advocates for the infidel dogmata of the philosophizers of the day; who . . . vilify the omnipotence of the Creator, and the operations of his hand. . . . We are shocked to see men so perversely degrading the noblest energies of their nature, as to voluntarily rank themselves with the beasts that perish.

Although its "natural consequence is violation of the laws of GOD and man . . . and the overthrow of all established good," the "monstrous system" allegedly propounded by Villers and Cuvier had "not wanted advocates among the inexperienced young disciples of science."[91] Such concerns with the ethical and social implications of the novel doctrines imbibed by impressionable students were particularly acute in the Scottish capital. In 1823 the *Edinburgh Magazine* published the confessions—albeit seemingly fictional—of a repentant reprobate who recalls how, two decades earlier, his youthful "study of medicine" in the city helped "lead to materialism . . . which had rather a malignant influence on my future character and happiness." This ruinous "creed of materialism," he laments, "was only riveted more strongly, by perusing the writings of *Cuvier*."[92] With Cuvier scorned as a dangerous materialist, Allen, only a year after he published his *Introduction*, abruptly quit Edinburgh to become the physician to Lord and Lady Holland, the most prominent Whig aristocrats in London.

In the same city, another, more substantial, English edition of Cuvier's *Leçons*, translated as *Lectures on Comparative Anatomy* (1802) by William Ross and the Irish surgeon James Macartney, was reassuringly expensive, costing a guinea for two volumes. The cost, format, and physical qualities of a book all have significant implications for its repute and meanings, and while Ross and Macartney's *Lectures* was targeted at London's medical community and replicated much of the same material given a radical or even materialist slant in Allen's cheap *Introduction*, the *Monthly Magazine* considered it sufficiently respectable to be "noticed in an account of the progress of Domestic Literature," even though the "work is not of British growth."[93] The rival *Monthly Review*, though, took umbrage, as noted earlier, at the "very ill-founded aspersion against English philosophers . . . coming from a Frenchman" disloyally repeated in Ross and Macartney's translation.[94] This residual distrust of the French only increased as the carefully negotiated Treaty of Amiens, which had facilitated both translations of *Leçons*, began to break down at the end of 1802. Although Cuvier's new approach to animal structure had reached British shores in the brief interlude of peace, it remained confined to medical publications, and, in Edinburgh at least, became tainted with seditious political and theological connotations. While, as seen earlier, Tilloch's *Philosophical Magazine* preempted even Cuvier himself in ascribing his renowned identification of the Parisian fossil opossum to the anatomical laws first rendered into English in Allen's and Ross and Macartney's translations, the same laws otherwise went largely unheeded in Britain after the resumption of war and the subsequent disruption of cross-Channel trade during the protracted commercial blockades.

The Consequence of This Principle

On the other side of the blockade, meanwhile, Cuvier spent increasing amounts of time away from Paris on official business for the Napoleonic government, and, by 1810, had mostly completed the great sequence of papers on fossil quadrupeds he contributed to the *Annales du Muséum*. When he returned to the French capital two years later, these perpetual absences and the slackening of his rate of publication had, according to Outram, "weakened his hold on patronage and the development of ideas" among the Parisian scientific elite.[95] At the Muséum d'histoire naturelle, Cuvier's principal colleagues were formulating new understandings of animal structure that were sharply at odds with the purported laws articulated in *Leçons*. Étienne Geoffroy Saint-Hilaire, beginning in a series of papers in 1807 and 1808, proposed that all vertebrates were formed on a single plan, and it was the modifications of their anatomical form that determined the functions they performed.[96] Still worse, Jean-Baptiste Lamarck's *Philosophie zoologique* (1809) reasserted his long-standing conviction that organisms were continuously transformed over time as new habits and functions were taken on, and others forsaken, in response to alterations in the environment.[97] Lamarck's transformism maintained that individual organs were subject to continual modifications that rendered any absolute or necessary relations between them entirely impossible.

As his successive papers appeared in the *Annales du Muséum* during the opening decade of the nineteenth century, Cuvier had retained a stock of separately paginated offprints that, in accordance with a long-held plan, he now published in four quarto volumes as *Recherches sur les ossemens fossiles*. In the lengthy "Discours préliminaire" included in the opening volume of this monumental work, Cuvier rebutted the alternative theories of animal structure recently advanced by his rivals.[98] Lamarck's transformism implied that the divergences between fossil and living species were not the result of sudden extinctions but merely the accretion of small modifications over enormous periods of time. There was, however, a conspicuous objection to this that Cuvier gleefully pointed out: the complete absence of any remains of intermediate forms, even among the ancient mummified animals brought back from Egypt after Napoleon's invasion in 1798.

Although it was less obvious, there was a still more fundamental impediment to Lamarck's conception of perpetual evolutionary modifications. In *Leçons* Cuvier had observed of the conditions necessary for existence that if only one of an organism's "functions were modified in a manner incompatible with the modifications of the other functions, that being could not exist," and now, in the "Discours préliminaire," he applied the same reasoning to the correlation of parts in an organism, noting tersely: "None of its parts can change without the others changing too."[99] Cuvier pointedly refrained from spelling out the full implications of this laconic observation, perhaps not wishing to break the resounding silence with which Lamarck's *Philosophie zoologique* was first received.[100] The surrounding text nonetheless made it clear that the perfect and harmonious correlation of the animal frame, without which a creature could not continue to exist, would require holistically coordinated alterations that were so complex as to render evolutionary change untenable. After all, even Lamarck could not conceive of a mechanism that would facilitate such intricately coordinated modifications across the entire organism.[101]

While Cuvier acknowledged that organic variations did occur, he insisted that, even in the case of domesticated dogs, these remained strictly limited, and, in effect, the inconceivability of the holistic and composite alterations necessitated by correlation ensured the fixity of species. It was left coyly implicit in the "Discours préliminaire," but the ramifications were certainly apparent to Cuvier's contemporaries in Paris. As Serres later recalled of the "principle of the correlation of forms":

> This principle, being based on the idea that the organs of an animal form an entire whole . . . it resulted that there could be no modification in any of these organs without producing corresponding and analogous results in all the others. . . . The consequence of this principle . . . was . . . to stop in its development the theory of evolutions, according to which slight transformations are allowed to take place in the organisms.

For Serres, working on embryological recapitulation, this impediment to progressive development represented the decisive moment when "new principles" became necessary, by which he meant Geoffroy's "principle of organic analogies" and its postulate of a unity of composition that could be readily accommodated with Lamarckian transformism.[102]

But to others, and particularly in Britain, it was precisely the same anti-evolutionary "consequence of this principle" that rendered it so peculiarly attractive. Cuvierian correlation, as E. S. Russell has contended, was a "regulative and conservative principle which lays down limits beyond which variation may not stray."[103] This seems to have been recognized almost as soon as the first copies of Cuvier's *Recherches* reached British shores. By 1814 the Scottish evangelical clergyman Thomas Chalmers, in a review of the initial English translation of the "Discours préliminaire," was noting approvingly of Cuvier: "The transition of the genera into one another is most ably and conclusively contended against by the author before us, who proves them to be separated by permanent and invincible barriers."[104] It would be almost another fifty years before Charles Darwin's *On the Origin of Species* (1859) examined the possibility of nonadaptational correlative changes being effected by natural selection modifying any one part. During the intervening half century, Cuvier's laconic riposte to Lamarck in the "Discours préliminaire" was, as will be seen in chapters 2, 3, and 7, regularly invoked, and augmented, in Britain as the most effective bulwark against domestic versions of transformism.

The Almighty Trumpet

The sentence in which Cuvier contended that no part could change without all the others being altered too concluded by asserting: "and consequently each of them . . . indicates and gives all the others." His imperiously succinct refutation of Lamarck segued into yet another reassertion of the now famous claim that a single part could yield the whole. In fact, the "Discours préliminaire" was not simply concerned with countering the claims of Cuvier's rivals, and it also elaborated the grounds for his distinctive method of examining fossil remains in unparalleled detail and with new rhetorical force. Although, as noted earlier, Cuvier was generally disdainful about addressing audiences beyond his own colleagues and peers, Corsi has proposed that the "Discours préliminaire" was based on a course of lectures Cuvier had given at the Athénée de Paris, a private institution funded by subscriptions, that were attended by "high-society people"—women as well as men—rather than the usual scientific specialists.[105] This atypical audience is certainly indicated by the new, more demonstrative manner Cuvier adopted in the "Discours préliminaire." He addressed his readers directly in the first person (even ironically acknowledging that he would test the "steadfastness of the reader" with the "arduous paths on which I am obliged to take him!"), cultivating a style that was deeply personal as well as accessible and engaging. In describing his technical pursuits, Cuvier's language was vivid and immediate, and often disarmingly colloquial. The study of fossil creatures, for instance, "bristles with . . . difficulties," but the law of correlation could make "these obstacles vanish," allowing an animal to be "recognized, at a pinch, from any fragment . . . of its parts." Indeed, with only "isolated bones, scattered higgledy-piggledy," the anatomist could "reach details that are astonishing."[106] With such descriptions, Cuvier invested the act of discovering new fossil creatures from just fragmentary parts of their remains with an excitement and energy, as well as a seeming simplicity and spontaneity, which ensured that the "Discours préliminaire" attracted the attention of audiences far beyond the anatomical specialists who read the more somber technical papers in the subsequent volumes of *Recherches*.[107]

Within the first few pages of the "Discours préliminaire," readers encountered a rehearsal of its argument concerning "fossil bones" in which Cuvier pledged:

> I shall expound the principles underlying the art of identifying these bones, or, in other words, of recognizing a genus and distinguishing a species from a single fragment of bone: on the certitude of this art rests that of the whole work.[108]

The worth of the entire four volumes of *Recherches*, which in Britain cost the colossal sum of eight guineas, was, in a gesture of extraordinary bravado, wagered on the dependability of the very understanding of animal structure that, over the preceding few years, had been contested by several of Cuvier's most prominent scientific colleagues. In fact, the veracity of all the other revolutionary scientific findings announced in Cuvier's magnum opus— including extinction, the vast extent of the earth's geohistory, and the rebuttal of Lamarckian transformism—was, from the start, made dependent on his vaunted ability to identify unknown creatures from single fragments of bone, something he described as an "art" rather than an exact procedure.

Hitherto, Cuvier had refrained, as in his 1804 paper on the Parisian opossum, from investing the anatomical doctrines he applied to fossil remains with the status of laws, instead referring to them simply as "zoological theories" (the invocation of the "laws of coexistence" in *Leçons* had not referred to fossils). Similarly, he had previously proposed, in his 1798 lecture at the Institut national, that a creature could be restored or reconstructed, rather than merely identified, only if a large proportion of its bones were well known. In the "Discours préliminaire," Cuvier now cast these erstwhile reservations aside, boldly proclaiming: "The claw, the shoulder blade, the condyle, the femur, and all the other bones taken separately, determine the teeth, and each other reciprocally. Beginning with each of them in isolation, he who possesses rationally the laws of organic economy would be able to reconstruct the whole animal."[109] Adopting the terminology of both laws and reconstructing (*"lois"* and *"refaire"* in the original French), he insisted that knowledge of the a priori laws of animal structure enabled a reconstruction to be made from a single bone in isolation.

It was this particular statement, where "Cuvier's prose style tended towards the hyperbolic," that ensured, as Rudwick has observed, "that he was later misunderstood to be claiming an almost magical ability to *reconstruct* an entire animal from a single bone."[110] Stephen Jay Gould has similarly remarked that it was "Cuvier's overenthusiasm in the *Discours Préliminaire*" that "spawned the legend . . . that paleontologists can reconstruct entire skeletons from single bits of bone."[111] With Cuvier relinquishing his earlier reluctance to indulge in brash assertions of the anatomist's purported capacity to make hypothetical reconstructions, such putative legends and misunderstandings may not, strictly speaking, actually constitute misapprehensions of what he claimed. In any case, myths and misconceptions are not without their own significance, and the shift in Cuvier's rhetorical style that Rudwick and Gould scorn as mere hyperbole and overenthusiasm had a decisive impact on the future fate of the law of correlation. Indeed, Cuvier's new ebullience, as subsequent chapters will show, helped transform it from a specialist concern of the anatomical experts and medical students who read *Leçons* or the papers in *Annales du Muséum* to one of the most celebrated—or at least widely invoked—scientific axioms of the nineteenth century.

Mindful of addressing a nonspecialist readership akin to the cultivated "high-society people" who attended his lectures at the *athénée*, Cuvier, in the "Discours préliminaire," also imbued his technical method of reconstructing extinct creatures with literary resonances. The laws that facilitated fossil reconstructions were so constant and enlightening that, in certain circumstances, tangible osseous remains were not actually required. Even when there was not the merest fragment of bone, Cuvier proposed, just the "imprint" or "track of a cloven hoof" allowed a "conclusion" that it had been made by a ruminant that was "quite as certain as any other in physics." With the same mathematical certainty that Cuvier had attributed to the relations between bones, this "single track," he insisted, was actually "a more certain mark than all those of Zadig."[112] In Voltaire's philosophical novel *Zadig* (1748), the eponymous ancient Babylonian sage is able to identify the precise characteristics of the queen's pet spaniel and the king's finest horse, both of which have escaped, having seen only their respective tracks. This remarkable feat, after an initial misunderstanding, eventually wins Zadig the respect of members of the court, who appoint him prime minister. Cuvier's putative powers were, of course, even greater than the fictional abilities of Zadig, enabling him to identify hitherto unknown extinct animals rather than just domestic pets, and then to actually reconstruct them from the most fragmentary traces of their past existence. Cuvier's scientific claims in the "Discours préliminaire" surpassed even the imaginative possibilities of Voltaire's orientalist fantasy.

Later in *Zadig*, the protagonist is chosen, above all other mortals, to hear the providential mysteries of fate from an angel who has been disguised as a hermit (even if, ironically, the same angel warns that "men were wrong to pass judgement on a whole of which they perceive only the smallest part").[113] Cuvier's avowal of his own access to "details that are astonishing" had already hinted at something no less preternatural in his anatomical powers, and readers who progressed from the first volume's "Discours préliminaire" to the third tome of *Recherches* were regaled with a still more prophetic description of how, with the extinct mammals of Montmartre, Cuvier's reconstructive abilities had enacted "almost a resurrection in miniature, and I did not have the almighty trumpet at my disposal." Eliding the eschatological trumpet that would raise the dead incorruptible at the day of judgment (1 Corinthians 15:52) with the inspired word of God that entered the prophet Ezekiel and enabled him to resurrect the dry bones of the Israelites (Ezekiel 37:1–10), Cuvier proclaimed that, rather than any divine agency sanctioned by Scripture, it was his "immutable laws" alone that ensured that "at the voice of comparative anatomy each bone—each fragment of bone—took its place again."[114]

In Britain these allusions to the prophetic language of Scripture were often assumed to signify Cuvier's theological orthodoxy. In Paris, however, his early reputation was actually more for religious irreverence.[115] Cuvier, who was born into the Lutheran denomination of Protestantism, generally maintained a scrupulous silence as to his own beliefs, only finally making religious references in his lectures during the spiritual revival in the later years of the empire, and in his published work following the restoration of the Bourbon monarchy.[116] Even then, his self-conscious invocations of the Creator remained conspicuous by their rarity, and, at most, evinced only a nominal deism. In fact, in 1808 official sponsorship for the printing of Cuvier's report on scientific education was withheld because of its lack of specific references to God and religion, and by 1824 he was still adhering to what he called his "principal rule to never go beyond the facts" and refusing to follow British geologists in equating the most geologically recent marine inundation with the Noahic Flood.[117] Despite assumptions about his intrinsic conservatism and orthodoxy (among historians as much as some of his contemporaries), Cuvier's theological and political outlook, according to Outram, was that of a "cosmopolitan liberal."[118] His rather motley fusion of passages from both Old and New Testaments certainly implied that the anatomist could, with an exclusively scientific understanding of natural laws, supplant the Deity and assume an almost divine potency for recalling the dead to their original forms. It was no less noticeable that, in the "Discours préliminaire," the Bible's account of the universal Deluge was treated as merely one of many textual sources from the Near East recounting stories of a great flood, none of which were privileged. The same relativism was also apparent in Cuvier's use of tropes both from Scripture and the literary archetype of *Zadig*—whose author was a notorious Enlightenment deist—to depict the powers of the comparative anatomist. Like Voltaire's fantastic novel, the Bible was another resonant literary source that invested the act of reconstruction with a supernatural mystique and vivid sense of drama.

These same literary qualities were also recognized by the novelist Honoré de Balzac. In his philosophical romance *La peau de chagrin* (1831), the narrator poses the rhetorical question "Is not Cuvier the greatest poet of our century?" before avowing:

> Certainly Lord Byron has expressed in words some aspects of our spiritual turmoil; but our immortal natural historian has reconstructed worlds from bleached bones. . . . He digs out a fragment of gypsum, descries a footprint in it and cries out: "Behold!" And suddenly marble turns into animals, dead things live anew and lost worlds are unfolded before us! . . . This awesome resurrection, due to the voice of a single man . . . seems pitiable.

Drawing directly on *Recherches*, which he calls the "geological treatises of Cuvier," Balzac's narrator explicitly conflates the language of the separate passages invoking *Zadig* ("a footprint") and Corinthians and Ezekiel ("resurrection, due to the voice"), making Cuvier's own relativism even more strikingly manifest.[119] After all, the inspired voice that, akin to the prophet Ezekiel, enacts this awesome resurrection evokes a pity comparable to that elicited by the poetry of Byron, who, a decade before Balzac's novel, had transformed what he called the "notion of Cuvier . . . derived from the . . . bones of enormous and unknown animals . . . that the world had been destroyed several times before the creation of man" into a demonic "assertion of Lucifer" in his controversial verse-drama *Cain* (1821).[120] Byron, of course, was also infamous for the sensuousness of his poetry as well as his scandalous personal life, and in *La peau de chagrin* the propriety of Cuvier's invocation of Ezekiel becomes still more questionable when, during a dissipated banquet, Balzac's narrator observes of an especially voluptuous courtesan who resembles a "sybil possessed by a demon" that a "glance from her might well bring dry bones to life" (the rather ribald joke seems deliberate). Intriguingly, later in the novel its aristocratic antihero, Raphael de Valentin, remarks that if a "new Messiah" were to appear in this "century of enlightenment," the Parisian authorities would "refer his miracles to the Académie des Sciences," where, presumably, they would pale in comparison with the secular—and perhaps immoral or even diabolic—resurrections of Cuvier, who was the *académie*'s perpetual secretary.[121]

Where Theory Fails

Alongside the overwrought rhetoric of the "Discours préliminaire," Cuvier had actually acknowledged that the reasons for some structural correlations were "less clear" than for others, especially when dealing with the numerous subdivisions of a particular class. The pragmatic solution was that "where theory fails observation must provide," as it could establish "empirical laws that are almost as certain as rational ones, when . . . based on sufficiently repeated observations." In "adopting the method of observation as a supplementary means," Cuvier proposed, an anatomist could recognize "a specific constancy . . . between a certain form of a certain organ, and another form of a different organ" even where the particular function that this constant correlation performed was not known. By collecting such observations together in a "general system of these relations," the same anatomist could establish a "standard constancy" that was almost as reliable as one determined by "actual reasoning."[122] The assiduous notation of customary correspondences among different elements in the animal economy could, for instance, ascertain that cloven hoofs almost invariably accompany ruminant dentition, and thus might even allow limited inferences to be made from certain representative bones.

What such a posteriori empirical observations, no matter how many times they were repeated, could never establish, though, was the rational a priori anatomical principles that Cuvier had continually compared with the laws of mathematics. And it was the mastery of these invariable organic laws, rather than the more prosaic alternative of endlessly repeated observations, that, in the often fervid prose of *Recherches*, afforded the almost preternatural capacity to reconstruct long-vanished and hitherto unknown animal structures from just a single fragmentary part. After all, an empirical method that relied on observing the organization of known living creatures would not necessarily be applicable to the as yet unspecified organic structures of the prehistoric past. In any case, the success of what Cuvier termed the "method of observation" often depended upon what he rather coyly called the "aid of a little appeal to analogy and effective comparison."[123] This required access to a vast array of cognate skeletal parts such as Cuvier himself enjoyed at the Muséum d'histoire naturelle, where he received the cadavers commandeered by his brother Frédéric from the neighboring menagerie, but which would not be so readily available to less well-connected naturalists.

The necessity of comparing numerous osteological specimens to establish recurrent morphological characteristics that Cuvier had previously acknowledged in his private letter to Autenrieth was now publicly avowed, in the "Discours préliminaire," as an essential means of corroborating his purportedly a priori scientific laws. In this light, as Coleman has contended, Cuvier's "'principles' were really rationalizations after the fact," and they actually "depended upon experience" to be successfully utilized.[124] Notably, though, these concessions that pragmatic observation was a necessary component of his method for identifying and reconstructing fragmentary fossil remains appeared alongside Cuvier's most ebullient and hyperbolic assertions of the same method's infallible powers. As will be seen in chapter 6, this curious combination of posturing and pragmatism in *Recherches* would have significant implications for its reception in Britain throughout the rest of the nineteenth century.

Armour to Defend His Faith

Although Britain and France remained at war, in 1810 the French authorities, after four years of the cross-Channel commercial blockade, agreed to once more accept importations of British colonial products, on the condition that French luxury items of commensurate value were exported to Britain. The expensive books that had been stockpiled in French warehouses during the British embargo were the perfect means to effect this balance of trade, and by 1812 the weight of French publications imported into Britain had returned to prewar levels.[125] Few French books were as luxurious as Cuvier's *Recherches*, and, having been published in Paris at the end of 1812, copies had crossed the Channel by the early summer of the following year. Its huge price, especially for what was effectively a collection of reprinted papers, restricted its circulation even among wealthy francophone men of science, and, unlike the earlier *Leçons*, the original edition of *Recherches* was not reviewed or extracted in British periodicals. By November 1813, however, an English translation of the "Discours préliminaire" had made Cuvier's introduction to his four-volume work considerably more accessible to British readers than to those in France or anywhere else.

This translation, as a notice in the *Edinburgh Review* observed, was "made with singular expedition," although the translator, the Scottish surgeon Robert Kerr, had died suddenly before it reached the printers.[126] The work was seen through the press by Robert Jameson, who added a preface, notes, and other supplementary materials and also gave it a contentious new title, *Essay on the Theory of the Earth*, which implicated the text with the very genre of speculative geotheory that Cuvier had long regarded with derision.[127] A decade earlier, Jameson had been appointed as Regius Professor of Natural History at Edinburgh with the support of the city's Tory lord advocate and the university's conservative patrons. The defeated rival candidate, who had been backed by the Scottish capital's Whig reformers, was John Allen, the extramural lecturer whose translation of *Leçons* had discerned parallels between Cuvier's understanding of animal structure and the radical political agenda of Edinburgh's liberals.[128] As with Allen's abridgment of only the initial tome of Cuvier's two-volume original in his *An Introduction to the Study of Animal Economy*, Jameson's edition of the "Discours préliminaire" also detached—and thereby recontextualized—the part of *Recherches* that was considered of most significance. Priced at six shillings, it similarly presented this extracted text to anglophone readers at just a fraction of the complete work's cost, ensuring, as Jameson noted approvingly, that it was "readily accessible not only to the naturalist, but also to the general reader."[129] As their respective supporters in the contest for the Regius Professorship might suggest, though, the purposes of Jameson's *Essay* were otherwise entirely different from Allen's radical appropriation of Cuvier's earlier work.

Advertisements in Scottish newspapers announcing the publication of the *Essay* advised potential purchasers that the "Christian may furnish himself from this production of a Parisian philosopher, with armour to defend his faith against those writers who have endeavoured to overturn it."[130] In his preface Jameson pledged that the translation would indeed "admonish the sceptic, and afford the highest pleasure to those who delight in illustrating the truth of the Sacred Writings, by an appeal to the facts and reasoning of natural history." As Byron mockingly demonstrated by having Lucifer ventriloquize Cuvier's words in *Cain*, many Christian denominations were increasingly apprehensive at modern geology's emphasis on the vast age of the earth and the reality of mass extinction.[131] Tailoring his *Essay* for precisely such conservative audiences, Jameson reassuringly urged that Cuvier's consideration of the fossils of extinct species "naturally leads our author to state the proofs . . . of the deluge," ensuring that the historical truth of "one of the grandest natural events described in the Bible, is . . . confirmed." In the French original, of course, Cuvier had treated the biblical account of the flood as merely one of many historical records of a recent geological catastrophe, but for Jameson the issue of "what regards the *deluge*" and the "many direct evidences of the truth of . . . scripture" became the most significant feature of the translated text.[132]

Although Jameson himself belonged to the moderate faction in the Presbyterian Church of Scotland, who relied on natural theological proofs of providential intention, his Regius Professorship required him to publicly avow his support for the Westminster Confession of Faith, which began by asserting the primacy of scriptural revelation.[133] As one of his patrons had noted during the contest for the post, "supporting by his doctrines the interests of piety & public virtue" was a necessary duty of "a Professor of Natural History," and Jameson's successful candidacy had benefited from his "sound . . . principles in both morals & Religion."[134] (Allen, on the other hand, was considered scientifically superior, although religiously far from appropriate.) During Jameson's first decade in the job, the need for such public demonstrations of "piety & . . . virtue" had only increased amid the continuing national emergency of the Napoleonic Wars, and the growing authority of the evangelical faction, with its emphasis on revealed theology, within the Presbyterian Kirk. While the "late Mr Kerr" had already added his own footnotes concerning sacred chronology before being "snatched from this transitory scene," Jameson went still further and repackaged Cuvier's geological evidence for a great marine inundation as a scientific verification of the Bible's infallible authority.[135] In crossing the Channel, Cuvier's religious relativism and irreverence were thrown overboard, rendering his "Discours préliminaire" compatible with the pious sensibilities of the Edinburgh establishment.

When Jameson's preface to the *Essay* finally moved on from the "enquiries" into the "*deluge*," which, he claimed, "form a principal object of the Essay of Cuvier, now presented to the English reader," it acknowledged that the translated text also "explains the principles on which is founded the art of . . . discovering a genus, and of distinguishing a species, by a single fragment of bone."[136] This, however, curiously underplayed the audacious claims that Cuvier had actually made in the "Discours préliminaire." Where Cuvier's narration insisted that knowledge of the rational laws of organic structure enabled the entire animal to be reconstructed from just a single bone, Jameson's prefatory observations, which, coming before any of Cuvier's own words, would have shaped how Kerr's ensuing translation was read, never alluded to this vaunted capacity for reconstruction. He instead proposed only that isolated fragments could facilitate identifications of species and genus. This divergence between what Jameson claimed for Cuvier and what the French savant had actually said was certainly apparent to readers of the *Essay*, with a reviewer for the *Philosophical Magazine* complaining, in regard to the "precise points of agreement" between geology and Scripture touted in the preface, "I have read the performance of Cuvier with much interest, but without being able to discover the promised agreement."[137] The same discrepancy between the preface and Kerr's translated text was also evident in Jameson's preference for terming what the deceased translator had rendered as "laws" of organic structure as merely "principles."

An influential review of Jameson's *Essay* in the *Edinburgh Christian Instructor* perhaps indicates why what, as Kerr's translation maintained, was "an art on the certainty of which depends that of the whole work," remained of relatively little concern to its editor.[138] The anonymous review, an early work of the evangelical theologian Thomas Chalmers, acknowledged the "qualifications of M. Cuvier as a comparative anatomist," and diverged from Jameson's preface in proposing that the "most amusing, and perhaps the soundest argument in the whole book, is that by which he unfolds his method of constructing the entire animal from some small and solitary fragment of its skeleton." Having been "highly gratified" with the "discussion upon this subject," Chalmers could not "resist the desire of imparting the same gratification to our readers" by quoting a lengthy portion from Kerr's translation. Chalmers immediately conceded, however, that it would have little impact on the "geological infidels of the day," who would simply "exclaim, that, though M. Cuvier be a good anatomist, it does not follow that he is a geologist." The "knife and demonstrations of the anatomist," no matter how ingenious or compelling, could do nothing to convince those whom the "hammer of the mineralogist and the reveries of the geologian" had induced to question their faith.[139]

For Chalmers, it was the Bible's "revealed history of God's administrations in the world" and the "authority of the sacred historian" that afforded the most effective refutation of the "fanciful and ever-varying interpretations" of the "antimosaical philosophers." While differing on some matters of interpretation, his review of the *Essay* deigned to "thank the respectable editor of this work, Mr. Jameson, for his becoming deference to the authority of the Jewish legislator [i.e., Moses], and his no less becoming and manly expression of it." Chalmers also accepted that, as Jameson contended, some of Cuvier's geological evidence regarding the replacement of extinct species provided the "very precious fruit" of a further "argument for the exercise of a creative power," although without yielding any knowledge of God's actual attributes. Even such limited arguments for the existence of a deity, by contrast, could never be "drawn" from what Chalmers dismissed as the "slender resources of natural theism."[140] Such natural theism could not afford a foundation for

Christian belief, which could come only from Scripture and the revelation of Christ's atonement, and instead its slender resources actually risked opening the way for mere deism.[141] South of the River Tweed, and principally in Oxford and Cambridge, Cuvier's view of the perfect correlation of animal structures would, as the following chapter will show, soon become integral to precisely the natural theological arguments that Chalmers considered so ineffectual. In Presbyterian Edinburgh, where the rational verification of scriptural revelation took precedence over the English universities' argument from design, Cuvier's purported capacity to reconstruct extinct creatures from just fragmentary bones might, as it was in Chalmers's review, be deemed quaintly amusing, gratifying, and interesting. It was nonetheless theologically inconsequential and thus of little significance to Jameson.

Translated, Copied, and Appropriated

While Jameson's preface to the *Essay* moderated Cuvier's hyperbolic rhetoric, it still ensured that his renowned "art of ascertaining . . . the fossil bones of land animals"—some of which were putatively drowned during the Deluge—was included in an argument for the accuracy of the "Mosaic account of the creation of the world" that, within fifteen years, had sold more than six thousand copies.[142] The ability to infer the integrated whole from a single part was therefore denuded of some of the seditious and irreligious connotations that Allen's *Introduction* had ascribed to it a decade earlier (although not all, as will be seen in the next chapter). As an obituary of Jameson in the *Edinburgh New Philosophical Journal* observed of the *Essay*, "This elegant and popular volume produced an excellent effect in our country. Cuvier, with all his genius and fame, was but partially heard of in Britain till this essay appeared, and it made him more familiar to our libraries than all his own invaluable writings put together."[143] Having vanquished Allen in the ideologically charged contest for the Regius Professorship back in 1804, Jameson, who had attended Allen's lectures in the 1790s, now presented a rival version of Cuvier that once again effaced his opponent's earlier scientific and political standing in Edinburgh.[144]

By the 1820s another of Edinburgh's extramural lecturers, Robert Knox, was complaining that the "works of the great Cuvier" were "studied, read, and got by rote" and "so many had translated, copied, and appropriated his views to themselves."[145] In London the publisher George Byrom Whittaker felt confident there was a market for ambitious serialized translations of both *Recherches* and *Le règne animal* (albeit with only the latter eventually reaching completion), and back in Edinburgh even Knox's closest colleague Robert

Edmond Grant translated *Le règne animal*, although it remained unpublished and its author was already switching his loyalty to Lamarck's transformism.[146] Knox had taken over John Barclay's anatomy school on Surgeons' Square, but his protest against the "Cuvierian school and its followers" shared none of his predecessor's revulsion at Cuvier's alleged materialism.[147] Rather, Knox was more radical even than Allen, and it was the appropriation of Cuvier by what he called "theologico-geologists" that prompted his ire.[148] His acerbic censure of Cuvier's British advocates, beginning in the extramural classrooms of Edinburgh, would continue for a further forty years and become hugely significant in later disputes over correlation. As such, Knox's enduring enmity for Cuvier and his followers will be considered in detail in chapter 8.

Knox's complaints, however, did little to hinder the continued success of Jameson's theologically inflected edition of the "Discours préliminaire." As Cuvier himself commented with evident satisfaction, "This discourse was in England the subject of special favour; it was reprinted four times in English and twice in America."[149] The editor of the American reprint of the *Essay*, Samuel Latham Mitchill, advised readers that this work "by a happy genius in France, comes to us, recommended and improved by the talents of a leading naturalist in Scotland."[150] Mitchill, whose claim that Cuvier had boasted "Give me the bone, and I will describe the animal" was discussed in the introduction, also added his own "brief memorandum concerning American fossils" discovered in the "boundless field for investigation" opened up by the young republic's "westward" expansion.[151] The version of Cuvier that, having crossed the Channel in the final years of the Napoleonic Wars, Jameson tailored to the particular religious sensibilities of Edinburgh, was now spreading across the Atlantic and to the farthest frontiers of Western civilization.

Conclusion

When Jameson was preparing a second edition of the *Essay* early in 1815, the protracted conflict between Britain and France reached a climax after Napoleon escaped from confinement on the island of Elba and marched a hastily assembled army into Belgium. The resumption of the Napoleonic Wars prompted Jameson to amend his preface, strategically repatriating the author of the "Discours préliminaire" as the "illustrious German naturalist Cuvier." A footnote explained that "Cuvier is a native of Mumpelgardt in Germany," although, as Jameson was surely aware, the city of Cuvier's birth had been annexed by France, and renamed Montbéliard, more than two decades before.[152] After Napoleon's final defeat at the Battle of Waterloo, Jameson could silently delete this specious addendum from the 1817 third edition.

Cuvier's liminal nationality, as well as Montbéliard's status as a Lutheran en-
clave within the nominally Catholic French state, continued to be valuable
resources for those who remained anxious at his close associations with the
defeated French emperor.

He was a "conspicuous dissenter from the national religion," as the Nonconformist *London Medical and Surgical Journal* noted after Cuvier's death in
1832, and its obituary closed by insisting proudly: "Baron Cuvier lived and
died a Protestant."[153] As a Protestant born outside of France's prerevolutionary borders, Cuvier could be domesticated as a kind of British—or at least
non-French—hero. Indeed, the most fulsome of the posthumous memorials to the internationally renowned savant, Sarah Bowdich Lee's *Memoirs of
Baron Cuvier* (1833), was initially published in English before being translated into French. Cuvier, Lee insisted, was "always the mediator . . . between
France and other nations; he resisted the antipathy of his countrymen against
those whom they chose to call barbarous; and with his whole force tried to
stem the torrent which their vanity . . . occasionally poured over that which
was wise and useful." Not only were the Gallic excesses that had precipitated
two decades of war not shared by Cuvier, but, despite his administrative efforts on behalf of the Napoleonic government, he had actually restrained the
national fervor for a "constant change of systems . . . calculated rather to
destroy than to improve."[154]

By the fifth and final edition of Jameson's *Essay* in 1827, its editor could
reflect: "In this country CUVIER was first made known as a geologist by the
publication of the present essay, which, from its unexampled popularity,
has made his name as familiar to us as that of the most distinguished of our
own writers."[155] Journals such as the *Monthly Magazine*, as was noted earlier,
had already justified including reviews of the translations of *Leçons* under
the heading "Domestic Literature." The domesticated version of Cuvier that
had become so familiar to British readers did not, of course, necessarily resemble the savant who, across the Channel, remained ensconced at the heart
of the French establishment during the regimes of both Napoleon and then
the restored Bourbon monarchy. Rather, Cuvier's works were successively repackaged for divergent British audiences during the opening decades of the
nineteenth century, from the medical students and radical Edinburgh liberals
at whom Allen aimed his *Introduction* to the godly readers anxious at Continental infidelity who were offered reassurance by Jameson's *Essay*.

Ironically, Jameson himself had, by the mid-1820s, altered his view of
Cuvier's Parisian rival Lamarck, and begun to secretly endorse the very theories of species change that were so brusquely rebuked in the "Discours préliminaire."[156] The most vociferous advocates of Cuvier's theological creden-

tials were now to be found south of the Tweed in the conservative Anglican seminaries of Oxford and Cambridge, as well as among educationalists in London committed to diffusing religiously reputable forms of science to new working-class readerships. What they had in common was an adherence to the argument from design that Chalmers, in Edinburgh, had considered so ineffectual (although he would later become more amenable to natural theological arguments concerning human conscience).[157] While the Scottish Presbyterians Chalmers and Jameson both marginalized Cuvier's law of correlation, the English proponents of natural theology embraced it as an irrefutable proof of the divine design of animal structures. With the endorsement of such influential supporters, who switched from producing translations of Cuvier's writings to appropriating his distinctive methods for their own researches, the purported capacity to reconstruct extinct creatures from just a single bone became, as the next chapter will show, more firmly established in Britain than in its native France.

Fragments of Design

In the first week of April 1829, Edinburgh's Wernerian Natural History Society conducted an "experiment" in which, as one of its members recorded, the "common domestic pig was the subject of our observations." In order to "ascertain the impression made by the feet of these animals," a specially appointed committee "caused them to walk across a board spread over with soft clay." However, "owing to the unruly nature of the animals," the committee could not "succeed in making them walk" on the board, and instead the pigs were "let out into a court" that already contained sufficient "soil" (a term that in the nineteenth century often implied the presence of excrement) to record the porcine footprints.[1] This grubby and unruly experiment, which established that pig's feet occasionally left bisulcated—or cloven—prints, had been prompted by one of the most pressing issues in early nineteenth-century British science: the reliability of Georges Cuvier's law of correlation. It was, as a Scottish commentator later remarked, "a strange and humbling accident that had reduced the settlement of such an important theory to a question of pettitoes."[2] What was soon dubbed the "pig's-foot controversy" was sparked by the naturalist and evangelical clergyman John Fleming, who, in March 1829, told the Wernerian Society, and subsequently the readers of the *Edinburgh New Philosophical Journal*, that "Baron Cuvier . . . under the influence of prejudices which few can avoid, has stated his confidence in the certainty of his deductions, with a boldness, which is the more astonishing, as it is equally at war with his own admissions and well-known facts."[3] Having intuited the ambivalences in Cuvier's hyperbolic assertions of his capacity to identify creatures from only isolated parts of their anatomy discussed in the previous chapter, Fleming gave a specific example where the French savant's vaunted method would lead naturalists astray.

Whereas Cuvier had claimed that the imprint of a cloven hoof enabled a conclusion that it had been made by a ruminant that was as certain as any deduction in physics, Fleming insisted that actual experience invalidated such generalizations. He averred:

> Observation had discovered many animals with cloven hoofs which ruminated; but, in such circumstances, would it be safe to infer that all cloven-hoofed animals ruminate? Conceive ourselves contemplating the footmarks of a sheep and sow. Under the guidance of Cuvier's demonstrations, we would conclude that both ruminated,—an inference true in the one case, and false in the other. Observation here warns us against the employment of a guide so liable to deceive us.[4]

Fleming did not attribute much significance to such "silly gasconading" over the ability to predict animal structures, and he contemptuously asserted: "I dare to call nonsense by its true name, even when uttered by a Cuvier."[5] South of the River Tweed, however, Cuvierian correlation was taken much more seriously. The geologist and Anglican clergyman William Daniel Conybeare could not "pass, without notice, one of Dr. Fleming's illustrations, where he taxes Cuvier with inaccuracy (assuredly using sufficient philosophical boldness)." As Conybeare observed with patrician condescension:

> Living in the country, I myself am in the habit of keeping some . . . "residuary legatees of all other animals", as a friend calls them. Now, my pigs are not bisulcous. . . . The impression of their feet in walking may, if carefully examined (as Cuvier says), be distinguished from the genuine bisulca. I take it for granted, however, that Dr Fleming possesses bisulcous pigs.[6]

Unlike the gentlemanly Conybeare, Fleming did not keep his own pigs. It was this which had necessitated the Wernerian Society's experiment in a squalid Edinburgh courtyard.

If this particular dispute between "Scottish naturalists" who "had not been wont so to study pigs' feet" and southern "country parsons" who "kept porkers" was never satisfactorily resolved, then Fleming was keen to remind Conybeare of a previous controversy that had decisively gone his way.[7] Three years earlier, it was another of Fleming's contributions to the *Edinburgh New Philosophical Journal* that, he claimed, "gave the death-blow to the diluvian hypothesis." In fact, Fleming proposed, the "period is probably not far distant, when the 'Reliquiæ diluvianæ' of the Oxonian geologist will be quoted as an example of the *idola specus* [idols of the cave]."[8] The target of Fleming's punning invective was William Buckland, whose *Reliquiæ Diluvianæ* (1823) had used the fossil remains washed into caves to augment Robert Jameson's contention, in *Essay on the Theory of the Earth* (1813), that Cuvier's account of a recent catastrophic marine inundation was compatible with the Noahic Flood recorded in the Bible.

Fleming had studied under Jameson in Edinburgh, but, by the early 1820s, it was south of the border, and in Oxford especially, that his erstwhile mentor's theologically orthodox interpretation of Cuvierian geology was most favorably received. From his remote Scottish parish, Fleming "carried the war into the enemy's country," incurring the "hatred of the Bucklandian school" for his insistence that past events should be examined only in relation to natural processes that were currently observable.[9] As editor of the *Edinburgh New Philosophical Journal*, Jameson was quick to recognize that "Buckland has been frightened out of his Deluge—by Fleming," although he himself was still "not disposed to give up the Mosaic Deluge—even geologically considered." His former student, by contrast, was exultant that the "Oxonian will have a hard morsel to chew," especially as a planned third edition of *Reliquiæ Diluvianæ* now had to be scrapped.[10]

In a footnote to his *Geology and Mineralogy Considered with Reference to Natural Theology* (1836), Buckland quietly retracted his earlier claim that Cuvier's "violent inundation" could be equated with the more "tranquil inundation described in the Inspired Narrative."[11] It was nonetheless still imperative for Buckland, amid the hallowed purlieus of Oxford, to maintain the religious credentials of his scientific pursuits. With Fleming, and then Charles Lyell, repudiating both Cuvier's catastrophism and its putative endorsement of the biblical Deluge, Buckland instead turned from the French naturalist's geology to the functionalist anatomy he first adumbrated in *Leçons d'anatomie comparée* (1800–1805). The perfect harmony of animal structures revealed by the law of correlation afforded a new impetus to the long-standing natural theological argument from design, and, with the identification of even hitherto unknown extinct species facilitated by just fragmentary fossils, extended a teleological understanding of organization into the prehistoric past. Among Edinburgh Presbyterians such as the evangelical Thomas Chalmers and even the more moderate Jameson, Cuvierian correlation was, as noted in the previous chapter, theologically inconsequential and therefore marginalized. Fleming too agreed, in a review of Jameson's *Essay*, to "pass over the very slender sketch here given of the methods of ascertaining species and genera of quadrupeds from detached parts."[12] For liberal Anglicans such as Buckland, as well as for his clerical counterparts at Cambridge, Cuvier's allegedly "very slender sketch" of his anatomical methods became increasingly vital to the reconciliation of science and religion, and with it the maintenance of the entire social order at a time of growing protest and upheaval.

As a graduate of Christ Church College, Conybeare apologized ironically to Fleming for the "narrow system of Oxford logic in which I have unfortunately been trained."[13] It was the ancient university's distinctive logic that compelled him to rush to Cuvier's defense over the particular configuration of porcine footprints, although he and Buckland soon realized that their Scottish adversary's fiery impertinence toward the renowned French savant had inadvertently handed them a strategic advantage. Sharing his compatriot's uniformitarian inclinations, Lyell relished the "sport" of Fleming's "anti-Bucklandite warfare." All the same, he warned that he "might have spared Cuvier," whose legions of "worshippers" in England meant "it would be a good policy to be more courteous towards" him. Lyell had heard that the "Oxonians made just the same remark, congratulating Buckland that Monsieur le Baron had been put on the pillory with him, and had come off with perhaps the greatest expressions of contempt."[14] While also eliciting laughter and bemusement, the bathetic "pig's-foot controversy" served to further identify Buckland with the most esteemed naturalist in Europe. He was soon hailed as "the English Cuvier," a mantle that was subsequently, and more enduringly, assumed by his brilliant protégé Richard Owen.[15] Although Owen had not been a student of Buckland's, his anatomical skills, honed in the medical schools of Edinburgh and London, would, by the mid-1830s, become integral to the natural theological science taught at Oxford, as well as that promoted by William Whewell at Cambridge.

As in Edinburgh during the initial two decades of the nineteenth century, though, English responses to Cuvier's law of correlation diverged dramatically in the 1820s and 1830s. While Buckland and other proponents of natural theology embraced it as an irrefutable proof of divine design, the same law was condemned as inherently irreligious by literalist interpreters of the Bible, for whom it was inconsistent with scriptural accounts of animal forms and habits. Notably, these biblical literalists actually preferred the transformist views of Cuvier's colleague and rival Jean-Baptiste Lamarck, which they considered more compatible with Scripture. Meanwhile, the materialist implications of correlation that were first discerned in Edinburgh at the turn of the century ensured that Cuvier, despite his own growing conservatism in Paris during the Bourbon restoration, was frequently invoked in the cheap newspapers and piratical reprints of proletarian radicals. The surprising and seemingly anomalous allegiances of such literalist Lamarckians and seditious Cuvierian materialists demonstrate the complexity and volatility of both British science and society in the early decades of the nineteenth century. They also cast significant doubt on attempts to align particular scientific approaches, especially those being imported from France in the early 1800s, with fixed ideological positions.[16]

By the close of the 1830s, Cuvier's law of correlation had more supporters in Britain than in its native France, and its now deceased discoverer was lionized as an icon of the conservative establishment by the influential gentlemen of science who controlled the elite forums of the scientific community. But during the tumultuous preceding two decades, this was by no means inevitable. The ownership of Cuvier's celebrated law of correlation, as this chapter will show, was fought over by radical advocates of materialism and clerical upholders of natural theology, with the battle extending from Oxford colleges and genteel lecture theaters to cheap pamphlets sold on street corners.

The Clumsy Fabric of Modern Materialism

Seven years before the "pig's-foot controversy," Fleming had sent a copy of his *The Philosophy of Zoology* (1822) to Cuvier in Paris. Although his understanding of English was only limited, the book's recipient soon realized that it was not intended as merely a respectful gift.[17] In a characteristically abrasive passage, Fleming avowed that the "whole history of the animal kingdom contradicts . . . expectations of *co-existing* characters," adding a lengthy footnote in which he remarked: "It is truly surprising to find such an observer as CUVIER, in the face of observations and his own experience, asserting the existence of this mutual dependence of the different organs." After quoting passages from *Leçons d'anatomie comparée*, Fleming continued:

> This specious reasoning, would certainly lead to the admission of these necessary laws of co-existence, were the statements advanced correct in all their bearings. But the operations of Nature are not restrained by such trammels. Quadrupeds possessing the common quality of being carnivorous, have not all the same number of teeth, nor of the same shape, neither the same kind of stomach. . . . Indeed, the number of varieties included under one species . . . intimate the variableness of the conditions of co-existence, and the absence of those supposed laws of relation, the belief in the *mathematical necessity* of which, has contributed to augment the clumsy fabric of modern Materialism.[18]

The aspiration to make his anatomical axioms as absolute and reliable as the laws of mathematics that had informed both *Leçons* and *Recherches sur les ossemens fossiles* (1812), which was dedicated to the renowned mathematician Pierre-Simon Laplace, was, Cuvier now read in the tome Fleming had sent him, instrumental in the rise of the materialist philosophy that had become synonymous with atheism, sedition, and amorality.

Cuvier quickly acknowledged Fleming's "testimony . . . of esteem," but adamantly refuted the insinuations of his footnote. He told him:

I might have wished, however, that you had more deeply or thoroughly understood my theory of the co-existences of organization, and the numerous applications which I have made of them in my work on fossil bones; you would then probably have recognized that my ideas on the subject are less separate from your own way of thinking, and especially you would have avoided representing them as supporting materialism.[19]

In another missive twenty years earlier, Cuvier had joked with his friend Charles de Villers about the attitude toward transubstantiation of those of his Parisian contemporaries he genially labeled "our materialists."[20] But with the horrors of the French Revolution increasingly attributed to the materialist doctrines of Enlightenment philosophers such as Baron d'Holbach, it became necessary for Cuvier, especially once he became a titled counselor of state to the restored Bourbon monarchy, to rigorously disassociate the scientific theories he had formulated in the immediate postrevolutionary period from the same pejorative designation. Cuvier had visited England in 1818, and, having met Buckland and several other prominent men of science, was aware that materialism carried similarly damaging connotations—with the added threat of foreign heterodoxy—on that side of the Channel too.

Although it had been the fulcrum of the initial English translations of his work, Cuvier did not travel north to Scotland, where Fleming was far from alone in ascribing such troubling views to him. An anonymous notice of *The Philosophy of Zoology* in the *New Edinburgh Review* reprinted the footnoted insinuation of augmenting materialism that had so alarmed Cuvier in Paris, with the reviewer approvingly remarking: "We are glad to see . . . a criticism on Cuvier, which we had made ourselves. We say glad, because of the dangerous weight which authority so high adds to error."[21] Fleming also discussed his anxieties over Cuvier's understanding of animal structure with many of his peers in the Scottish capital. He confided to Chalmers that he had "ever thought Cuvier a strange mixture of anatomical knowledge and geological ignorance seasoned with infidelity."[22] In 1822 the extramural anatomy lecturer John Barclay read the proofs of *The Philosophy of Zoology* and agreed with Fleming that Cuvier "seems to have placed too much confidence in structure. . . . Comparative anatomy falls here short of its object. . . . It can hardly predict from the structure and its organs the different instincts of the hare and the rabbit."[23] Barclay added that he had doubts about "what he [i.e., Cuvier] calls the phenomena of organic life," and, as seen in the previous chapter, he had scorned his fellow private anatomy teacher John Allen's abridged translation of *Leçons* for showing that "were any person capable of supporting . . . the doctrines of materialism, it would certainly be Cuvier."[24]

The perception that Cuvier advanced a view of life and structure that was manifestly materialistic was not confined to Edinburgh, though.

A Conspiring Band of Sceptics and French Physiologists

Concern over materialism had been heightened throughout Britain by a vituperative public debate at the Royal College of Surgeons in London's Lincoln's Inn Fields that had resulted in blasphemy charges at the neighboring Court of Chancery.[25] Unlike in Edinburgh, however, for some in this dispute it was the putative materialism of Cuvier's conception of organic structure that constituted its most compelling feature. The outspoken surgeon William Lawrence, in an introductory lecture at the Royal College of Surgeons in March 1816, proclaimed: "Our first step in the study of life is to examine the organs, that are its material instruments." What such an examination revealed, he continued, was that "all the various parts of an animal are so closely connected to each other, that this relation may be traced even in the minutest particulars." In the published form of his lecture, Lawrence added a footnote stating:

> The mutual relations of the organs, and the laws of co-existence, to which their combinations are subjected, are so well pointed out by Cuvier, that I quote his words, as the work in which they occur, is not so generally known in this country, as the other productions of this justly celebrated zoologist.[26]

The words that Lawrence quoted were his own translation of a passage from the "Discours préliminaire" to *Recherches*, which, despite the success of Jameson's *Essay*, was apparently not as familiar as Cuvier's other works, at least to members of the medical profession in London.

Lawrence was confident that his "audience" of the "venerable elders of our profession," as well as the "general body of London surgeons" and "students of several schools of medicine," would be better acquainted with *Leçons*.[27] It was to this "most comprehensive and philosophic production" that he turned in his next lecture a week later, in which he contended that it was simply the "functions of the living animal body" that "distinguish its history from that of dead matter." As Lawrence told his medical auditors: "It is justly observed by Cuvier that the idea of life is one of those general and obscure notions produced in us by observing a certain series of phenomena, possessing mutual relations."[28] Like the Teutonic "teleomechanism" and "vital materialism" that, as noted in the previous chapter, Cuvier's approach to the question of life closely resembled, such statements could be interpreted, with equal validity, in either teleological or materialist terms, with Lawrence clearly opting for the latter.[29] Rather than a superadded immaterial or spiritual essence, life was no more than a correlate of physical organization, which, as Cuvier seemed to suggest, was produced by the harmonious cooperation of all the body's constituent parts. Lawrence additionally proposed that consciousness itself was dependent on the material organization of the nervous system, a conclusion that Cuvier had explicitly disavowed in his criticisms of phrenology in 1808.[30] Whatever reservations might be expressed in Paris, Lawrence's lectures at the Royal College of Surgeons implicated Cuvier and his law of correlation in a conspicuous endorsement of materialism that would soon reverberate far beyond Lincoln's Inn Fields.

In 1817 Lawrence's former tutor John Abernethy angrily exclaimed: "That in England . . . the mere opinions of some French anatomists, with respect to the nature of life, should be extracted from their general writings, translated, and extolled, cannot, I think, but excite the surprise and indignation of any one fully apprized of their pernicious tendency." By invoking such Gallic heterodoxy, Abernethy opined, his erstwhile student "inculcates opinions tending to subvert morality, benevolence, and the social interests of mankind" and therefore "deserves the severest reprobation." Abernethy even insinuated that Cuvier "collected information by reading," and had actually plagiarized the more meaningful elements of his understanding of animal structure from John Hunter's eighteenth-century notion of the sympathy of every part of the body for all the others. While Cuvier had invested his own law of correlation with the offensively materialist tendency typical of a "nation where the writings of its philosophers . . . have greatly contributed to demoralize the people," the British Hunter had included the sympathy of parts in a morally uplifting argument for the existence of a vital principle distinct from matter. Accepting the "indispensable duty publicly to answer . . . the Modern Sceptics," Abernethy proposed himself as a "voluntary advocate in the cause of Hunter versus Cuvier," and was confident of persuading his audience—or "Gentlemen, (of the Jury)," as he legalistically addressed them—of the worth of "my client and his country."[31] The end of the Napoleonic Wars two years earlier had done nothing to temper Abernethy's Francophobia nor to absolve Cuvier of the suspicions aroused by his tainted nationality.

Lawrence was bemused that the "French . . . seem to be considered our natural enemies in science, as well as in politics," and mocked Abernethy's hysterical fears of a "conspiring band of sceptics and French physiologists . . . a nest of plotters . . . threatening, in the noise and alarm which preceded their discovery . . . to eclipse even the green bag conspiracy of another place."[32] The so-called green-bag conspiracy involved a secret cache of parliamentary papers exaggeratedly alleging widespread public disorder that, at the beginning of 1817, were exploited by reactionary aristocrats in an unsuccessful attempt to compel the government to revive the political restrictions of the 1790s.[33] Abernethy's allegations of a Cuvierian conspiracy of seditious materialists, Lawrence implied, were no less extravagant and overblown. Yet the final years of the 1810s were marked by acute economic privations among the working classes, and a resultant upsurge of political protest, culminating in the mass reform meeting at Manchester in August 1819 that, following its violent suppression, became known as the Peterloo Massacre. In the following year, a very real conspiracy to murder the Tory prime minister Lord Liverpool, hatched in London's Cato Street, was exposed by a government spy and its perpetrators hanged.[34] The materialistic understanding of life that Lawrence claimed to have derived from Cuvier had irreligious implications that accorded with the subversive atheistic agenda of many of the protesters (although often they maintained a deist or even a quasi-Christian orientation). By the early 1820s, the lectures Lawrence had originally delivered within the grand "portals of the great edifice in Lincoln's-Inn-Fields" were being sold in cheap unauthorized reprints on working-class street corners.[35]

From his cell in Dorchester jail, the radical publisher Richard Carlile, who was serving a three-year sentence for blasphemy and sedition, arranged his own edition of "all the Lectures ever delivered by Mr. Lawrence in his character of Professor in the Royal College of Surgeons." This piratical reprint, as Carlile's weekly newspaper the *Republican* advertised, was "sold in numbers at 3*d*. each, for the convenience of those who cannot purchase the whole at once" and cost "but half the price" of the original authorized editions.[36] Lawrence, threatened by the loss of his professional positions at leading London hospitals, had already withdrawn his *Lectures on Physiology, Zoology, and the Natural History of Man* (1819) from sale, and he was now aghast at such militant appropriation of his materialist approach to life. He was unable to obtain an injunction against the radical reprinters, though, as the Court of Chancery ruled that there could be no copyright in a blasphemous book. This judicial denunciation only made Lawrence's unprotected lectures even more attractive to piratical publishers eager for printed weapons that could be used to attack the government and the established Church, and Carlile's edition of the lectures was ironically dedicated to the lord chancellor in honor of his "INJUSTICE IN REFUSING TO ESTABLISH THE AUTHOR'S RIGHT OF PROPERTY IN THEM."[37]

Carlile's similarly unauthorized reprint of Lord Byron's controversial verse-drama *Cain* (1821) had already revealed the potential of Cuvierian geology to undermine conventional religious opinions of the earth's creation.[38] Lawrence's lectures, officially anathematized and issued in cheap installments, now helped introduce Cuvier's allegedly materialist understanding of animal structure to radicalized working-class audiences (fig. 2.1). As a plebeian correspondent to the *Lion*, Carlile's monthly successor to the *Republican*, reflected in 1828: "M. Cuvier . . . I then considered as a person not entirely destitute of truth or talent, especially as he had been so highly extolled by Mr. Lawrence in his Lectures."[39] While the same correspondent subsequently changed his mind, his initial response suggests that Carlile's piratical reprinting of Lawrence's praise for Cuvier, including his translation of passages from the "Discours préliminaire" outlining the law of correlation, helped secure the famous French savant a favorable reception among proletarian radicals.

As well as atheism and materialism, Carlile also held radical views on female emancipation, and his paramour, Eliza Sharples, edited her own militant journal *Isis*. After Cuvier's death in May 1832, many periodicals, as noted in the last chapter, ran obituaries eulogizing his purportedly orthodox attitudes, especially on religion. *Isis* reprinted a brief tribute from the liberal *Foreign Quarterly Review*, although its editor balked at one particular paean. An aside in the original claiming that "he was an enlightened and liberal Protestant, and watched over the interests of his co-religionists" was accompanied in *Isis* by a footnoted intervention from Sharples protesting: "Cuvier, the geologist and comparative anatomist a religionist! A Frenchman would shrug up his shoulders and say 'Bah!' to that."[40] Drawing attention to the scrupulous silence with regard to religion that Cuvier's British obituarists strategically ignored, radicals such as Sharples and Carlile created a further domesticated version of the French savant that accorded with their own antiestablishment stance. With Cuvier himself having prospered, and even ascended to the rank of baron, by acquiescing to the growing conservatism of the Ultra-royalist French government, this insurrectionary incarnation was no more legitimate than any other British appropriation of his scientific authority.[41]

The cheap weekly newspaper the *Poor Man's Guardian* also marked Cuvier's passing with a reprinted obituary, but it looked further than the *Foreign Quarterly Review* and instead offered its plebeian readers a translation of the effusive *éloge* for his former mentor that Marie-Jean-Pierre Flourens delivered at the Académie des sciences in Paris. Flourens, as the *Poor Man's Guardian* reported, had eulogized Cuvier's "law of the co-relation of the organs," which meant that it was "often sufficient alone to know a single bone, or a part of a single bone, to acquire the most extended notions of the structure and habits of an animal."[42] While the *Poor Man's Guardian* did not share Sharples and Carlile's avowedly atheistic and materialist agenda, it led the campaign for the repeal of newspaper taxes, and championed working-class demands for suffrage and trade union rights.[43] Its respectful tribute to the deceased

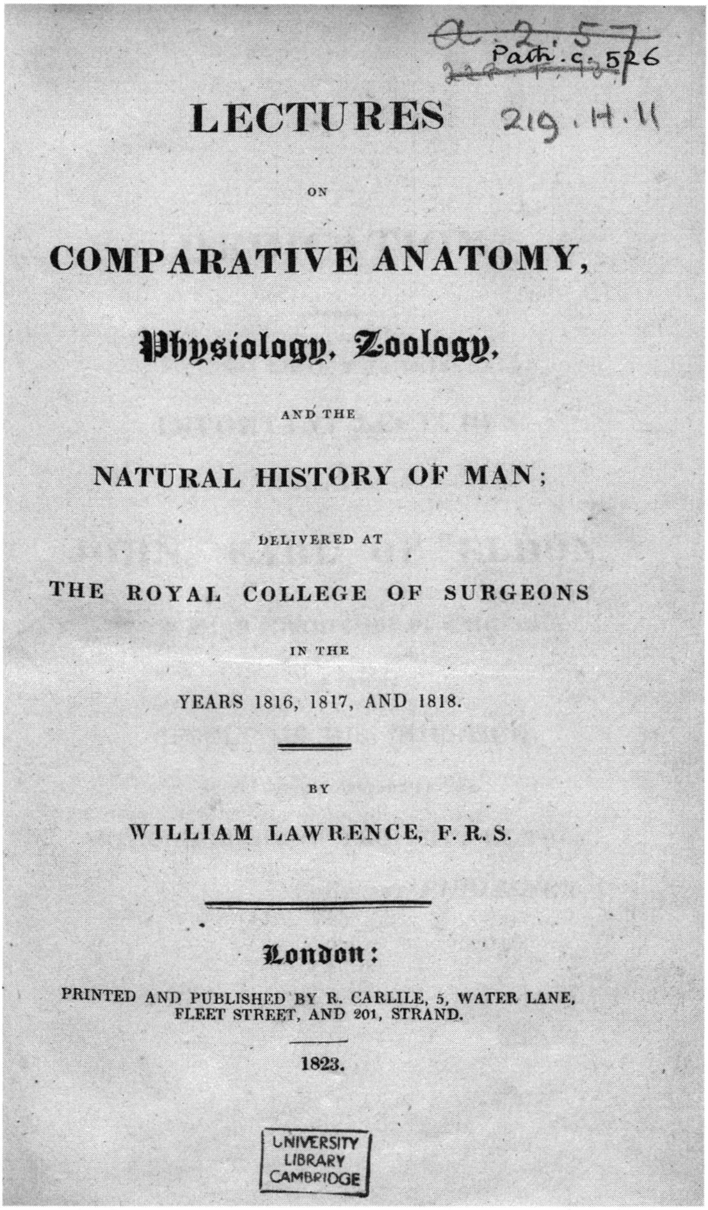

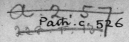

LECTURES

ON

COMPARATIVE ANATOMY,

𝔓𝔥𝔶𝔰𝔦𝔬𝔩𝔬𝔤𝔶, 𝔝𝔬𝔬𝔩𝔬𝔤𝔶,

AND THE

NATURAL HISTORY OF MAN;

DELIVERED AT

THE ROYAL COLLEGE OF SURGEONS

IN THE

YEARS 1816, 1817, AND 1818.

BY

WILLIAM LAWRENCE, F. R. S.

London:

PRINTED AND PUBLISHED BY R. CARLILE, 5, WATER LANE,
FLEET STREET, AND 201, STRAND.

1823.

FIGURE 2.1. Title page of William Lawrence, *Lectures on Comparative Anatomy, Physiology, Zoology, and the Natural History of Man* (London: R. Carlile, 1823). This pirated reprint of Lawrence's lectures, sold in cheap installments costing threepence, brought Cuvier's law of correlation to radicalized working-class audiences, presenting it as an endorsement of a materialist understanding of life. Reproduced by the kind permission of the Syndics of Cambridge University Library.

French naturalist indicates that, as John Allen had perceived in Edinburgh three decades earlier, Cuvier's conception of the body's physiological inter-dependence was by no means antithetical to the political arguments of those protesting for democratic reform.

A Subject Unacceptable to an Academical Audience

Nowhere was the threat of radical protest and the moral dangers of atheism and materialism felt more acutely than at Oxford, the ancient bastion of clas-sical learning and Anglican orthodoxy. In 1821 the antiquary Rowley Lascelles lamented the "mad philosophies . . . nothing but the fruits of materialism and atheism" that had recently gained adherents in the "world abroad." He urged the necessity of "leaving other *concerns* to the Church and University," and anxiously counseled: "I would throw out this warning to Oxford, against following that down-hill road (or rather precipice), the assimilating its opin-ions, science, and studies . . . to those fashionable in the world."[44] But spurred by criticism, particularly from Edinburgh, Oxford's classical and clerical cur-riculum had already begun to incorporate some of the new sciences that were taught in the Scottish capital and debated in London's scientific institutions, even if they remained optional and did not lead to a degree.[45] In 1813, the same year he joined the Geological Society of London, Buckland assumed Oxford's readership in mineralogy, and soon added the newly established readership in geology. His inaugural address for the latter appointment, two years before Lascelles's jeremiad on fashionable atheism and materialism, acknowledged that within the "Metropolis" there was now a "general taste" for geology "even amongst the imperfectly educated classes of society."[46] Throughout this lecture, which in its published form was entitled *Vindiciæ Geologicæ* (1820), Buckland insisted that, far from negating the Bible and Oxford's rigorous theological orthodoxy, geology had all the religious and scholarly attributes necessary for the university's ancient pedagogic traditions.

Importing from Edinburgh the conclusions of "Cuvier's admirable Essay on the Theory of the Earth," Buckland assured his Oxonian auditors that, as Jameson's preface made clear, "geological investigations can be shewn to be in no way inconsistent with the true spirit of the Mosaic cosmogony." Indeed, the "grand fact of *an universal deluge*," Buckland pledged, is "proved on grounds . . . decisive and incontrovertible." Unlike Jameson, however, in *Vindiciæ Geologicæ* Buckland also invoked a separate, rational argument for divine providence. Reflecting on the "importance of Geology to the sci-ence of Anatomy," he discerned in the "organization of animals" and the "delicately proportioned laws of coexistence . . . such undeniable proofs of a nicely balanced adaptation of means to ends . . . that he must be blind indeed, who refuses to recognize in them proofs of the most exalted attributes of the Creator."[47] The form of the animal body was the means that was adapted to facilitate the performance of the functions that were its end, or, in teleological language, its final cause. This fit between form and function was so perfect that the structure entailed by a creature's adaptation to its environment could not have resulted from mere heredity or in accordance with a general unified type, and instead testified to the intelligent choices made by a divine designer.[48] While Lawrence, addressing London surgeons, had ridiculed "the very idea of resorting . . . amid the blood and filth of the dissecting-room . . . to this low and dirty source for a proof of so exalted and refined a truth," Buckland's own proofs of the perfect adaptations of animal anatomy were the decorous, unsoiled fossils—"Antiquities of the Globe itself"—that could be displayed in the Ashmolean Museum alongside artifacts from the classical world.[49]

In *Vindiciæ Geologicæ* the unmistakably Cuvierian terminology of the "laws of coexistence" was yoked to the "beautiful and energetic language" in which the "acute and learned Paley sums up . . . the results of the minute and elegant investigations pursued in his invaluable volume on Natural Theology."[50] Three years before his death in 1805, William Paley, the archdeacon of Carlisle, had revived the long-standing teleological argument that the evident design of the material world confirmed the existence of a deity, whose wisdom and benevolence could be inferred from the order, beauty, and happiness of his creations. In *Natural Theology* (1802), he employed instances from human anatomy, natural history, and astronomy, as well as analogies with complex mechanical contrivances such as the watch, to demonstrate that the structure and organization of a variety of natural phenomena were so intricate and perfect that they were explicable only by a belief in providential design.[51] As Buckland recognized, the language of *Natural Theology*, which brimmed with metaphors and other rhetorical figures derived from classical authors such as Paley's favorite Cicero, played an important role in rendering design and Deity a seemingly irrefutable pairing.[52]

Although Paley's book did not include any of the remarkable prehistoric creatures whose discovery by Cuvier began to be reported in British periodicals during the late 1790s, his particular vocabulary in *Natural Theology*, and especially his regular allusions to the "correlation" of the "animal structure" and its "correlation of parts," became interchangeable with Cuvier's own terminology regarding fossil species in anglophone renderings of the French savant's writings.[53] In fact, what Cuvier designated "la corrélation des formes" in his "Discours préliminaire" was regularly translated imprecisely into English as the "correlation of parts."[54] With Paley insisting that such structural "correlation . . . fixes intention somewhere," the conflation of Paleyite and Cuvierian terminology inevitably entailed teleological assumptions about divine foresight that, as seen in the last chapter, were largely absent from the original versions of Cuvier's texts.[55]

In Oxford, Paley's argument from design, and the emphasis on perfect adaptation that it shared with Cuvier's law of correlation, was a valuable means of validating the religious credentials of the new sciences that liberal reformers endeavored to include in the traditional curriculum.[56] Buckland's predecessor as reader in mineralogy John Kidd, in his own inaugural address on becoming Regius Professor of Medicine in 1822, recognized that his audience was "composed chiefly of individuals to whom the pursuit of physical science both is and ought to be a relaxation rather than a study." Kidd remained confident, though, that the "natural theology of Dr. Paley is so generally recommended and read in this University" that he "need not fear, in employing this argument as the groundwork" for his lecture on comparative anatomy, that he had "selected a subject unacceptable to an academical audience." At the very time that Carlile was reprinting Lawrence's materialistic lectures for radicalized plebeian readers in London, Kidd assured his Oxonian auditors that, in relation to anatomy, "Paley's argument is important . . . as tending to counteract the influence of those who would inculcate atheistical opinions." And while Lawrence, and then Carlile, recruited Cuvier as an exponent of materialism, Kidd instead showed how the contention of "Cuvier, one of the most philosophical physiologists of the present day," that animals' feet have a constant "relation to the character of their teeth" afforded "evidence of consummate skill in the structure and functions of natural objects."[57]

By the mid-1820s, as the keeper of the Ashmolean Museum Philip Bury Duncan later reflected, a "taste for the study of natural history had been excited in the University by Dr. Paley's very interesting work on Natural Theology, and the very popular lectures of Dr. Kidd on Comparative Anatomy, and Dr. Buckland on Geology."[58] Cuvier, who was acclaimed in both Kidd's and Buckland's lectures, also contributed to Oxford's new taste for natural history, although, as his proponents at the university were careful to ensure, his distinctive understanding of animal structure was divested of any of the hints of materialism or irreligion that were scorned by Presbyterians in Edinburgh and embraced by radicals in London. Its apparent parallels with Paleyite teleology meant that Cuvierian correlation, when employed in a natural theological argument for divine providence, could gain admittance to Oxford's ancient sanctuaries of Anglican orthodoxy.

That Ancient and Venerable Stock of Classical Literature

The urbane hauteur with which Conybeare regarded the dour Edinburgh graduate Fleming during the "pig's-foot controversy" reflected his own training in classical literature at Oxford. The Greek and Latin texts that dominated the traditional *literæ humaniores* curriculum instilled the genteel patrician values that putatively characterized Oxford's alumni, and that Conybeare had maintained even as his adversary resorted to abuse and impertinence. In his *Outlines of the Geology of England and Wales* (1822), Conybeare lamented the "injurious effects" on the "zealous partisans" of the "Edinburgh school . . . produced by . . . excessive addiction to theoretical speculations," averring that an "ignorance of the rules of classical composition, and of the languages, and philosophy of polished antiquity, are by no means essential advantages in research of this nature."[59] Buckland, in his inaugural address, similarly assured his Oxonian audience that, while geology and other sciences had long "formed a leading subject of education" both on the Continent and north of the border, it was not necessary to "surrender a single particle of our own peculiar, and, as we think, better system of Classical Education." Instead, he proposed "ingrafting (if I may so call it) . . . the study of the new and curious sciences of Geology and Mineralogy, on that ancient and venerable stock of classical literature from which the English system of education has imparted to its followers a refinement of taste peculiarly their own."[60]

Buckland's rooms in Corpus Christi College were notoriously strewn with miscellaneous disarticulated bones and fossil remains, which he interpreted in the approved Cuvierian manner.[61] In February 1824 he presented some "detached bones" found in quarries north of Oxford at a meeting of the Geological Society, asserting: "Although the known parts of the skeleton are at present very limited, they are yet sufficient to determine the place of the animal in the zoological system." This particular creature, he proposed, was a colossal reptile that, at Conybeare's suggestion, he named the megalosaurus. Buckland conceded, however, that the proportions of even this extraordinarily large saurian "fall short of those which we cannot but deduce from a thigh-bone of another of the same species."[62] The prehistoric monster whose unparalleled size was discernible from just its femur was the iguanodon, which had first been identified in the early 1820s by the Sussex surgeon Gideon Mantell. Like Buckland, Mantell, as will be seen in chapter 5, also employed what he called "Cuvierian sagacity" in his celebrated discovery of the iguanodon, and the French savant's scientific authority helped confirm his controversial classification of its teeth as herbivorous but nonetheless reptilian.[63] When, at the end of 1829, Buckland received new specimens of this remarkable saurian, he once more inferred its gigantic proportions using the paleontological technique of extrapolating the whole from a single part, although this time, and in contrast to Mantell with his humble medical apprenticeship, imbuing the well-known Cuvierian method with his own Oxonian training in classical literature.

Presenting the "largest metacarpal bone which has yet been discovered" to the fellows of the Geological Society in December 1829, Buckland suggested:

> If we apply to the extinct animal from which it was derived, the scale by which the ancients measured Hercules (*"ex pede Herculem"*), we must conclude that the individual of whose body it formed a part, was the most gigantic of all quadrupeds that have ever trod upon the surface of our planet.[64]

The proposition that the size of the Greek divinity Heracles, whom the Romans renamed Hercules, could be measured solely by his foot was attributed to Pythagoras. The Latin maxim expressing this ancient method of proportional reasoning had already become proverbial, especially among Oxford's classicists, and newspaper reports of Buckland's paper at the Geological Society intimated that his use of it was both extempore and archly humorous. As the *Literary Gazette* reported: "The Chairman inquired of the professor if the latter could form any correct estimate of the size of the reptile: the professor, smiling, answered '*ex pede Herculem!*'"[65] Such droll showmanship—in which he would switch rapidly from the sublime to the ridiculous, thereby enabling him to distance himself from his own speculative theorizing—was a familiar trait of Buckland's lecturing, whether to undergraduates in Oxford or to scientific gentlemen in London, and did not necessarily imply any disparagement of the Pythagorean technique (fig. 2.2). In fact, while Buckland generally used such humor only in his lectures before excising it from his more sober published work, he opted to retain the *ex pede Herculem* aside (even if in parentheses and denuded of some of the archness imparted by his genial demeanor) in the version of his paper published in the Geological Society's *Transactions*.[66]

The same Latin maxim had long been a refrain of discerning reviewers anxious, as the *Edinburgh Literary Journal* expressed it, to avoid "multiplying quotations of a similar kind," who could instead observe urbanely "but *ex pede Herculem*."[67] Such literary overtones were evidently appealing to Buckland in his endeavor to engraft new sciences such as geology onto Oxford's traditional *literæ humaniores* curriculum. His close friend, and fellow Oxonian, William John Broderip, who subsequently trained in the law but retained the passion for natural history he cultivated while a student at Oriel

Drawn from Nature & on Stone by Geo. Rowe.   Printed by C. Hullmandel

GUL. BUCKLAND B.D . F.R. S. MIN., ET GEOL. OXON PROFF. 1823.

FIGURE 2.2. George Rowe, *William Buckland.* Lithograph, 1823. National Portrait Gallery, London. Buckland, evidently addressing an audience, discusses the hyena jaw in his hand surrounded by the accoutrements of his engaging lectures, including, in the background, a pictorial restoration of two extinct hyenas. © National Portrait Gallery, London.

College, applied the same ancient axiom directly to Cuvier's reconstructive techniques. Broderip related how a solitary "ungueal phalanx of an Edentate quadruped" was sent to Paris, and then proclaimed: "Taking this ungueal bone for his basis, with far less material than he had for his opinion who ventured to pronounce 'ex pede Herculem', Cuvier measured the animal to which it must have belonged as having extended to the length of twenty-four feet."[68] In the hands of Buckland and Broderip, the Cuvierian method for inferring the size and appearance of extinct creatures from merely a single bone became an undertaking that was appropriate even for the gentlemanly classical scholars of Oxford.

The Truth of the *Word of God*

Despite Buckland's efforts to classicize Cuvierian correlation, Oxford's traditional emphasis on the evidential value of written texts rather than physical phenomena also informed many of the criticisms of both him and Cuvier emanating from the English provinces. The Leicestershire clergyman George Bugg lamented that the "Translation of M. Cuvier's 'Theory of the Earth' into our language," which was then "propagated by Professor Buckland, at Oxford," had induced an "amazing readiness to relinquish the plain and literal meaning of divine truth," which he considered "both contagious and alarming." In his anonymous *Scriptural Geology* (1826–27), Bugg, an evangelical Trinitarian who believed the earth had been created in six solar days no more than six thousand years ago, insisted that it was not "necessary to impute to M. Cuvier *motives* of an hostile nature against the Bible," as whether or not the French savant "had studiously set himself to do mischief to the Volume of Inspiration, he could not have adopted a more effectual method." But it was not just Cuvier's conception of the vast extent of the earth's geohistory that, in Bugg's opinion, "cannot possibly exist consistently with a fair and literal construction of the *Word of God*." The "peculiar, and almost miraculous talent of M. Cuvier" that supposedly enabled him to "distinguish . . . species by half a bone" was, notwithstanding Cuvier's own allusions to Corinthians and Ezekiel, no less inconsistent with biblical accounts of animal forms and habits.[69]

After quoting the passage from Jameson's *Essay on the Theory of the Earth* in which Cuvier confidently claimed that a cloven-footed imprint could only be left by a ruminant, Bugg curtly opined: "Here I would offer a remark or two." His principal cavil with the quotation was that

> *Moses*, whom M. Cuvier's system tends to accuse of unusual ignorance and error, can correct the celebrated Author in the application of the first principles of his peculiar science. It is quite incorrect to say that ruminants are "the *only* animals having" "cloven hoofs": For Moses informs us, from the mouth of an infallible instructor, that,—"The *swine*, though he divide the hoof and be *cloven-footed*, yet he *cheweth not the cud*"! (Levit. xi. 7).

The very same issue that, three years later, the members of the Wernerian Natural History Society would attempt to resolve by experimenting with pigs in a squalid Edinburgh courtyard could be determined for Bugg simply by reference to Scripture. The Bible was equally infallible on arcane matters of natural history as it was on religion and morality. The scrutiny of such scriptural minutiae also had very significant implications, for Bugg had evidently noticed Cuvier's extraordinary wager, in the opening pages of the "Discours préliminaire," of the veracity of all his other revolutionary scientific findings on the reliability of the law of correlation. Bugg coolly laid his own bet:

> I shall leave the reader here to make his own reflection, only asking him one question.— Can an Author so obviously incautious as M. Cuvier unquestionably is, in the very principles of his instruction, be a safe guide to follow in matters not only affecting the foundation of geological *science*, but, by consequence, the truth of the *word of God*?

The entire edifice of the catastrophist historical geology in which "M. Cuvier proves revolutions from the extinct animals, and proves animals to be extinct from his skill in comparative anatomy" was rendered invalid by the discrepancies between Cuvier's "theory and speculation" regarding animal structure and the inspired truths of the Bible.[70]

Although Bugg's stridently italicized denunciation of virtually every aspect of Cuvierian geology was notably shrill and uncompromising, his insistence on the Bible's infallible authority, even in scientific matters, was far from unusual in this period.[71] And others who adhered to the literal truth of Scripture similarly had little respect for Cuvier's understanding of animal anatomy, regardless of the reverence for it evinced by Oxford dons and London surgeons. Mary Roberts, a popular writer on the natural history of her native Gloucestershire who in the 1820s renounced her Quaker roots for evangelical Millenarianism, acknowledged the problems "in accounting for the remains of various non-descript species, for the Mosaic records are brief," but she still proposed "walking by the light of revelation, and merely considering facts as they occur." It was its absence from the Bible's enumeration of the creatures on Noah's ark, rather than any actual skeletal evidence, that persuaded Roberts of the extinction of what she called the "carnivorous elephant, or Mastodon," although she was also confident of her ability to challenge even internationally renowned authorities on the fossil remains. As she boldly asserted in *The Progress of Creation* (1837):

> Cuvier describes this animal as herbivorous, but surely without reason. We can judge of its nature, only by its remains; and as the most striking characteristic is found in the enormous grinding teeth, which resemble those of a carnivorous species, there is good reason to believe that the creature preyed on animal food.[72]

With apparently no awareness of the hierarchies of scientific expertise that the Geological Society and other metropolitan institutions were then busily constructing, Roberts assumed that her opinion of the mastodon's dentition was every bit as valid as that of Cuvier.

Roberts's insistence that the extinct mastodon could not have been herbivorous related to her scriptural interpretation that "carnivorous creatures were not admitted within the ark." Her explanation for the problematic fact of Noah's protection, as recorded in Genesis, of "lions and tigers, bears and wolves, hyænas and wild cats" was to question whether such "carnivora, which now thirst for blood . . . were always so inclined." Instead, she proposed that the "carnivorous propensity" could be "drawn forth by circumstances." In order to maintain the literal truth of the biblical account of the zoological consequences of the Noahic Flood, Roberts concluded that the "carnivora of the present day may . . . have been rendered such by various existing causes, by the effect of climate, or by the facility of obtaining animal food."[73] While Roberts was able to accommodate the extinction of creatures such as the mastodon that were not permitted to enter the ark with her literal reading of Scripture, Bugg was adamant that it contravened the Bible's inspired teachings. For him, it was far more conceivable that Cuvier's assertion of the "supposed unchangableness of the form of animals" was erroneous, and that many purportedly extinct creatures had simply "changed their *form* as well as their *diet*, from vegetable to flesh." Bugg ridiculed the Cuvierian contention that the "form and habits of animals are stationary, and that they do not transgress certain known boundaries" as an "anomaly in the philosophy of nature," and demanded: "Animals, he supposes, have a peculiar conformation of parts, from which they never materially depart, either by change of food, time, or place. I ask for *evidence* of this."[74] With the inspired words of Moses negating the supposedly invariable relation between teeth and hooves, Bugg had no confidence in the perfect and harmonious correlation of the animal frame, which, for Cuvier, rendered extensive organic variations untenable.

Remarkably, biblical literalists such as Bugg and Roberts appeared to endorse the transformist view of animal structure that Cuvier's colleague Jean-Baptiste Lamarck had long espoused in Paris. In fact, such views were certainly not unknown in provincial evangelical circles in England. In 1833 another scriptural literalist, the Essex clergyman Frederick Nolan, directly invoked Lamarck's transformism in attempting to reconcile the limited size of Noah's ark with the vast number of different animals now known.[75] Lamarck himself had died blind and in poverty in December 1829, but his conviction that, rather than becoming extinct, organisms were continuously modified over time as alterations in the environment required new habits and functions was, as Pietro Corsi has shown, "known and commented upon in Britain during the 1820s and early 1830s," and it occasionally received support.[76] As well as Jameson's clandestine endorsement noted in the last chapter, and the more overt advocacy of another of Edinburgh's erstwhile Cuvierians Robert Edmond Grant, Lamarck's theories of species change were also embraced by working-class atheists who quickly recognized their subversive potential.[77] Although such radical Lamarckians were numerically negligible, their raucous assertions that nature was both self-creating and self-perpetuating, and did not require any divine superintendence, were no less threatening to the Anglican establishment than the cheap unauthorized editions of Lawrence's blasphemously materialist lectures.

Adrian Desmond has argued that in Britain "Lamarckianism was taken up by groups that flatly rejected aristocratic authority," with such "fierce materialists" forging a "left-wing Lamarckian science" that, crucially, contrasted with "Cuvier's more conservative . . . and safe science," which appealed to his admirers among the British establishment.[78] Such a simple dichotomy, however, obscures the diversity in both radical and conservative opinion in the opening decades of the nineteenth century.[79] It also fails to recognize the divergent political and theological meanings that could be—and indeed were—ascribed to the same scientific theories. That the materialism embraced by irreligious reprinters such as Carlile was ostensibly inspired by Cuvier's approach to comparative anatomy, while literalist interpreters of Scripture such as Bugg and Roberts scorned Cuvierian correlation and instead advocated Lamarck's rival transformism, indicates that the affiliations between particular scientific theories, especially those imported from France, and the diverse religious and political orientations that characterized early nineteenth-century Britain were considerably more complex than historians have hitherto acknowledged.

Painful Recollections of Mortality

Although without concurring with the remarkable Lamarckian transformism of Bugg and Roberts, Oxford's own emphasis on tradition and textual authority had, by the early 1830s, become noticeably more strident. The influential clique of Anglo-Catholic clergymen who began the Oxford Movement considered the church doctrines set down by the apostles more meaningful than the Old Testament Scriptures favored by evangelical biblical literalists, but they were no less critical of Buckland's attempts to reconcile his liberal Anglicanism with Cuvierian science.[80] Their leader John Henry Newman, who had attended Buckland's geology lectures in the 1820s, condemned the "educated men" who, "uneasy, impatient, and irritated" by "original and apostolic doctrine," "praise Newton or Cuvier" and asserted "their Christian liberty . . . of thinking rightly or wrongly, as they please."[81] What had particularly provoked the ire of Buckland's High Church colleagues was his invitation to the new British Association for the Advancement of Science to hold its second annual meeting, in the summer of 1832, in Oxford's hallowed halls.

Founded a year earlier, the British Association was intended to bring men of science together at peripatetic meetings that were free from the sectarian and denominational divisions that beset other areas of national life. The generalities of natural theology were ideally suited to this atmosphere of latitudinarian tolerance in which Anglicans and dissenters could agree on the existence of divine providence while eschewing more contentious discussions of its precise characteristics.[82] Buckland, as both the British Association's president as well as the local host, delivered the closing lecture of the Oxford meeting on 23 June. This was barely a month since Cuvier's sudden death during an outbreak of cholera in Paris, and Buckland, having begun his lecture by alluding to the "discoveries of the immortal Cuvier," broke off to acknowledge that he could not "utter the name . . . and associate it with the term 'immortal', without being at once arrested and overwhelmed by melancholy and painful recollections of mortality." Lamenting the loss of a "man, whose friendship I have ever counted among the most distinguished honours of my life," Buckland informed his interdenominational auditors that

> the genius of Cuvier . . . has shown that the frame and mechanism of every animal present an uniformity of design and a simplicity of purpose, which prove to demonstration that every individual, not only of existing species, but of those numerous and still more curious races which have lived and perished in distant ages . . . were framed and fashioned by the same Almighty mind. Gentlemen, to this great and good man not only are the sciences of natural history profoundly indebted, but the higher science of morals also owes a debt of deep and everlasting obligation, for that he has proved to demonstration the high and solemn truth to which I have alluded, viz.—the unity and universal goodness of the Great Creator.[83]

No mention was made of the erstwhile diluvian hypothesis that had taken precedence in *Vindiciæ Geologicæ*, and Buckland instead ascribed Cuvier's contribution to religion and morality solely to his exposition of the perfect mechanisms of animal structure.

The recently deceased savant had occasionally referred to the mechanical qualities of organic forms, as when he proposed that each of the body's component parts "can be considered as a machine part" in *Leçons d'anatomie comparée*.[84] Such parallels, however, were relatively unusual in Cuvier's voluminous writings, and, as Dorinda Outram has contended, he always resisted the "full-scale use of the machine analogy."[85] Buckland's eulogy nevertheless made Cuvier's distinctive conception of animal organization seem indistinguishable from the mechanistic metaphors of divine contrivance that dominated Paley's *Natural Theology*. The reason for Buckland's mournful allusion to Cuvier's immortality had been his identification, back in 1796, of the extinct South American edentate he named the megatherium. This enormous sloth-like creature, Buckland remarked, possessed "gigantic claws . . . forming most powerful instruments for scraping roots out of the ground," which combined the attributes of a "spade . . . a hoe and a shovel" with such perfection that he hailed it as the "Prince of sappers and miners" (fig. 2.3). Noting that "I speak in the presence of Mr. [Marc Isambard] Brunel, the Prince of diggers," Buckland jovially proposed that the engineer, then engaged in an innovative attempt to excavate under the Thames, "eyes him and says, 'I should like to employ him in my tunnel.'"[86] As well as maintaining the Paleyite concern with providential contrivance, Buckland's whimsical insistence on the mechanical capacities of the megatherium's structure also aligned the exotic subject of his closing lecture with the more practical concerns of his mixed audience, among whom Oxonian gowns and mortarboards were outnumbered by the frock coats of the professional classes.

Describing the colossal dimensions of this "animal apparently the most monstrous of the monster kind," Buckland conceded that, with its diminutive head and elephantine "posterior portions," the megatherium might appear awkward and ungainly. He assured his audience, though, that "the object of this apparently incongruous admixture of proportions was to enable the creature to stand at ease on three legs," thereby leaving "one of its fore paws at liberty to be exercised without fatigue in the constant operation of digging roots." The megatherium's seemingly discordant structure was in fact perfectly suited to its own fossorial feeding habits, and, as such, it revealed the "inexhaustible richness of contrivances whereby Nature has adapted every animal to a comfortable and happy existence in that state where it was destined to move."[87] This, as Buckland finally concluded after speaking for almost three hours,

> afforded also to natural theology a powerful auxiliary, showing from the unity of design and unity of structure, and from the symmetry and harmony that pervade all organic beings in the fossil world, as well as in the present, that all have derived their existence from one and the same Almighty and Everlasting Creator.[88]

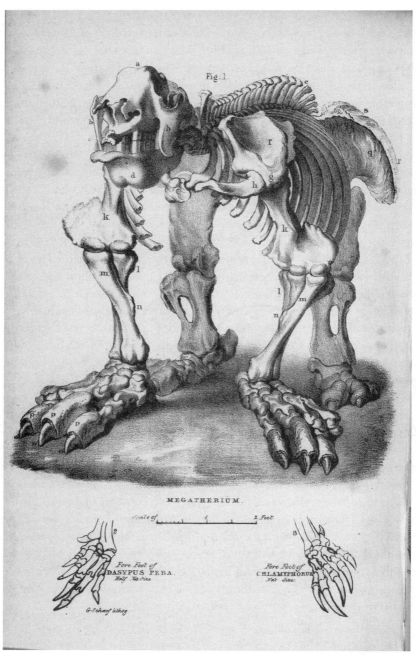

FIGURE 2.3. *Megatherium*. William Buckland, *Geology and Mineralogy Considered with Reference to Natural Theology*, 2 vols. (London: William Pickering, 1836), vol. 2, plate 5. This frontal view of the megatherium emphasizes its large and powerful claws, which Buckland compared to the industrial tools of miners and sappers. © The Trustees of the Natural History Museum, London.

The scientific legacy so recently bequeathed by Cuvier ensured that animal organization was the principal manifestation of the natural theological argument from design that was integral to both the ecumenical aspirations of the British Association as well as Oxford's own institutional requirements for incorporating scientific subjects into its curriculum. Indeed, Buckland's grandiloquent lecture on the megatherium, according to John Brooke, represents the "high point of natural theology in England."[89]

The only problem was that Cuvier himself had actually considered the megatherium anything but perfectly designed. In the revised second edition of *Recherches sur les ossemens fossiles* in 1823, he had observed that, like the closely related sloth, its ill-proportioned body parts were "inconsistent with the laws of co-existence which we find established throughout the animal kingdom."[90] When Buckland delivered another version of his British Association lecture at the Linnean Society in London in March 1833, he acknowledged that this was "almost the only passage in the works of Cuvier that I do not read with entire assent and admiration," and explicitly criticized the tendency of the "learned author . . . to view the structure of this animal, as Buffon had done before him, in relation only to its defects." In order for the sloth-like megatherium to add "another striking case to the endless instances of perfect mechanism and contrivance . . . so admirably illustrated by the judicious Paley," it was necessary for Buckland to rebuke Cuvier's fallacious "imputation of feebleness or imperfection."[91] A handwritten transcript of his closing lecture in Oxford the previous June, made by his wife, Mary, indicates that Buckland had at least hinted at his disagreement with his renowned friend even then. The unjustly maligned sloth family, as he told his audience, had "been misunderstood by almost every naturalist, including Buffon, and even the immortal Cuvier himself."[92] This gentle censure of the two French savants certainly appeared in local newspaper reports of Buckland's lecture, which were soon reprinted in the national press.[93]

As the British Association's president, Buckland arranged with the publisher John Murray to compile a report of the Oxford meeting that would be incorporated with a hitherto unpublished account of the initial meeting at York in 1831. When this combined *Report of the First and Second Meetings* was published in 1833, however, those who had attended Buckland's closing lecture might have struggled to reconcile the official record of it with their own recall of events. Not only was the speaker's whimsical humor—including comparing the megatherium's bulky posterior to similar afflictions in humans—entirely expurgated, with Buckland seemingly taking heed of a complaint in *Fraser's Magazine* that his "warm and affectionate tribute to the memory of his immortal friend Cuvier . . . ill accorded with an immediate

transition to the levity of a mountebank."[94] Any hint of criticism of the same
deceased naturalist was also strategically concealed. The line in the handwritten transcript regarding previous misapprehensions of the sloth family was
amended, removing Cuvier's name from Buckland's observation that they
were "considered by Buffon and other naturalists to afford the greatest deviations from the ordinary structure of quadrupeds." The closing remarks of the
Marquis of Northampton, which were seemingly reported verbatim, likewise
confirmed that Buckland had informed his audience about an "animal which
has been represented even by Buffon as imperfect in its constitution."[95] While
this report was drafted by William Vernon Harcourt, the British Association's
general secretary, Buckland had seen the printer's "corrected sheet," and, as
he admitted to Harcourt, he had "struck out the name of Cuvier which stood
in Lord Northampton's speech associated with that of Buffon under the imputation of superficial knowledge."[96] With Cuvier's scientific authority helping to ratify his own anatomical demonstrations of divine design, Buckland
evidently considered it necessary to censor even the most trivial criticisms of
his late friend.[97]

Stronger Proof of the Unity of Design

Aware that neither biblical literalists nor his High Church colleagues at Oxford required any further encouragement to decry Cuvier, Buckland continued to elide direct criticism of him and instead peddle the same circumspect
generalization that "it has been the constant practice of naturalists, to follow
Buffon in misrepresenting the Sloths, as . . . imperfectly constructed" when he
reused material from his lecture to the British Association in a contribution
to the series of books on natural theology commissioned in accordance with a
bequest from the Earl of Bridgewater.[98] The eight *Bridgewater Treatises on the
Power, Wisdom and Goodness of God as Manifested in the Creation* (1833–36)
were written by leading men of science and theological commentators, and
endeavored to elucidate the evidence for design in recent developments in
several scientific fields. These natural theological primers were to be written
in an accessible style, with Buckland planning to provide a "popular general
view of the whole subject [of geology] avoiding technical detail as much as
possible." Exactly what Buckland considered to be popular, however, is suggested by his proposal to restrict detailed descriptions to footnotes "which
the Country Gentleman may skip," and, more significantly, the expensive
cost that his contribution shared with the rest of the series.[99] Although ranging in price from nine shillings and sixpence to one pound and fifteen shillings, the eight *Bridgewater Treatises*, as Jonathan Topham has shown, were

widely reproduced in more accessible formats and reached middle-class and even plebeian audiences who often valued their nonspecialist scientific exposition more than their overtly theological agenda.[100] It was in this way that the particular version of Cuvierian anatomy that Buckland had been developing over the last two decades reached a readership far beyond the privileged purlieus of Oxford.

In the footnotes to *Geology and Mineralogy Considered with Reference to Natural Theology* that apathetic country gentlemen were expected to skip, Buckland finally disavowed the diluvian hypothesis and its support for scriptural revelation. The principal aim of his two-volume *Bridgewater Treatise*, whose publication was continually delayed and eventually came out in 1836 as the last in the series, instead entailed "extend[ing] into the Organic Remains of a former World, the same kind of investigation, which Paley has pursued with so much success in his examination of the evidences of Design in the mechanical structure of the corporeal frame of Man." These "petrified Remains" of "extinct species," Buckland averred, were "made up . . . of 'Clusters of Contrivances', which demonstrate the exercise of stupendous Intelligence and Power," and therefore "afford an argument of surpassing force, against the doctrines of the Atheist."[101] With Lawrence's advocacy of Cuvier in his materialist lectures at the Royal College of Surgeons having been appropriated by avowed atheists such as Carlile, Buckland insisted that the French savant was unequivocally on the side of the Church, and Paleyite natural theology in particular.

In fact, he proclaimed:

> We can hardly imagine any stronger proof of the Unity of Design and Harmony of Organizations that have ever pervaded all animated nature, than we find in the fact established by Cuvier, that from the character of a single limb, and even of a single tooth or bone, the form and proportions of the other bones, and condition of the entire Animal may be inferred.[102]

Notably, this emphasis on the "Unity" of divine design actually resembled the transcendentalist anatomy of Cuvier's Parisian colleague and rival Étienne Geoffroy Saint-Hilaire, who, as seen in the last chapter, posited that all vertebrates were formed on a single plan of creation and shared homological components that indicated a unity of composition. The pervasiveness of this unity, for Buckland, confirmed the unitary nature of the Deity.[103] Even in hailing Cuvierian correlation as the most effective weapon in the armory of natural theology, Buckland subtly departed from Cuvier's own long-standing opposition to Geoffroy's transcendentalism, which had grown ever more vehement in the last years of his life.[104]

Similarly, it was, once again, the ostensibly "imperfectly contrived" mega-therium whose fossil bones, for Buckland, afforded "imperishable monuments of the consummate skill with which they were constructed. Each limb, and fragment of a limb, forming co-ordinate parts of a well-adjusted and perfect whole."[105] Cuvier's particular approach to animal organization was not only co-opted to reinforce both a natural theological argument from design with which he had little sympathy and a notion of compositional unity that he detested, but the creature chosen as its principal exemplar was one that, as Buckland had acknowledged at the Linnean Society but otherwise scrupulously censored, he himself considered an ignominious exception to the law of correlation. The version of his late friend that Buckland presented with an assurance arising from their personal intimacy was actually hardly any more authentic than other appropriations of Cuvier by those who had not known him.

A chapter of Buckland's book was devoted to fossil fish, on which Cuvier had long planned an exhaustive monograph as a sequel to *Recherches*. Having failed to complete the enormous project by the early 1830s, Cuvier, only months before his death, presented his ichthyological specimens to the young Swiss naturalist Louis Agassiz. The work that Agassiz soon after began to publish, *Recherches sur les poissons fossiles* (1833–43), not only replicated the format of his late mentor's magnum opus—albeit appearing in irregular serial installments, as will be discussed in chapter 4—but also deployed the same methods for identifying and reconstructing prehistoric marine fauna. As Agassiz told his father in a letter from 1832, often with the "fishes I try to restore . . . I have only a single tooth, a scale, a spine, as my guide in the reconstruction."[106] In a public lecture given in New York a decade later, he explained that, just as "Cuvier . . . made . . . restorations from single bones," so he too could "'restore' a fish from isolated scales" by using his prior "knowledge" of the "relations of the scales to other portions of the animal."[107] Like Cuvier, Agassiz was adamant that merely a single disarticulated part could facilitate a complete reconstruction of the extinct creature from which it originally came, an announcement that elicited rapturous applause from his American audience.

Long before he traveled to America, Agassiz had made several visits to Britain, becoming an honorary member of the British Association in 1835. In the following year, Buckland's *Geology and Mineralogy* acclaimed the recent investigations of "M. Agassiz . . . on the character of the external coverings, or Scales," of fossil fish. This "character," Buckland avowed, "is so sure and constant, that the preservation, even of a single scale, will often announce the genus and even the species of the animal from which it was derived." The individual scales that, like Cuvier's single fossil bones, yielded such detailed information of the whole, together formed a "kind of external skeleton," which, according to Buckland, was fashioned in accordance with the fish's "adaptation to the medium in which they lived."[108] No less than extinct quadrupeds such as the megatherium, fossil fish, Buckland proposed, were perfectly adapted to their particular circumstances and habits, thereby revealing the guiding hand of divine design. Agassiz, however, had no truck with such teleological providentialism, and instead considered that the perfection of animal forms could only be determined by their resemblance to man. And unlike the deceased Cuvier, Agassiz could resist the Paleyite appropriation of his work. In his 1839 German translation of Buckland's *Bridgewater Treatise*, Agassiz announced in the preface that he was "somewhat embarrassed by the theological-teleological interpretation of many facts."[109]

Despite Agassiz's protests, the success of Buckland's much-discussed *Geology and Mineralogy* helped align both the young Swiss ichthyologist and, more significantly, his French mentor with particular religious and ideological positions.[110] At a fashionable metropolitan dinner party in the spring of 1837, Charles Lyell encountered the radical Whig aristocrats Lord and Lady Holland as well as a "Mr. Allen," who, as he told his sister, "has been for thirty years physician to the Holland family" and "lives with them on a very independent footing." This, of course, was the erstwhile Edinburgh radical John Allen, who, only a year before joining the Hollands' entourage in 1802, had published the first English translation of Cuvier's *Leçons d'anatomie comparée*. Now, more than three decades later, Lyell noted that during the conversation, "Lord Holland asked me about Buckland's book. . . . He seemed not to have formed a high estimate of the said Bridgewater." Later in the evening, "Lady Holland . . . took her share in the talk," and, when Lyell "spoke of . . . Cuvier," she reflected "how much she lamented his having abandoned the line in which he was so great, to meddle with politics (in which he played so inferior and, in her opinion, unworthy a part)."[111] The Hollands' physician and trusted confidant seemingly did not dispute the criticisms of either of his benefactors. Whereas Allen, as seen in the previous chapter, had once discerned democratic implications in Cuvierian correlation, the French savant, having been ennobled by the reactionary Bourbons in Paris and now decisively appropriated by Buckland in Britain, was clearly no longer to the taste of those of a radical Whig persuasion. This consolidation of the position of conservative gentlemen of science such as Buckland as the sole custodians of Cuvier's legacy was part of a "new consensus" between science and the liberal Tory and moderate Whig governing classes, from which erstwhile radicals and the working classes were excluded, that James Secord has identified as a key development of the 1830s.[112]

A Kind of Conservative Energy

In June 1832 Buckland had delivered his closing lecture to the British Association's meeting at Oxford only days after the Reform Bill extending the franchise by more than a half was finally given royal assent. When one of the audience asked if the megatherium was an "animal likely to have made holes to burrow in," Buckland replied, with a topical pun, that "he was convinced *his friend was no boroughmonger*, but rather, from the appearance of his claws, *a radical reformer*,—a joke which," as *Fraser's Magazine* reported, "brought forth much laughter and applause."[113] Buckland's characteristic joviality notwithstanding, the actual political meanings that many of his fellow *Bridgewater Treatise* authors ascribed to the natural world, and Cuvier's view of animal structures especially, were far from radical or reformist.

The elderly entomologist William Kirby perceived in nature a "kind of conservative energy" that could "prevent any dislocations in the vast machine," and even parasitic insects that weakened their predatory hosts, he proposed, "may, with great propriety, be called *conservatives*, since they keep those under that would destroy us." Throughout his *On the History, Habits and Instincts of Animals* (1835), Kirby employed the politically charged terminology of "rank and station" to express the particular environmental niches for which providence had adapted different animals. Drawing on Buckland's lecture to the British Association, he avowed that putatively disorderly creatures such as the megatherium, "instead of being an abortion, imperfect, misshapen, and monstrous, are exactly, and in every respect, adapted for the station which God has assigned to them."[114] Although Buckland did not share Kirby's high Tory paternalism, his more liberal Paleyite appropriation of Cuvier, by demonstrating the unerring wisdom and beneficence of the Deity, similarly implied that the created order, including the current social order, was perfect, even if, like the enigmatic megatherium, its flawless design was not always immediately discernible. This "*providential* natural theology," as Topham has termed it in distinction to a more philosophical "*prophetic*" strain that looked to things as they may yet become, presented class hierarchy and social inequality as ineluctably part of the divinely ordained natural order.[115]

Buckland's Oxford colleague John Kidd, in his own *Bridgewater Treatise*, reused much of the material from his inaugural lecture as Regius Professor of Medicine in 1822, although he did find one new reason for commending Cuvier's approach to anatomy. In *On the Adaptation of External Nature to the Physical Condition of Man* (1833), Kidd recounted the Cuvierian mantra that in every animal the several parts have such a mutual relation . . . that if any part were to undergo an alteration, in even a slight degree, it would be rendered incompatible with the rest; so that if any part were to be changed, all the other parts must undergo a corresponding change: and thus any part, taken separately, is an index of the character of all the rest.

Kidd had in mind the terse sentence in the "Discours préliminaire" in which Cuvier implied that, because of the perfect correlation of parts in the animal frame, Lamarck's perpetual evolutionary modifications were simply untenable and species were fixed. Lamarckian transformism, as noted earlier, had already been enthusiastically received by radical working-class atheists (as well as some biblical literalists), and its ultimate implication, as Lyell had warned only a year before Kidd's *Bridgewater Treatise*, was a "progressive scheme" in which the "orang-outang . . . is made slowly to attain the attributes and dignity of man."[116] The prospect of such a radical leveling of traditional hierarchies was made all the more fearsome by the clamor for democratic political reforms during the protracted passage of the Reform Bill. In this anxious atmosphere, the impediment to Lamarck's transformist theories that Cuvier had divulged so laconically was elaborated on in several of the *Bridgewater Treatises* and, in the process, made both politically conservative and overtly providential.

Precisely such a conservative version of Cuvierian correlation was used to quell Lamarck's radical British adherents by the surgeon Charles Bell. In his *Bridgewater Treatise, The Hand* (1833), Bell exemplified, in regard to that especially complex and intricate organ, the "curious relation of parts which has been so successfully employed by Paley to prove design, and from which the genius of Baron Cuvier has brought out some of the finest examples of inductive reasoning."[117] Bell's was the shortest of all the *Bridgewater Treatises*, and, like both Buckland and Kidd, he had augmented the meager but nonetheless expensive volume with material from earlier publications. This included a passage from *Animal Mechanics* (1827–29) in which he extolled the "wonders disclosed through the knowledge of a thing so despised as the fragment of a bone," proclaiming that, in the "masterly manner of Cuvier," such a partial relic could enable an anatomist to "estimate not merely the size of the animal, as well as if he saw the print of its foot, but the form and joints of the skeleton, the structure of its jaws and teeth, the nature of its food, and its internal economy."[118] Although they fitted seamlessly into Bell's costly exposition of the workings of *The Hand*, the original format in which these laudatory sentiments had appeared was a cheap anonymous pamphlet that, as part of the Library of Useful Knowledge published by the Society for the Diffusion of Useful Knowledge (SDUK), was explicitly targeted at plebeian readers. The London-based SDUK, whose activities will be considered in more detail in the next two chapters, endeavored to stymie radical publishers such as Carlile by supplanting their atheistic wares with inexpensive but morally wholesome reading matter tailored to the proletarian masses. Although the SDUK generally maintained a neutral stance on religion, Bell's *Animal Mechanics* employed exactly the same combination of Cuvierian functionalism and the Paleyite argument from design as *The Hand*, and, as Topham has suggested, "might almost be called a working-man's *Bridgewater Treatise.*"[119]

When the SDUK also proposed publishing a short work on *Animal Physiology* (1829) in its Library of Useful Knowledge, Bell, having read the proofs, complained at the

> opinion expressed in the first page. I do not think that the Society should identify itself with these opinions about *life*. *Life is nothing but organization in action* . . . is the line of argument which Mr. Lawrence followed, borrowing from the French, and is objected to by the physiologists of this country.[120]

Lawrence's principal Gallic borrowing, of course, was of Cuvier's allegedly materialist understanding of life as a correlate of the harmonious cooperation of the body's various organs, which had been eagerly appropriated in Carlile's piratical reprints of Lawrence's lectures. Bell, who ensured that the SDUK had the offending lines suitably amended, recognized the necessity of contesting Carlile's radical usurpation of Cuvierian correlation with the safer, more conservative, and explicitly natural theological version of the French savant's approach to anatomy that subsequently became a mainstay of the *Bridgewater Treatises*.[121] This militant purpose was certainly evident to readers of Bell's widely disseminated pamphlet. As Carlile later reflected in the *Lion*, when he was in Manchester in 1827, a "Scottish gentleman" had brought a copy of the initial part of *Animal Mechanics* to "my lodgings there, as something that I could not master. It was left with me in challenge."[122]

The Cuvierian Cultivators of Comparative Anatomy

The conservative Cuvierian functionalism that came to dominate the *Bridgewater Treatises* had not been evident in the first of the series to be published, William Whewell's *Astronomy and General Physics Considered with Reference to Natural Theology* (1833). Whewell's *Bridgewater Treatise* departed from the standard Paleyite argument that divine design was demonstrated by the mechanical contrivances of organic structures, and instead he maintained that overarching and harmonious natural laws helped reveal the mind of the Deity, an approach suited to the emphasis at Cambridge, where Whewell was a fellow of Trinity College, on mathematics and Newtonian mechanics.[123] In his *History of the Inductive Sciences* (1837), Whewell proposed that "Cuvier's . . . master-principle" of correlation, which enabled him to "reconstruct the whole of many animals of which parts only were given," was another of those "beautiful and orderly laws by which the universe is governed." Whewell, though, acknowledged that such an "opinion . . . may appear presumptuous in a writer who has only a general knowledge of the subject."[124] In fact, having "followed Cuvier principally thinking him the fairest and wisest writer I had seen," Whewell sent a copy of his *History* to Richard Owen in London, admitting that he had "always been afraid of physiology as a branch of my understanding" and asking for comments on "anything which you think defective or erroneous."[125]

Owen, by his early thirties, had already risen to the Hunterian Professorship of Comparative Anatomy and Physiology at the Royal College of Surgeons, and his skillful investigations of the functional anatomy of creatures such as the pearly nautilus were the talk of the metropolitan scientific scene (fig. 2.4). The self-possessed tyro assured Whewell of the "benefits and pleasure which I have derived from your History."[126] He was nevertheless more proactive in the preparation of its sequel *The Philosophy of the Inductive Sciences* (1840), frequently assisting Whewell with advice and information. After checking the proofs, Owen could not resist "expressing how much we—i.e. the Cuvierian cultivators of Comparative Anatomy—are, and always must be, indebted to you for the clear statement of the scientific character of teleological reasoning."[127] The youthful Hunterian Professor was no less accommodating to Buckland at Oxford, telling him, in July 1832, that "since the decease of the lamented Cuvier, there is no one whose opinion . . . I look for with more anxiety than your own."[128] Five years later, however, the shoe was firmly on the other foot, and Owen, having made a "careful reading," as he informed Buckland, of "your Bridgewater Treatise," had several "notes and suggestions as arose therefrom," which, he gently urged, would be "worthy of your attention."[129] He did not tell Buckland, but Owen hinted to Whewell that he had even "found out teleologically the business of the Megatherium," and was "never satisfied with the diet of *roots* assigned to him by some of our friends."[130] This alternative interpretation of the megatherium's perplexing feeding habits, which will be discussed in chapter 4, would replace Buckland's own in subsequent editions of his *Bridgewater Treatise*. Although Owen had not been educated at either of the ancient universities, and was instead trained in the narrower medical curricula of Edinburgh and

FIGURE 2.4. Robert Lee, *Oxford Street with Placards Referring to Owen's Researches*. Pen and ink, undated. Caroline Owen Commonplace Book, MS0283, Royal College of Surgeons, London. This sketch, presumably made in 1832 or shortly thereafter, shows advertising bills for Owen's *Memoir on the Pearly Nautilus* (1832), his work on the reproductive system of the platypus, and the partial megatherium skeleton that had recently arrived at the Hunterian Museum. The bills have just been pasted onto a temporary hoarding on London's main commercial thoroughfare, while the bill-paster, standing to the right, holds several more sheets of the same advertisements. Reproduced by kind permission of the President and Council of the Royal College of Surgeons of England.

London (where he studied under both John Barclay and John Abernethy), his technical prowess and specialist expertise helped both sustain and actually shape the natural theological science taught, albeit with different inflections, at Oxford and Cambridge.

And it was not long before Buckland and Whewell required Owen's assistance once more. In the summer of 1838 Cuvier's successor at the Muséum d'histoire naturelle, Henri de Blainville, disputed his predecessor's classification of two fossil jawbones from Oxfordshire as mammalian, and, more perturbingly, cast doubt on the entire edifice of Cuvierian correlation. Initially published in the *Comptes rendus* of the Académie des sciences, Blainville's attacks on Cuvier, with whom he had been estranged since 1816, were translated for the *Magazine of Natural History* by its irascible editor Edward Charlesworth, who shared his concern with the "too rigid enforcement of the Cuvierian doctrines."[131] Blainville was eager to "draw the attention of English naturalists to a matter of such great importance in palæontology, and to show how questions of this nature ought to be treated." Detailing the ambivalences in Cuvier's identification of the tiny *Didelphis*, which, like the famous Parisian opossum he had discovered in 1804, seemingly exhibited marsupial characteristics, Blainville asserted:

> Without doubt persons who are little versed in the study of organic structures, and who place too implicit a reliance on perhaps rather a presumptuous assertion, that, by the aid of a single bone, or of a simple facette of a bone, the skeleton of an animal can be reconstructed . . . may very possibly think it strange that four or five half-jaws . . . should be insufficient to indicate promptly and with certainty to what class the animal to which they belonged should be referred.

While, as Blainville acknowledged, the "assertion above quoted . . . has almost passed into common phrase," he insisted that it "becomes strained, and even quite fallacious, when the forms in question are more or less isolated, whether recent or fossil." Cuvier's famous capacity to identify or even reconstruct extinct creatures from merely single bones was actually strictly limited, and only effective, according to Blainville, "when we apply it to known animals, or such as differ but little from them." This, he pledged, would soon "be placed beyond doubt in the continuation of my great paleontological work" *Ostéographie* (1839–64).[132]

Even before the initial installments of Blainville's serialized opus reached Britain, his own alternative classification of the Oxfordshire jawbones as reptilian, as well as his resultant critique of the Cuvierian method, were vehemently rebutted by Owen at the Geological Society. Owen began by conceding that "when the accuracy of one of Cuvier's interpretations is called in question by an anatomist so eminent as his successor in the chair of Comparative Anatomy, the confidence of geologists in their great guide in Palæontology becomes liable to be shaken." But Blainville's "conflicting opinion" of the *Didelphis*, he noted disdainfully, was "maintained on mere theoretical grounds, and without actual inspection of the fossil itself." Owen, on the other hand, had access to all the relevant specimens, and his detailed anatomical examination of them established further "strong evidence of the marsupial nature" originally identified by Cuvier when he visited Oxford in 1818. Notably, Owen acknowledged that the "two original fossils" had been specially "transmitted to me from Oxford," and that it was "in compliance chiefly with Dr. Buckland's request that I have undertaken to test the validity of these doubts."[133] A further fossil jawbone was also sent to Owen, at Buckland's behest, by the Oxford-educated magistrate William John Broderip. By authoritatively vindicating the reliability of Cuvier's interpretative techniques, Owen, although speaking at Somerset House in central London, was once again upholding the natural theological agenda of the ancient universities.

Owen's paper on the *Didelphis*, delivered on 21 November 1838, elicited a "protracted and brilliant discussion" among the Geological Society's fellows that, as the *Athenæum* reported, was actually "more favourable to the views of M. de Blainville than we were prepared to expect."[134] It was therefore necessary for Whewell, as the Geological Society's president, to override this unexpectedly volatile discussion and strategically enshrine Owen's response to Blainville as a resounding triumph for Cuvierian functionalism. In his annual presidential address, Whewell gave a detailed and highly partisan account of the dispute, before concluding:

> I have dwelt the longer on this controversy, since it involves considerations of the most comprehensive interest to geologists, and, we may add, of the most vital importance. . . . The battle was concerning the foundations of our philosophical constitution; concerning the validity of the great Cuvierian maxim,—that from the fragment of a bone we can reconstruct the skeleton of an animal. This doctrine of final causes in animal structures, as it is the guiding principle of the zoologist's reasonings, is the basis of the geologist's view of the organic history of the world; and, that destroyed, one half of his edifice crumbles into dust. If we cannot reason from the analogies of the existing, to the events of the past world, we have no foundation for our science; and you, Gentlemen, have all along been applying your vigorous talents, your persevering toil, your ardent aspirations, idly and in vain.[135]

More audaciously even than Cuvier's wager at the beginning of the "Discours préliminaire," Whewell staked the intellectual viability of geology as a science, as well as the long-standing labors of the Geological Society's fellows, on the accuracy of the French savant's vaunted ability to reconstruct from just fragmentary disarticulated bones, and the teleological corollaries it allegedly entailed. It was, after all, the basis of "our philosophical constitution," as Whewell grandiloquently avowed. As with the ancient universities, the more recent forums of the scientific community such as the British Association and the Geological Society, both of which were regularly led by either Buckland or Whewell, were likewise beholden to Cuvierian correlation. And once again it was Owen's anatomical prowess that, as he had shown in his riposte to Blainville, was integral to its continued success.

Conclusion

It was therefore something of a problem that Owen's own researches had, from the mid-1830s, increasingly veered away from Cuvierian functionalism and instead prioritized form over function in tracing the osteological homologies that revealed an underlying skeletal archetype. He assured Whewell, in a letter from October 1837, that he wished to establish a "harmonious theory combining the transcendental and the teleological views," but in reality, as Nicolaas Rupke has shown, Owen eschewed such a synthesis and maintained an epistemological duality in which Geoffroy's transcendentalism generally took precedence over Cuvier's putative teleology.[136] Yet it was at precisely this time that Owen began to assume Buckland's erstwhile mantle as "the *British* Cuvier," even if Thomas Henry Huxley later quipped that he "stands in exactly the same relation to the French as British brandy to cognac."[137] Despite his growing prestige, Owen's professorial salary at the Royal College of Surgeons was relatively meager, and he bitterly resented what he called his "present anomalous position, holding a Cuvierian rank without the means of doing it justice." Buckland regularly advised the Tory prime minister Robert Peel on scientific matters, and informed him that "since the death of Cuvier even France herself has looked up to Owen as the only worthy successor of that great man."[138] With such a recommendation, Peel duly awarded Owen more than half of the government's annual funds for scientific grants.[139] The remunerative support of his influential patrons at Oxford and Cambridge, however, required that Owen did not entirely relinquish his advocacy of Cuvier's functional anatomy. Rather, he simultaneously espoused a transcendental approach when addressing more radically inclined audiences in London, and, on other occasions, maintained the teleological perspective that he knew would satisfy the conservative gentlemen of science at the ancient universities.[140]

Owen was in fully Cuvierian mode when Blainville charged his predecessor with a further "gratuitous hypothesis" regarding an extinct species of pangolin in the *Comptes rendus* for February 1839.[141] Although not having actually read it, Owen assuaged Whewell's anxieties by apprising him of the likely tenor of "Blainville's paper" and "this mistake as he calls it of Cuvier." Even Owen, though, could not hide his exasperation at Blainville's continual quibbling over minor points of interpretation, and incessant proclivity for "dealing a blow at the principle of a part of the skeleton revealing the nature of the whole animal."[142] In Paris, Blainville's constant carping had proved so deleterious to his predecessor's posthumous reputation that Cuvier's former secretary, Charles-Léopold Laurillard, needed to consult Owen when given the "opportunity to express anew the principle of the identification of the bone as formulated by Cuvier." He asked him to "tell me honestly whether you consider M. de Blainville's attacks against that principle to be based on reason," before gingerly confiding his own opinion: "I do not know if it is my respect for Cuvier distorting my judgment, but I find so little logic in the reasoning of M. de Blainville that I cannot believe he is right."[143] Across the Channel, where Cuvierian correlation was now more firmly established than in its native France, Owen doubtless concurred with his diffident correspondent. Even in Britain, however, Blainville's insidious criticisms could not be dismissed so readily, especially once the serialized installments of his *Ostéographie*, whose impending anti-Cuvierian sentiments had already been touted in the *Magazine of Natural History*, began to arrive in the summer of 1839.

What was needed was an unequivocal vindication of the capacity to infer the precise characteristics of a prehistoric creature from only a single part of its structure: a British equivalent of Cuvier's spectacular demonstration of his predictive powers when anticipating the presence of a marsupial pelvis in the ruptured block of gypsum that, in 1804, yielded the famous Parisian opossum. Only such a definitive proof of the method's reliability could turn the tide of Blainville's escalating criticisms, and also finally denude Cuvierian correlation of the lingering taint of materialism and Gallic irreligion that it had accrued over the previous four decades. In the final months of 1839, as the following chapter will show, an opportunity for precisely such an indubitable demonstration of the potency of Cuvier's method arrived unexpectedly from the other side of the world, although it needed all the skills of Owen, Buckland, and Broderip—and not just their scientific abilities—to turn it into the nineteenth century's most celebrated exemplar of the law of correlation.

PART II

Triumph, 1839–54:
Bones, Serials, and Models

3

Discovering the Dinornis

On the evening of 3 August 1848, a mysterious "seafaring man" knocked on the door of a house in London's aristocratic suburb of Belgravia.[1] The "shabbily-dressed" traveler was not visiting a blue-blooded noble, though, and instead wished to see the surgeon and paleontologist Gideon Mantell.[2] Mantell's son Walter had, while in New Zealand, amassed a large collection of remains of the extinct struthious bird named the dinornis, or moa as it was called in indigenous traditions, whose discovery nine years earlier had been widely hailed as one of the greatest scientific accomplishments of the nineteenth century. Only six months before, in February 1848, Mantell himself had eulogized Richard Owen's famous identification of the dinornis from a single "fragment of bone" as the "most striking and triumphant instance of the sagacious application of the principles of the correlation of organic structure enunciated by the illustrious Cuvier,—the one that may be regarded as the *experimentum crucis* of the Cuvierian philosophy."[3] Mantell's unexpected visitor, however, had a very different interpretation of how Owen had been able to infer the existence of a hitherto unknown bird from merely a broken piece of femur bone.

Introducing himself as "Dr Rule," he informed Mantell that "he brought home the first fragment of Moa bone—that Owen first described: [and] that he himself pointed out to O. that it was a bird." Clearly shocked by the revelation, Mantell lamented: "This is worse and worse. . . . Is it possible Owen can have told me such audacious falsehoods, and which, in my ignorance, I have promulgated."[4] If Owen had already known that the fragment of bone came from a bird, then his successful inference of its precise characteristics would have been much less striking as an exemplification of Georges Cuvier's law of correlation. The celebrated paleontological feat that

Mantell himself had recently designated a "brilliant example of successful philosophical induction—the felicitous prediction of genius enlightened by profound scientific knowledge" was, the mysterious traveler alleged, merely a cynical fabrication.[5] Aghast at having inadvertently propagated such potential falsehoods, Mantell excised any hint of panegyric from his next published account of the discovery of the dinornis in *Petrifactions and Their Teachings* (1851). Where he had previously celebrated Owen's putative deployment of Cuvierian correlation, Mantell now instead emphasized that the "fragment of a thigh-bone of a bird . . . had been brought to England by a Mr. RULE, who lent the specimen to Professor Owen." In an accompanying footnote, he alluded conspiratorially to the private conversation in which "Mr. Rule informed me" of these events.[6]

The mysterious Rule, meanwhile, had himself already ensured that his own version of the events surrounding the dinornis's discovery reached an audience beyond just Mantell. Five years before his impromptu arrival at the paleontologist's Belgravia residence, Rule had published a signed article on "New Zealand" in the July 1843 number of the London-based *Polytechnic Journal*. After an impersonal account of the flora and fauna of the far-off colony, Rule alluded to his own intimate involvement in the identification of the most celebrated specimen of New Zealand's natural history. Explaining how a "large portion of femur was brought to me" when he was employed as a surgeon in the Antipodes before subsequently returning to London, Rule asserted that the "great care and circumspection with which Professor Owen described this interesting article has my warmest acknowledgements."[7] He nonetheless insisted that, following on from Owen's own account of his renowned inference earlier in the same year, "I have to offer the following brief remarks." Even before he "placed this rare specimen in natural history in the hands of Professor Owen," the "fragment of bone under consideration" was, according to Rule, "clearly that of a bird of very great strength. It evidently did not belong to beast, fish, nor amphibious creature." In attempting to "prove the ornithic character of the fragment," Rule avowed that it was he, rather than Owen, who had "direct[ed] attention to the characteristic cancellous structure so prominently developed on its external surface." In fact, before meeting Owen, he had already "sought, without finding, for the bone of some bird that might equal it in size . . . in the public and private rooms of the British Museum," and rather than then simply entrusting the fragment to Owen's specialist expertise, the two men had "together compared it with the largest thigh bones of the ostrich in the Museum of the College of Surgeons."[8] The much-heralded discovery of the dinornis was, as far as Rule was concerned, the result of a scientific collaboration in which his own contribution had been decisive.

Even if the circulation of the *Polytechnic Journal,* a short-lived monthly aimed at educated readers, remained limited, Rule was determined that his article would be seen by influential men of science. In July 1843 he wrote to William Buckland, who six months earlier had forwarded to Owen the more substantial set of bones from New Zealand that enabled him to triumphantly confirm the original inference made from just Rule's broken femur. Alluding to the "fragment of bone of an extinct bird found in new Zealand, and brought home by me," Rule informed Buckland that "I have given the tradition relative to the bird in the Polytechnic Journal of this day; a copy of which I shall be happy to forward through your bookseller."[9] It is unlikely that Buckland responded to Rule's offer, and instead he seems to have passed the missive on to Owen, who, as will be seen shortly, certainly did read the *Polytechnic Journal* article with close attention. But with Rule proactively touting his own alternative to Owen's version of events, and especially once his cause was taken up by the embittered Mantell at the end of the 1840s, the suggestion that it was the seafaring surgeon who first noticed the ornithic character of the fragmentary femur cast considerable doubt on Owen's insistence that his acclaimed inference instead relied solely on Cuvier's law of correlation.

Yet, with hardly any exceptions beyond Rule's article in the *Polytechnic Journal* and Mantell's *Petrifactions and Their Teachings,* the primacy of the Cuvierian law in the dinornis's identification remained uncontested throughout the nineteenth century. It was not until the 1960s that Rule's contribution was again brought to light, and Owen's putatively "brilliant anatomical prediction" revealed to be "less simple and more interesting than at first appeared."[10] This enduring and almost universal acceptance of the sanctioned version of the chain of events between Rule's bringing the fragmentary femur to Owen in October 1839 and his receiving, via Buckland, the more substantive remains in January 1843 was far from fortuitous. Rather, it is testament to the remarkably scrupulous control that Owen and his supporters, Buckland and William John Broderip most notably, exercised over how the dinornis's discovery, and its purported exemplification of Cuvierian correlation, was represented in all sectors of the Victorian print media, from the proceedings of specialist societies to mass-circulation magazines. Of course other nineteenth-century men of science similarly attempted to suppress the contributions of colonial collectors and collaborators, Roderick Murchison most notoriously, but Owen's more effective and enduring erasure of Rule reveals his exceptional mastery of the innovative publishing formats that began to appear in the 1830s and 1840s.[11] Far from being the obscurantist preserver of "nepotism and privilege . . . 'raised' to abhor the 'ignorant savages' of the press" that Owen has often been portrayed as by historians, he enthusiastically embraced the new opportunities presented by a publishing industry undergoing a radical overhaul following the advent of the steam-powered printing press, the consolidation of vast new reading audiences, and the development of syndication practices that extended even across the Atlantic.[12]

With the willing connivance of such influential editors as Charles Dickens and John Gibson Lockhart, Owen, along with Broderip, ensured that any doubts over the reliability of Cuvier's approach to paleontology, such as those continually raised by Henri de Blainville in Paris, were rapidly and robustly refuted. The prevailing conventions of anonymous journalism even enabled Owen to revise and correct—and sometimes actually write—the accounts of the dinornis usually attributed to other authors. While historians have previously noted that the alleged reconstruction of the gigantic New Zealand struthian from merely a single bone assumed a mythological status in the nineteenth century, this chapter will, for the very first time, reveal exactly how that myth was both constructed and sustained.[13]

The clandestine manipulation of how the dinornis's discovery was reported in the press was considered necessary because, even more than Buckland's own Cuvierian accomplishments considered in the previous chapter, it afforded an indisputable affirmation that only providential design could have produced such a perfectly integrated mechanism. Owen's striking paleontological feat, moreover, bolstered the conservative creeds of natural theology at precisely the moment that early Victorian Britain was descending into economic depression and violent social unrest. To make matters worse, a seductive epic of evolutionary development that captivated the public in the mid-1840s threatened to supplant divine design with self-sufficient natural laws. The dinornis, once it was shorn of Rule's awkward insinuations and fashioned as an exemplar of Cuvierian correlation, could be directly invoked as a bulwark against both the revolutionary threat of the Chartist movement and the controversial transmutationism of the best-selling *Vestiges of the Natural History of Creation* (1844).

The Greatest Zoological Discovery of Our Time

A decade before his unexpected arrival at Mantell's home in Belgravia, John Rule had similarly sought an entrée at the Royal College of Surgeons. Having taken lodgings in nearby Fetter Lane, he wrote to Owen, the college's Hunterian Professor of Comparative Anatomy and Physiology, on 18 October 1839 stating:

I desire to offer for sale a portion or fragment of a bone . . . part of the femor [sic] of a bird now considered to be wholly extinct. It is unchanged after a lapse of many years—it may be centuries of years, in which it has been im-bedded in the mud of a river that disembogues into one of the bays in New Zealand.

The letter included the "six inches" of fragmentary bone, which Rule val-ued at "ten guineas," with a request that "no injury be done to it, should it be rejected."[14] A further enclosed missive from February 1837 explained that the broken femur had originally been sent to Rule, who had not yet visited New Zealand, by a relation in the colony. Recently returned from New South Wales, Rule had initially offered the bone to the British Museum, where John Edward Gray, the keeper of the zoological collections, recommended that he take it to Owen. As an Edinburgh-trained surgeon and a former member of the Royal College of Surgeons (having been released in 1826 in order to join the Royal College of Physicians), Rule considered it unnecessary for an inter-mediary to contact the renowned Hunterian Professor on his behalf.[15]

The Royal College of Surgeons declined the purchase of the bone, as the "price asked . . . was deemed too high"—but not before Owen had made an "exhaustive comparison" of the "unpromising fragment," initially with "similar-sized portions of the skeletons of the various quadrupeds," and then with the "skeleton of the Ostrich." Here, it was at once apparent that, as Owen later recalled, "'the bone' tallied in point of size with the shaft of the thigh-bone in that bird," and also exhibited the same "reticulate impressions." Hav-ing been "stimulated to more minute and extended examinations," Owen eventually "arrived at the conviction that the specimen had come from a bird" as large as an ostrich, although with the "striking difference" that the "huge bird's bone had been filled with marrow, like that of a beast." While no other birds of comparable dimensions had hitherto been found in New Zealand, which had been settled by British traders and missionaries only since the end of the eighteenth century, it was clear that the fragmentary femur must have belonged to a flightless bird of extraordinary and unparalleled stature. Con-tradicting Rule's account in the *Polytechnic Journal,* Owen implied that these deliberations were undertaken alone, and it was only when the bone's "owner called the next day [that] I told him, with much pleasure, the result of my comparisons."[16]

On 12 November 1839, barely three weeks after receiving Rule's letter, Owen "exhibited the bone of an unknown struthious bird of large size, pre-sumed to be extinct" before the fellows of the Zoological Society. He began by acknowledging that it "had been placed in his hands for examination by Mr. Rule, with the statement that it was found in New Zealand, where the natives have a tradition that it belonged to a bird of the Eagle kind."[17] This was a relatively accurate précis of the section of Rule's letter in which he alluded to the indigenous beliefs that "it was a bird of flight, not a bird of passage, as it never quitted the New Zealand forests." Significantly, however, Owen made no mention of Rule's own empirical observations on the fragment in the same missive, where he noted: "That it was a bird of great size and strength, the bone is evidence. . . . Inspection shows that it belongs to a bird—and not a beast, nor a fish, nor amphibious animal" (statements later repeated in his *Polytechnic Journal* article).[18] Denuding him even of his Edinburgh MD, Owen presented "Mr. Rule" as merely the unwitting purveyor of a largely fallacious native tradition.

Instead emphasizing his own exhaustive process of comparison, Owen concluded:

> So far as a judgement can be formed of a single fragment . . . [and] so far as my skill in interpreting an osseous fragment may be credited, I am willing to risk the reputation for it on the statement that there has existed, if there does not now exist, in New Zealand, a Struthious bird nearly, if not quite, equal in size to the Ostrich.[19]

Although implicitly invoking Cuvier's celebrated extrapolations from similarly fragmentary remains, this bold conjecture and dramatically avowed risking of a burgeoning reputation was at first declined by the Zoological Society's publication committee and only appeared in its *Proceedings* early in the following year after Owen agreed that responsibility for the paper rested solely with himself. It was soon reprinted, including the opening allusion to Rule, in the *Annals of Natural History*, before also appearing in the Zoological Society's *Transactions* with a lithographic plate of the fragment (fig. 3.1).[20] Surprisingly, though, Owen's high-stakes quasi-Cuvierian gamble seems not to have attracted any attention elsewhere in the press, either in scientific publications or in more general periodicals.

This would change radically three years later, when, in January 1843, William Williams, a missionary stationed in New Zealand, sent a consignment of miscellaneous bones to the Oxford home of Buckland, who in turn passed them on to Owen. Some of the specimens that Williams sent had been received from William Colenso, another member of the Church Missionary Society as well as a resourceful colonial naturalist who would later claim, like Rule, to have also recognized the bones' avian characteristics without Owen's assistance. From these bones, Owen was able to reconstruct a giant struthi-

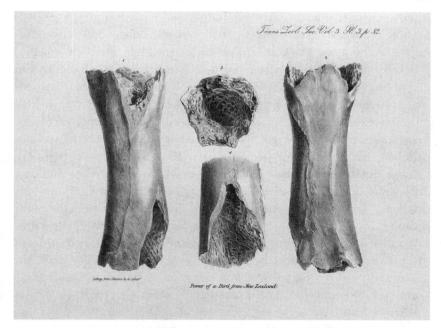

FIGURE 3.1. George Johann Scharf, *Femur of a Bird from New Zealand*. Richard Owen, "Notice of a Fragment of the Femur of a Gigantic Bird of New Zealand," *Transactions of the Zoological Society* 3 (1842): plate 3. This lithograph shows four views of the same fragment of femur bone, revealing the reticulate impressions on its inner surface that, Owen claimed, were the principal indication that it came from a bird. Reproduced by kind permission of the President and Council of the Royal College of Surgeons of England.

ous bird that appeared to confirm his earlier inference to a remarkable extent, although actually the dinornis, as Owen named it, was of considerably larger stature than an ostrich. Its size notwithstanding, it was soon apparent to Owen's closest friends and supporters that the most significant aspect of this colossal creature was its manifest vindication of the reliability of Cuvier's law of correlation.

On 20 January, barely a day after Owen began to articulate the "mysterious bones," Broderip wrote to Buckland, and quickly moved on from their jovial mutual congratulations:

> It is no joke and I look upon this as the greatest zoological discovery of our time. . . . All this not from any guess but from severe philosophical induction. This is not only another proof of the powers of our great physiological friend; but it comes well in aid against the smears that de Blainville . . . and other French physiologists have lately directed against the followers of Cuvier, and the principle of building up the whole skeleton from a bone or even the fragment of a bone.[21]

Although penned in haste, these impromptu comments became the template for how Owen and his supporters would present the discovery of the dinornis for several decades to come. In particular, the insistence on its status as a "severe philosophical induction" aligned Owen's procedures not only with Cuvier but also with the British tradition, enshrined in the revered methodology of Francis Bacon, of patient reasoning from established facts and the rejection of speculative theorizing.[22] The authority of such purportedly Baconian inductions had been regularly impugned by Blainville over the previous decade, and in the opening parts of his serialized *Ostéographie* (1839–64), Cuvier's embittered successor at the Muséum d'histoire naturelle completely denied the "degree of prevision" by which his predecessor could, from "only one bone," "reconstruct the entire skeleton, and consequently the rest of the organization of the animal to which it belonged." In fact, Blainville scornfully insisted: "In my opinion, no one has ever been able to provide proof of this pretension."[23] By November 1839, as will be discussed in chapter 8, the *Lancet* had printed a supportive English translation of this indignant skepticism toward correlation. The resounding corroboration of the prediction that Owen had made in the very same month afforded, as Broderip—with his legal training as a magistrate—immediately realized, precisely the kind of palpable "proof" for Cuvier's cherished axiom that Blainville claimed was wanting.

Reasons Obvious to the Student of Natural & Moral Science

Or at least it would if it were accepted that Owen's successful inference relied exclusively on the same rigorous philosophical induction undertaken by the great French savant, and remained untainted by any awkward insinuations that Owen had already been tipped off by Rule regarding the bone's avian characteristics. When, only four days after Broderip's letter to Buckland, Owen announced his discovery to the Zoological Society on 24 January 1843, he omitted all mention of Rule, even of the indigenous tradition that the bone belonged to an eagle he had previously been willing to acknowledge. While referring back to the original "femur . . . of which the shaft is described and figured in the Society's Transactions," Owen now avoided any allusion to what was already known about this bone. Instead, he concluded of the "*Dinornis of New Zealand*" that "it may not be saying too much to characterize it as one of the most remarkable acquisitions to zoology in general which the present century has produced."[24] Rule's de facto rejoinder in the *Polytechnic Journal* six months later, however, showed that his contribution to this extraordinary discovery could not be suppressed so easily.

It was likely Buckland who alerted Owen to the retaliatory article on "New Zealand" when, as noted earlier, he forwarded Rule's letter of July 1843 offering to send a copy of the *Polytechnic Journal*. Owen clearly read Rule's allegations with great scrutiny, and vigorously refuted them in successive articles on the dinornis for the Zoological Society's *Transactions*. In November 1843, for instance, he insisted on the necessity of "giving the rein to a too exuberant fancy," claiming to have avoided "laying undue stress on the native tradition of the gigantic Eagle . . . cited by Mr. Rule," and instead scrupulously maintained the "cautious scepticism due to second-hand testimony."[25] The indigenous beliefs that were the only aspect of Rule's input that Owen had hitherto acknowledged were now scornfully dismissed as simply whimsical hearsay that was irrelevant to the dinornis's reconstruction.

By 1846 Owen refused even to name Rule, alluding loftily to merely "a correspondent of the 'Polytechnic Journal.'"[26] In later years Rule was deper-sonalized still further, becoming only "an individual" in Owen's *Memoirs on the Extinct Wingless Birds of New Zealand* (1879).[27] This haughtiness toward Rule did not go unnoticed. In the 1930s the New Zealand politician Lindsay Buick remarked of Owen: "Curiously enough . . . he is singularly reticent and indefinite regarding his dealings with this gentleman. When he does refer to him, it is merely as 'Mr. Rule,' or as 'an individual,' or as the 'vendor' of the bone." Buick, as his puzzled tone makes evident, did not realize the precise reason for this reticence, and he suggested, rather implausibly, that Owen might have deferentially "regarded him [i.e., Rule] as a surgeon rather than as a doctor of medicine, and as surgeons affect the title of 'Mr.' and not that of 'Dr.,' Owen naturally gave him the surgeon's prefix."[28] The intimation that there was something "singularly . . . indefinite" about Owen's treatment of Rule nonetheless suggests an uneasiness about his conduct toward someone whom Buick considered a gentleman.

With the taint of Rule's purported contribution to the dinornis's discovery painstakingly eradicated, Owen could instead present his "inferences deduced from the structure of the shaft of a femur" as, in exact accordance with Broderip's suggestion to Buckland, a compelling "vindication of the fruitful principle of physiological correlations, the value of which . . . there has been a tendency to deprecate in an otherwise estimable osteological work." An accompanying footnote informed readers that this deprecatory work was none other than "The 'Osteographie' of Prof. De Blainville." The circumstances of Owen's initial hypothetical inference of the dinornis's existence and its resounding confirmation three years later afforded a renewed "confidence in the laws of correlation" that was especially needful in the wake of Blainville's venomous attacks.[29] But the credibility of these particular circumstances required that Rule remain hidden from view.

The awkward interloper does seem to have vanished for a number of years, possibly having returned to the Antipodes, but his reappearance, on Mantell's Belgravia doorstep in August 1848, required Owen's supporters to once again, and still more vigorously, refute any suggestions that Rule had anticipated the famous Cuvierian interpretation of the avian femur. Less than a year after Mantell disclosed details of his private conversation with Rule in *Petrifactions and Their Teachings*, Broderip, in an unsigned contribution to the *Quarterly Review*, strategically retold the story of the dinornis's discovery. He related how the "bone in question had been . . . offered for sale by an individual who averred that it came originally from New Zealand." Omitting any hint that this anonymized "individual" had also apprised Owen that the bone came from a bird, Broderip instead moved on to Owen's inductive inference and how the arrival of the bones in 1843 "fully confirmed the prevision" of the "great Struthious bird of New Zealand."[30] It was precisely this same "degree of prevision" that, as seen earlier, Blainville had denied to Cuvier.

Broderip, whose unwavering loyalty to Owen seems to have been motivated by genuine affection and admiration, then addressed Rule's cavils, indignantly avowing: "All criticisms and misgivings as to the original audacious induction from the fragment . . . being thus quashed, there remained only attempts at detraction from the merit of the discovery." Broderip's *Quarterly* article also rebutted the doubts cast on Owen's priority by "Dr. Mantell," who had asserted that, even if Rule's native tradition were disregarded, the New Zealand missionary "Mr. Colenso was the *first* observer that . . . determined the Struthious affinities of the birds . . . ere any intelligence could have reached him of the result of Professor Owen's examination."[31] As will be examined further shortly, Broderip's doughty defense of Owen in the *Quarterly* was secretly revised and corrected, and in some parts actually written, by Owen himself. Seeing Rule and himself so disingenuously stripped of credibility, Mantell fumed, in a private letter, that the "fulsome and dishonest panegyric on Owen" was "written by Owen and his bosom friend, Broderip." It was, he bemoaned, the "subserviency of the press" that had allowed "this eulogistic account of himself by O.!" Although the "scientific world" was apparently alert to the deceitful circumstances of this "scandalous attack," Mantell was concerned that the "public are gulled; for the great *We* is as potent as ever with the multitude."[32]

Having once more effectively expurgated Rule from the sanctioned version of events received by an unsuspecting public, the furtive collaborators could again insist—in a virtual paraphrase of Broderip's letter to Buckland— that the

> confidence of comparative anatomists and geologists in the fruitful principle of physiological correlations, if it had ever been shaken by the ingenious and active-minded but somewhat eccentric author of the "Ostéographie", must have been completely restored by the results of the palæontological labour of our distinguished countryman, carried on in the true spirit of Cuvierian inquiry.[33]

The discovery of the dinornis, however, could only be so strategically culti-vated as the nineteenth century's most exemplary instance of the "true spirit of Cuvierian inquiry" if Rule's contribution were entirely expunged from the historical record. With the aggrieved surgeon still awkwardly haunting the margins of London's scientific society with his scandalous secrets, this evi-dently required the use of some highly disreputable journalistic practices.

When, forty years later, Owen's grandson compiled his reverential *The Life of Richard Owen* (1894), he memorialized his late grandfather's celebrated identification of the New Zealand struthian by quoting extensive passages from "a writer in the 'Quarterly Review' (March 1852)," which, inevitably, reconfirmed that there had been no inkling of the famous femur's avian characteristics before it reached Owen.[34] Soon after the biography's publi-cation, though, the taxonomist Charles Davies Sherborn, who assisted with the book's scientific portions, sorted through Owen's remaining correspon-dence and came across the letter that Rule had sent him back in October 1839. Alarmed at what he found, Sherborn added a confidential note dated 15 November 1894:

> This is the original letter received by Owen along with the fragment of the femur of the Moa. It is of the greatest possible interest in connection with the subject & I wish it to be preserved. . . . The reason why it has not been pub-lished is that it was not found until after the "Life" was written, & for other reasons obvious to the student of natural & moral science.[35]

Sherborn's ambiguous final line suggests that even Owen's executors acknowl-edged that this letter impugned the Cuvierian credentials of the dinornis's discovery, and that the sustained campaign to deny Rule any credit in it had been, at best, morally questionable. Tellingly, Sherborn, although not willing to destroy Rule's problematic missive, recognized the need, even in the 1890s, to keep its contents hidden.

What at the Present Time Is Termed Popular Information

Despite the lingering shadow of Rule, the significance of Owen's achievement in identifying the dinornis from just a broken femur bone was immediately recognized at the elite forums of the scientific community. Roderick Murchison, in his presidential address to the Geological Society in November 1843, acclaimed the prediction arising from the "examination . . . of a single fragment of a bone" as "a discovery in natural history, which I do not hesitate in characterizing as one of the most remarkable of modern times."[36] At the following year's meeting of the British Association for the Advancement of Science, Hugh Strickland apprised his auditors of a "very imperfect fragment of a femur, which Professor Owen did not hesitate to assign to an extinct gigantic bird." In assessing recent progress in the science of ornithology, Strickland declared proudly: "This bold conclusion, which from the imperfection of the data seemed prophetic rather than inductive, was speedily confirmed by the arrival of fresh consignments of bones."[37] By the late summer of 1844, the successful prophecy of the dinornis's existence was the toast of the gentlemanly scientific specialists gathered in York for the British Association's annual meeting.

A few months earlier, the very same "discovery" was similarly commended as "unexpected and marvellous" by another eminent man of science. Mantell's *The Medals of Creation* (1844), however, was written expressly as a "familiar exposition of the elementary principles of Palæontology" for the "amateur collector," and its account of the dinornis instead dwelled on the bird's "enormous size" as established by the "bones . . . sent to England by Rev. W. Williams."[38] No mention was made of the original femur or Owen's daring inference. This was before Mantell's estrangement from Owen, when he was still willing to acclaim the dinornis's discovery before the Geological Society's members as the "*experimentum crucis* of the Cuvierian philosophy," and the only explanation for his reticence in *The Medals of Creation* was presumably that, as will be seen shortly with many journalists, he did not consider the details of a hypothetical induction, no matter how successful, of interest to the book's popular audience.[39]

By the early decades of the nineteenth century, the Enlightenment notion of a unified bourgeois public sphere had been replaced with a recognition that there were numerous discrete reading audiences each with its own specific interests and intellectual competences.[40] The most novel development was the emergence of new working- and lower-middle-class audiences rendered literate by religious schooling and now eager for scientific instruction, as well as publishers and authors alert to both the commercial and the philanthropic opportunities afforded by what was newly designated "popular science."[41] The effectiveness of Owen's striking identification of the dinornis could only be fully realized if it were accepted not just by scientific specialists but also by the burgeoning new readerships of works such as Mantell's *The Medals of Creation*. After all, Blainville's criticisms of Cuvierian correlation were already reaching just such popular audiences.

A notice of the *Ostéographie* in the *Foreign Quarterly Review* in 1840 began by noting how "geology [is] one of the most popular sciences of the day" and that the "celebrity Monsieur de Blainville has obtained as a naturalist and comparative anatomist, is of itself sufficient to attract attention to this work." It then printed, following the previous year's translation in the *Lancet*, its own English rendition of the book's assault on Cuvier's famous principle, stating: "The following observations are well worthy perusal, as they point out . . . the difficulties, if not impossibilities, of determining with certainty all the peculiarities of the entire skeleton of an animal, of which only a few bones are possessed." The review, by the Polish naturalist Jan Lhutsky, urged that it was necessary to promulgate Blainville's attempt to "throw . . . doubt on the correctness of many of what are termed the restorations of animals, that, in the course of the last few years, have been brought out rather hastily, and with too much presumption . . . to excite the wonder of the ignorant." With "geology . . . a branch of what at the present time is termed popular information," this was now more important than ever.[42]

The label "popular information" signified the rational instruction addressed to plebeian readers in inexpensive publications that eschewed the contentious subjects of politics and religion. This flood of reputable cheap reading matter, which by generating small margins on an extensive sale enabled publishers to realize the potential of a mass market, was intended to supplant the blasphemous, seditious, or just plain trashy unstamped press that had been proliferating since the end of the Napoleonic Wars.[43] The lucrative market for this so-called popular information had, by the 1840s, been effectively cornered by two rival organizations: the London-based Society for the Diffusion of Useful Knowledge (SDUK) and the philanthropic Edinburgh publishing firm established by the brothers William and Robert Chambers.[44] The SDUK's endorsement of Cuvierian functionalism in its Library of Useful Knowledge was noted in the previous chapter, but the Chambers brothers, like others in the Scottish capital, were more ambivalent about the renowned French savant. Two years after Lhutsky's review, *Chambers's Journal*, which offered its readers an engaging combination of fiction and

popular information for just one and a half pence, carried a sardonic account
of the more "credulous members" of the "class of philosophic and scientific
enquirers" that reflected:

> The happy inductions of Cuvier from a single fossil bone as to the gen-
> eral character of the animal to which it belonged, may be said to rank with
> the steam-engine and the Rosetta stone—the one good hit amongst many
> misses. . . . Even this eminent man was not incapable of being led too far. . . .
> From . . . only a few fragmentary and disconnected particulars, it is amazing
> how plausible a hypothesis is sometimes formed—perfect in itself, tallying
> perfectly with . . . every thing that could be wished—except the truth.[45]

Although Blainville was not named in this invective, it was exactly the same
criticisms of Cuvier made in his *Ostéographie* that were being repackaged for
the more than fifty thousand readers who purchased *Chambers's* each week.[46]
With such sentiments appearing in the house journal of W. & R. Chambers
in July 1842, Owen's supporters jumped at the chance, less than a year later,
to trumpet his discovery of the dinornis, with all its potential to refute Blainville, in a publication of Chambers's main rival as a purveyor of popular
information.

The Fireside of the Peasant and Artizan

Since the mid-1830s, Broderip had anonymously contributed short but authoritative articles on scientific subjects to the *Penny Cyclopædia* (1833–43).[47]
This ambitious reference work was first proposed by the publisher Charles
Knight to the SDUK following the success of their previous joint venture, the
Penny Magazine. As Knight calculated in a handwritten proposal, the "circulation of the Penny Magazine" had already revealed the enormous "number
of persons . . . who are desirous to acquire information, when presented to
them at a very low price."[48] In its early years Knight's forecast was vindicated,
and the *Penny Cyclopædia's* weekly numbers sold around seventy-five thousand copies.[49] *Hogg's Weekly Instructor* approved of Knight's plan for what
it called

> science stripped of its scholastic technicalities and obscurities, and rendered
> as delightful and instructive . . . presented to the masses in weekly penny-
> worths . . . [and] sent to the fireside of the peasant and artizan, to be their
> companions and instructors, at a rate more moderate than the companion-
> ship of an ale flagon for a few hours would be.[50]

Broderip's regular anonymous entries for the *Penny Cyclopædia* therefore
afforded the perfect opportunity to make Owen's inference regarding the di-

nornis, already well known among the gentlemen of science, a familiar topic even at the "fireside of the peasant and artizan."

In the *Penny Cyclopædia's* entry on "Struthionidæ," published in early 1842, Broderip concluded with a section on extinct specimens that dealt only with Owen's prediction, still unconfirmed at that stage, of the past existence an "unknown struthious bird of large size" in New Zealand. As well as reprinting Owen's paper from the *Proceedings of the Zoological Society*, Broderip also noted that "through the kindness of friends, we have been enabled to present our readers with some information not previously published."[51] In fact, as with many of Broderip's other contributions to the *Penny Cyclopædia*, Owen had been intimately involved in the composition of this entry. In February 1842 Broderip had asked him, "Will you tell me where to find . . . your account of the fossil struthian?" and then, later in the same month, sent him his "condensed treatise on Struthionidæ" with the request that Owen "be *very* critical and close in your reading and tell me honestly whether you think it will do."[52] As Nicolaas Rupke has shown, Owen "shaped" many of Broderip's contributions to the press during the 1850s by making "both minor and major alterations to the proofs," and it seems likely that the two men were already deploying similar collaborative practices a decade earlier with the *Penny Cyclopædia*.[53] Ironically, an American reader even assumed that another of Broderip's entries must actually have been penned by Owen, who responded sniffily, "I cannot permit myself to retain the credit of writings which are not mine, but which any zoologist might be proud to acknowledge," before insisting with disingenuous finality: "I have never contributed any article to that work."[54] Even if he was reluctant to admit it, Owen was directly involved in the attempt to enthuse the *Penny Cyclopædia's* plebeian readers about his Cuvierian accomplishments, with, tellingly, the same strategic silence regarding Rule being observed as in Owen's own published statements.

Owen's was not the only private information that appeared in Broderip's entries, though. When the bones confirming Owen's prediction arrived at the Royal College of Surgeons on 19 January 1843, Broderip, as noted earlier, was present. The following day he wrote to Buckland, relating how they had gone back to the original announcement in the Zoological Society's *Proceedings*: "We went over Owen's paper on the fragment of bone, the work of a man in the dark with the exception of the glimmering that he could collect from that fragment. Every word comes true to the letter."[55] Broderip, as will be discussed in the following chapter, was remarkably adept at exploiting the *Penny Cyclopædia's* serialized format, especially in interpolating an account of these exciting new developments into a timely but otherwise unrelated entry on the "Unau," an arboreal sloth from South America. What is even more remarkable is that, in this same entry for the SDUK's steam-printed endeavor to diffuse cheap useful knowledge among the proletarian masses, Broderip repeated exactly the same excitable rhetoric used in his handwritten letter to Buckland.

As the *Penny Cyclopædia*'s entry on the "Unau," published in the spring of 1843, proclaimed:

> It is curious and instructive, with these wonderful bones before one, to look back to Professor Owen's description of the fragment of bone which first came under his notice. . . . Entirely in the dark, with the exception of the glimmering light which he extracted from that fragment . . . every word that he then wrote comes true to the letter.[56]

The emotive and somewhat hyperbolic tone appropriate in a jubilant letter to a close friend, as well as the hackneyed trope of light overcoming darkness, were exactly replicated in the closely packed double-column quarto pages whose uninviting density was one of the principal reasons for the *Penny Cyclopædia*'s unprecedented cheapness (fig. 3.2). David Masson later remarked that, despite its intellectual virtues, many authors had a "cowardly shame at acknowledging obligations to a work of reference which had the unfortunate word 'Penny' as part of its name," and they considered that "to let the words 'Penny Cyc.' figure among [their] footnotes . . . would have been like walking down Regent Street . . . arm in arm with your uncle Hodge from the country."[57] Broderip, though, clearly saw nothing ignominious in replicating the very same words he had privately written to an Oxford academic and doyen of the Anglican establishment in the vulgar pages of the *Penny Cyc*.

The justification, of course, was that it afforded a means of disseminating, beyond just members of the scientific elite, the same lesson regarding the "doubts expressed by certain modern French physiologists as to the value of the method of Cuvier" that Broderip had recognized in his letter to Buckland and now replicated in his entry on the "Unau." Broderip did acknowledge the vastly different audiences for his private missive and the cheap weekly numbers from which the *Penny Cyclopædia*'s disreputable title was derived, adding a haughty caveat about the "value of the method of Cuvier" inevitably absent from the letter to Buckland: "Let us not be misunderstood: it is an instrument not to be wielded by every hand." Cuvier's method instead required "vast experience in all the phases of organic forms and a powerfully comprehensive mind."[58] While the "peasant and artizan" whom *Hogg's Weekly Instructor* identified as the *Penny Cyclopædia*'s prospective audience were instructed to have full confidence in Cuvierian correlation, regardless of Blainville's much-publicized criticisms, they were nevertheless exhorted

Teeth $\frac{5-5}{4-4}$, either contiguous or separated by equal intervals; upper ones trigonal; the anterior of the lower ones trigonal, the second and third subcompressed, the external face longitudinally sulcated; the last the greatest and bilobate.

Unau. (De Blainville.)

Head of the femur impressed by the ligamentum teres; tibia and fibula distinct. Astragalus with two excavations

Mylodon robustus. (Mus. Coll. Chir.)

anteriorly. Heel-bone long, thick. Falcular claw great and semiconical.

Species. *Scel. leptocephalum*, O.
    Scel. Cuvieri, O. (Syn. *Meg. Cuvieri*, Lund).
    Scel. Bucklandi, O. (Syn. *Meg. Bucklandi*, Lund).
    Scel. minutum, O. (Syn. *Meg. minutus*, Lund).

Genus 5. *Cælodon*, Lund.
Teeth $\frac{4-4}{3-3}$.

Genus 6. *Sphænodon*, Lund.*

In the Unau the number of cervical vertebræ is seven; in the Three-toed Sloth, *Bradypus tridactylus*, the number is nine.

The close approximation of the Sloths to the Birds in many parts of their organization calls upon us here to notice a discovery which will make the year in which we write (1843) a very remarkable one in the zoological calendar; and before we enter into the particulars of that discovery, we will just illustrate the close approximation above referred to, by observing that if nothing but the broken gigantic pelvis of the bird hereinafter noticed were laid before even an experienced eye, it might fairly enough be taken on a superficial view to have belonged to the genus Mylodon, though a closer examination would detect certain minute characters which show that it could not have belonged to a quadruped.

In the article STRUTHIONIDÆ will be found Professor Owen's descriptions of the fragment of a femur said to have been found in New Zealand, laid before the Zoological Society of London in 1839. [Vol. xxiii., p. 147.]

On the 17th of May, 1842, the Reverend William Williams wrote from Poverty Bay, New Zealand, to Dr. Buckland, and his letter contained an extract from another, sent by way of Port Nicholson, in February of the same year.

'It is about three years ago, on paying a visit to this coast south of the East Cape, that the natives told me of some extraordinary monster, which they said was in existence in an inaccessible cavern on the side of a hill near the river Wairoa, and they showed me at the same time

* Professor Owen thinks that both this genus and *Cælodon*, Lund, are indicated rather than satisfactorily established. The teeth of the sloth, he observes, are first developed in the form of hollow, obtuse cones, and do not assume the cylindrical form until worn down to the part which has acquired in the progress of growth the normal thickness; and this is afterwards maintained without appreciable alteration during the subsequent uninterrupted growth of the tooth. The compressed molars of the *Scelidotherium*, which, he remarks, doubtless follow the same law of development, would present in the young animal the form of hollow wedges, and such he suspects to be the nature of those teeth which are figured by Dr. Lund in the above-cited Danish memoir, plate xvii., figs. 5-10, and on which he has founded his genus *Sphænodon*.

to recognize that it could be deployed only by a few expert practitioners far removed from their own humble circumstances.

In reality, however, the *Penny Cyclopædia's* colossal size, as *Hogg's Weekly Instructor* subsequently realized after its completion in twenty-seven volumes, had "carried the aggregate work beyond the pecuniary limits at first intended, thus making it only purchasable by the middle class and better paid working-men."[59] Certainly, the bulk of the actual audience for the *Penny Cyclopædia*, as James Secord has shown with other SDUK publications, hailed from a "modest yet decidedly middle class background."[60] The intellectual quality of many of its entries also attracted even gentlemanly and specialist purchasers, and, ironically, Broderip's account of Owen's discovery of the dinornis seems to have gained most attention among readers already firmly in the camp of Cuvier's supporters, and at the same elite forums of the scientific community where correlation had by now been resolutely defended. Murchison, for instance, commended "a most graphic sketch of this monstrous bird and its analogies from the pen of my friend Mr. Broderip, Penny Cyclopædia (*Unau*)" to his auditors at the Geological Society, while Strickland recounted how Buckland, at the British Association in 1848, "commencing with the *Dinornis* . . . proceeded to remark that the 'Penny Cyclopaedia' was a work of great excellence."[61] Buckland, of course, had had the opportunity to ponder the excellence of the same account of the bird's reconstruction both in a handwritten missive and in the *Penny Cyclopædia's* printed pages.

Even when Broderip's entry did reach readers previously unacquainted with the precise details of the confirmation of Owen's inference, as when Walter Mantell reported that the "account in the Penny Cyclopædia, art. *Unau*, has interested us greatly" after he had traveled to New Zealand and discovered with "much regret that no copy of Prof. Owen's paper on the Dinornis [from the Zoological Society's *Proceedings*] is in this colony," they were hardly the autodidactic plebeians the SDUK's massive reference work was originally targeted at.[62] With Rule soon divulging a no less interesting account of the same events to Walter's outraged father, the necessity of ensuring that it was Owen, Broderip, and Buckland's own interpretation of the discovery of the dinornis that appeared in other new formats of popular science publishing remained as urgent as ever.

Burked by the London Press

While Owen's original hypothetical induction had, as noted earlier, generated surprisingly little interest in the press, the subsequent arrival of the bones confirming his inference was reported widely. What was considered most noteworthy about the discovery of the dinornis in the early press coverage, however, was the bird's remarkable size. Owen's ability to infer its existence from merely a fragment of bone, on the other hand, seemed to be considered no more alluring to readers of newspapers and periodicals than it was to the audience for Mantell's avowedly popular *The Medals of Creation*. Indeed, throughout the 1840s *Chambers's Journal* carried excited reports of what it termed the "*Dinornis*, a bird, by the side of which the ostrich is a dwarf," avowing that it "must have been not less than sixteen feet high!," without making any reference to the circumstances of its original identification.[63] By October 1843, though, the "height of Dinornis," as Broderip reported to Buckland, had to "yield to the severity of comparative measurement, and Owen has been compelled to cut down the largest species . . . to ten feet." Although themselves disappointed at the "reduced stature" of this nonetheless "gigantic species," Broderip and Buckland, unlike the journalists whose curiosity had been piqued by the bird's initially gargantuan proportions, were more concerned with the comparative methods that had subsequently shrunk its size.[64] The problem was inducing the correspondents of mass-circulation periodicals such as *Chambers's*, which in the previous year had actually re-packaged Blainville's criticisms of Cuvier, to share the same priorities.

Commercial periodicals that attracted plebeian readers by their strikingly cheap cover prices were often unable to fill their pages with costly original copy, especially if they were dependent on the contributions of just one or two correspondents, as was the case with *Chambers's*, which was largely written by Robert Chambers.[65] It was therefore common practice to supplement leading articles with reprinted passages extracted from noteworthy recent books, lectures, or reports in other journals.[66] This "scissars-and-paste school of authorship," as Charles Knight dismissively termed it, originated in the eighteenth century, although it became more prominent with the weekly miscellanies that began in the 1820s and sought to keep readers abreast of the latest intellectual fashions and tastes, including developments in science.[67] By the 1840s the practice was used more discreetly as particularly egregious instances drew widespread criticism, with even *Chambers's* condemning the "loose 'scissors and paste' style of publication . . . of the present day."[68] But it was not itself entirely averse to filling occasional columns with intriguing excerpts from other publisher's wares.

In the early 1850s, with Rule still at large and Mantell now divulging details of their private conversations, accounts of the dinornis's discovery that bore Owen's distinctive imprint began to appear in books by his close friends. In a philosophical treatise entitled *The Intellectual and Moral Development of the Present Age* (1853), the Welsh lawyer Samuel Warren insisted that, back in "the year 1839," an unnamed "shabbily-dressed man" had announced only that "he had got a great curiosity, which he had brought from New Zealand." There was no question, according to Warren, that "Professor Owen considered" the "fragment" without any prior indications as to "whatever animal it might have belonged." It was only by not even deigning to acknowledge the existence of Rule's insinuations that Warren could maintain, as Broderip and Buckland had likewise appreciated, that his friend had undertaken a rigorously philosophical induction that "carried comparative anatomy much beyond the point at which it had been left by his illustrious predecessor Cuvier."[69] The strategic benefits Owen accrued from his friendship with Warren will be examined further in chapter 5, and the latter's account of the dinornis certainly included particular details of the disputed encounter with Rule—such as the spitefully specific avowal that the surgeon was "shabbily-dressed"—that only Owen could have told him.

The same source of inside information also enabled Warren to proffer an exclusive tidbit to readers eager to actually encounter the famous fragment: "Any one in London can now see the article in question, for it is deposited in the Museum of the College of Surgeons in Lincoln's Inn Fields." It was, paradoxically, the very drabness and diminutive size of "this old shapeless bone," which, by making tangible the audacious and dramatic character of Owen's "immense induction of particulars," rendered it one of the exciting spectacles of metropolitan science, rivaling the more exceptional visual pleasures available at the capital's other sites of scientific instruction and entertainment (fig. 3.3).[70] This humble but captivating "article in question" had presumably been lent to the Hunterian Museum by Benjamin Bright, the friend of Broderip who had stumped up ten guineas for the now fabled femur after the Royal College of Surgeons declined to purchase it. It certainly remained the property of the Bright family for another two decades until it was finally presented to the British Museum in 1873.[71]

When couched in such vivid language and accompanied by confidential details of the famous bone's whereabouts, Warren's partisan rendition of Owen's discovery of the dinornis proved irresistible to periodicals with spare column inches to fill, even those that had hitherto been indifferent to the circumstances of the bird's initial identification. Soon after the publication of his *Intellectual and Moral Development*, Warren wrote to Owen relating how during a railway journey

a stranger in the carriage was reading the monthly no. of "Chambers' Journal":—& he called my attention to an article entitled "*A Wonderful Bone*"—& behold, there appears *my* account (duly acknowledged) of *your* old bone!! It's

FIGURE 3.3. Thomas Hosmer Shepherd, *The Hunterian Museum*. Watercolor, circa 1842. Royal College of Surgeons, London. By the early 1850s the famous fragment of the dinornis's femur bone could be viewed by the public in the large gallery of the Hunterian Museum, with visitors guided personally by Owen, as here, or by one of his staff. Reproduced by kind permission of the President and Council of the Royal College of Surgeons of England.

in the no. for 12 March 1853. The gentleman was quite enchanted & said "I'm going tomorrow to Lincoln's Inn Fields to see the bone", & he added that he supposed thousands would now go to see it. Chambers' Journal sells they say (I believe) 60,000 if not more, a week.

Later in the letter, Warren complained that his book had been "*burked* by the London press," suggesting that it had suffered the same dismemberment as the cadavers supplied to anatomists by the notorious murderer William Burke (see chapter 8).[72] His evident exhilaration at the no less brutal severing of a particular section from the book in an Edinburgh-based periodical, especially one notable for its large circulation but, like the *Penny Cyclopædia*, also able to attract gentlemanly readers, suggests that, notwithstanding his fears of being "burked," Warren recognized the advantages of the "scissors and paste" system of reprinting extracts for generating publicity and increased sales.

With *Chambers's* titling its extract "A Wonderful Bone," a phrase more hyperbolic even than those actually used by Warren, and prefacing it "We gladly make room for the following," which "touches on the subject of comparative anatomy, and the pitch to which a study of it has been carried in this country," the same advantages would have been no less apparent to Owen and his fellow advocates of correlation.[73] After all, the passage from *Intellectual and Moral Development* reprinted in "A Wonderful Bone" retained and perpetuated the denigration of Rule, based on private information supplied by Owen, in Warren's book. This was especially significant, as *Chambers's* had earlier printed an "outline" of a "lecture, delivered . . . by Dr Mantell," which repeated the contentious claim that Rule had declared his bone to be from a "large bird" when he first met Owen.[74] The war of words over Rule's exact role in the famous discovery was perpetuated by means of "scissors and paste" journalism.

The gentlemanly periodical reader in Warren's railway carriage certainly seemed convinced, even "enchanted," by the integrity of Owen's Cuvierian induction, and, along with thousands of other potential visitors, planned to come and witness for himself the spectacle of the diminutive bone from which the inference was made. Being "burked" by the press enabled the carefully honed version of the events surrounding the dinornis's identification to permeate into mass-circulation periodicals, including those, such as *Chambers's*, that initially considered that questions of paleontological method would hold little interest for their predominantly plebeian readers.

Our Friend Dickens

Surveying the popular press in the decade after the completion of his own *Penny Cyclopædia*, Knight nominated as the two "most successful periodical works above a penny—'Chambers' Journal', 'Household Words.'"[75] The first of these best-selling journals had, after years of merely remarking on the dinornis's size, finally been induced to notice the audacious prediction of its past existence only via the convoluted "scissors and paste" system of reprinting extracts from books. With the second of Knight's nominations, the process was considerably smoother and much more direct. A year before "A Wonderful Bone" appeared in *Chambers's*, Owen received an urgent request from the office of *Household Words*, a new weekly costing twopence and with the considerable selling point of being edited by Charles Dickens. Owen had been a friend of the famous novelist since the mid-1840s, and on 1 July 1852 the subeditor of *Household Words*, William Henry Wills, sent him the incomplete proofs of a prospective article, stating:

> I have bottled up as well as I was able what you were good enough to tell me. Will you kindly fill up the blank I have left & correct the errors which I fear I have made. The messenger will wait. The passage occurs in the last two paragraphs.[76]

The text that Owen hurriedly corrected and completed as the messenger waited was published nine days later as "A Flight with the Birds," and, notably, its final paragraphs, precisely where Wills had left the blank for Owen to fill, featured yet another account of the dinornis, beginning indicatively: "About this gigantic bird we have a good deal to say."[77]

With license to amend and supplement what Wills had already gleaned from their conversations, Owen could surreptitiously enforce his own version of the contentious story of the struthian's discovery. The still unnamed Rule was relegated from a surgeon to a mere "sailor . . . dealing in . . . marine stores" (he was defamed still further in other periodical accounts, which labeled him an "illiterate seaman").[78] Owen's own social status, on the other hand, was emphasized as much as his scientific integrity. He asserted, in the third person, that "a gentleman — one Professor Owen — who had a remarkable predilection for old bones . . . concluded from certain structural evidences, that this bone belonged to a bird."[79] The circulation of *Household Words* was only slightly smaller — at about forty thousand — than that of *Chambers's*, which had so impressed Warren.[80] Wills's rushed request therefore represented an unmissable opportunity to once again ensure that it was the sanctioned Cuvierian interpretation of events that reached such a large readership.

Access to this increasingly important popular audience was by no means a given for men of science in the 1850s. When, for instance, Thomas Henry Huxley wished to influence the press coverage of a proposal for dispersing the British Museum's natural history collections, he told a correspondent: "I have written to Rob. Chambers requesting he will give us an article in Chambers Journal to show the advantages of our plan for the people — Can you get at the 'Household Words'? If one only knew that snob Dickens."[81] Huxley, like Knight, recognized which were the most influential of the cheap periodicals, but his lack of personal contact with the haughty novelist meant that *Household Words* remained beyond his reach (and nor was he any more successful with *Chambers's*). Owen himself conceded, in private, that "Dickens has grown so arrogant," although he was prepared to overlook this exasperating pomposity while he was afforded the kind of access to *Household Words* that Huxley could only yearn after.[82] In fact, within three months of Wills's request, Dickens had asked Owen to "write some familiar papers on Natural History, yourself, for this Journal."[83] Unsurprisingly, he once more availed himself of the chance to coyly acclaim that "eminent gentleman in charge of the unrivalled museum of . . . the College of Surgeons," and also endorsed the validity of the method by which the "anatomist . . . is able to divine from a mere fragment of the skeleton, the nature and affinities of the animal of which it has formed part."[84]

Even if he did not realize they had actually been penned by his own collaborator with the *Penny Cyclopædia*, the final paragraphs of "A Flight with the Birds" certainly attracted the attention of Broderip. He wrote gleefully to Owen on 12 July 1852: "Look in the Times of to-day—paragraph headed 'a bird *twenty* feet high' with a voucher by our friend Dickens in 'Household Words.'"[85] Just as *Chambers's* reprinted passages from Warren's *Intellectual and Moral Development* under its own improvised title "A Wonderful Bone," so the *Times*, deploying the same practices of "scissors and paste" journalism, had extracted the particular paragraphs from *Household Words* relating the story of the dinornis's discovery.[86] Although not knowing of Owen's direct involvement with the original article, Broderip, who as a magistrate helped Dickens to find female miscreants for his reformatory for fallen women at Urania Cottage, intimated that it was their mutual friendship with the campaigning novelist that had facilitated the dissemination of their particular version of the events of 1839, denigrating Rule and affirming the Cuvierian credentials of Owen's inference, into the pages of Britain's most influential daily newspaper.[87]

The Pages of the Am. Jour. Are at Your Service

Owen's friendly relations with editors extended even across the Atlantic, where such informal contacts similarly enabled him to exert control over how the story of the dinornis's discovery was represented when it was reprinted in the American press. At the end of a long, convivial letter to Owen from July 1843, Benjamin Silliman, the Yale professor who both edited and financed the *American Journal of Science*, promised: "We shall always be happy to do any thing in our power to promote your important researches & the pages of the Am. Jour. are at your service."[88] In the light of such an offer, it is hardly surprising that when an annotated abstract of Owen's next article on the dinornis, an offprint of which he had himself forwarded to Silliman, was carried in the *American Journal*, its readers were assured that Owen originally "received from New Zealand the single shaft of a femur" and made his inference "on this evidence alone."[89] With Rule expurgated even more abruptly than he was in British accounts, the abstract was categorical in its assessment of the article's most salient contribution to science: "*This paper . . . affords us some beautiful and instructive examples of the wonderful principle of the correlation of structure in animals.*"[90] Notably, the *American Journal*'s anonymous commentary was written by Edward Hitchcock, whose controversial interpretation of fossil footprints found in the Connecticut valley as the only surviving remnants of unprecedentedly gigantic birds was, as will be examined in chapter 9, dramatically corroborated by the advent of the dinornis.[91] Hitchcock's partisan glossing of Owen's article ensured that readers across the Atlantic received precisely the same sanctioned account of the dinornis's identification as their British counterparts.

The nineteenth-century press's perpetual reprinting, on both sides of the Atlantic, of extracts, abstracts, and other forms of syndicated text has been termed "literary replication" by James Secord. Emphasizing how, in this complex process, "textual stability" is "difficult to achieve," Secord proposes that reprinted texts often become radically unstable and are denuded of the same range of meanings they previously held.[92] Such instability might even be assumed to be a necessary component of Secord's model of literary replication, with, for instance, Leslie Howsam suggesting the "metaphor" that, "like cells, texts replicate themselves, but with variants."[93] However, the *American Journal*'s annotated abstract of Owen's paper reveals that, even when the replication occurs on a transatlantic scale, the original meanings of reprinted texts do not inevitably mutate and instead can, with careful supervision, be retained. The private correspondence of both Warren and Broderip similarly shows how they assiduously tracked the reprinting of texts relating to the dinornis in *Chambers's* and the *Times*, evidently approving of the replicated versions that continued to propagate their own particular interpretation of the bird's disputed discovery. The original texts, moreover, had, unbeknownst even to Broderip himself, actually either been written by Owen or, as with Warren's book, prepared with his confidential advice.

Owen and his supporters engaged in a carefully orchestrated attempt to exercise control over the burgeoning range of meanings entailed in the process of literary replication, and, notably, they were successful in inducing publications as influential and widely read as the *Times* and *Chambers's*, or as geographically distant as the *American Journal*, to reprint what were, in effect, highly partisan and even deliberately distorted accounts of the dinornis as if they were merely exciting reports from impartial sources. The elaborate mechanisms of puffery and self-promotion facilitated by the conventions of "scissors and paste" journalism, anonymous authorship, and the covert bonds of personal friendship ensured that certain forms of literary replication could be designed and deliberate as much as variable and volatile.

Mantell's indignation, seen earlier, at Broderip and Owen's clandestine collaboration on an account of the latter's achievements for the *Quarterly Review* showed that there were still limits to how far such conventions could be exploited. In fact, although the *Quarterly's* editor, John Gibson Lockhart, was, like Dickens, a friend of both men and seemingly aware of their suspect working practices, he too was concerned they might have contravened the accepted standards even of anonymous puffery. As Broderip told Owen in May 1852 after having received his amended manuscript of the article's second installment:

> Your copious insertions have produced the enclosed letter from L. at which I am not surprised, as you may suppose from what I said to you. The additions cover ten additional pages of *slips* and . . . are very lengthy filling many consecutive pages and I think L. is not wrong about "minute vindications". You are far above all that.[94]

The surviving proof sheets of this *Quarterly* article show that Owen continued to add still further insertions, especially in the sections concerning the dinornis.[95] Nor, despite Broderip's reassurances, were any of the "minute vindications" of Owen's priority over Rule and Colenso removed from the published version. Ironically, the anonymous article that became so engorged with Owen's "copious insertions" began with an earnest assurance that its subject (and secret author) was "not one of those who delight either in talking or writing about themselves."[96]

When Owen's friend George Henry Lewes inquired about the author of the *Quarterly's* glowing appraisal of his work, he disingenuously replied: "I believe my reviewer is a barrister, a beak . . . the author of most of the Nat. Hist. articles in the Penny Cyclopædia."[97] Owen evidently recognized that his furtive collaboration with Broderip, which had raised the hackles even of Lockhart, could not be acknowledged beyond a close circle of his most loyal supporters. This nefarious joint effort was nevertheless regularly reprinted, as an ostensibly impartial account of the dinornis, in a variety of new formats, including, ironically, his own grandson's *Life of Richard Owen*. The prevalent "scissors and paste" system of reprinting ensured that the underhand puffery that had tested even the *Quarterly's* indulgent standards, along with Owen's no less clandestine contributions to *Household Words* and Warren's *Intellectual and Moral Development*, shaped how the story of the dinornis's discovery was received—with Cuvierian correlation foregrounded and Rule a conspicuous absence—for the remainder of the nineteenth century.

That Insatiable Animal the Many-Headed Monster

The reason why Owen and Broderip were willing to so extensively employ journalistic practices that, as even they realized, might be construed as dishonest and reprehensible was that the dinornis, and especially its putative identification by means of Cuvier's law of correlation, had particular political and theological resonances in the tumultuous context of early Victorian Britain. Owen's original inspection of Rule's femur bone occurred amid the violent response to Parliament's rejection, in July 1839, of the People's Charter, which, following the anticlimactic compromises of the Reform Act seven years earlier, demanded universal male suffrage and a secret ballot. With Cuvierian correlation having long been tainted by connotations of materialism and irreligion, it was imperative that the story of Owen's induction of the dinornis be perpetuated, through the processes of "scissors and paste" journalism and literary replication, in ways that upheld the values of the political and religious establishment. Even the reprinting of the story in the ostensibly apolitical genre of popular information had ideological implications, with the SDUK's radical opponents accusing it of attempting "to stop our mouths with *kangaroos*" (and Broderip did contribute an exhaustive account of "Marsupialia" to the *Penny Cyclopædia*).[98] Unlike kangaroos, however, the specific circumstances of the dinornis's discovery gave it a political potency far greater than merely diverting the attention of plebeian readers from sedition and blasphemy to arcane Antipodean animals.

In May 1842, as a second wave of Chartist protests demanding more radical reforms of Parliament gained the support of over three million workingmen, the militant medical journal the *Lancet* similarly called for the reform of the no less moribund and unrepresentative Royal College of Surgeons.[99] The diversion of college funds from surgical matters to Owen's paleontological researches had long irked the *Lancet's* irascible editor Thomas Wakley, and he now satirically invoked the Hunterian Professor's celebrated method of drawing inferences from fragmentary fossils:

> The Council of the London College of Surgeons remains an irresponsible, unreformed monstrosity in the midst of English institutions—an antediluvian

relic of all, in human institutions, that is most despotic and revolting. . . .
According to the law of correlation, Mr. OWEN, treading in the steps of
CUVIER . . . divined the whole structure of an animal from the inspection
of a single part. The same law will apply to political structures; and when
we put this single fact—"irresponsibility in the management of a public
institution"—before the eyes of the student, it gives him the key to the elaborate evils which have marked the course . . . of the College of Surgeons.

The political application of Cuvier's law meant that just one indication of
institutional venality could expose "how ruthless and fierce [were] the instincts which governed the whole organisation." Far from being intrinsically
conservative, Cuvierian correlation, as seen in chapters 1 and 2, had been
regularly appropriated on behalf of radical and democratic causes in the initial decades of the nineteenth century, from Edinburgh Whigs demanding
political change in the wake of the French Revolution to the rabble-rousing
reprints of William Lawrence's materialist lectures in the 1820s. It was once
more co-opted in the *Lancet*'s satirical squib, this time as a radical scourge of
corruption and vested interests, revealing unreformed institutions, whether
the Royal College of Surgeons or Parliament itself, to be monstrous beasts
with "teeth . . . sharp, and set like scissors, to cut and tear."[100]

Although it had recently introduced significant fiscal reforms and progressive legislation on employment conditions, Robert Peel's Tory adminis-
tration resolutely resisted, throughout the turbulent summer of 1842, both
the Chartists' petition calling for Parliamentary reform and the *Lancet*'s
demands for wholesale medical reform.[101] At the beginning of the following year, it was Peel whom Broderip first contacted, once he had written to
Buckland, when the bones that confirmed Owen's inference of the dinornis's
past existence were unpacked, although, embarrassingly, he misled the prime
minister over the bird's size. Once the accurate dimensions had finally been
established, Broderip ruefully told Buckland:

> As I have been the innocent cause of giving the impression entertained by
> Owen and myself at the unpacking to Sir Robert Peel, take, I pray, the first op-
> portunity of letting him know the reduced stature. Sir Robert, of all men, has
> a right to the earliest intelligence of this.[102]

In November 1842, two months before the arrival of the bones from New
Zealand, Peel had awarded Owen a lavish civil list pension of two hundred
pounds per year. In bestowing this, he insisted: "I have not inquired what are
your political opinions, and am wholly unaware of them. My only object . . .
is . . . to encourage that devotion to science for which you are so eminently
distinguished."[103] Broderip's anxious haste to inform Peel of both the dinor-

nis's reconstruction and then its unfortunate shrinkage was not merely an apolitical gesture of gratitude, though.

Owen's politics were in fact distinctly Peelite on the vexed issue of Chartism, and, having volunteered for the Honorary Artillery Company in 1834, he spent much of the following decade supplementing the government's efforts to quell the threat of proletarian revolt. On 24 January 1840, barely two months after he announced his prediction of the New Zealand struthian to the Zoological Society, his wife, Caroline, recorded in her diary: "R. at H.A.C. on guard. . . . The order to keep guard originated with the Home Office. No alarm of Chartists, however, disturbed the tranquillity of . . . the night."[104] By involving Peel so intimately in the subsequent confirmation, exactly three years later, of the same prediction, Broderip's eager communications to the prime minister aligned the discovery of the dinornis with the staunch resistance to Chartism that he embodied, denuding correlation of the taint of the radical potential that the *Lancet* had satirically ascribed to it. Peel himself soon after commissioned a portrait of Owen "in the act of lecturing, holding the dinornis bone" as a "pendant to that of Cuvier" already hanging in the gallery of his country home (fig. 3.4).[105]

Owen later reflected admiringly that, in postrevolutionary Paris, "Cuvier, had the advantage of subserving the prepossessions of the 'party of order' and the needs of theology."[106] The putatively conservative connotations of the French savant's law of correlation, as exemplified by Owen's identification of the dinornis, became still more important as the economic depression of the 1840s persisted and the example of the revolutionary movements that shook Continental Europe in 1848 prompted a third and final wave of Chartist insurrection. Broderip, correcting the proofs of Owen's latest examination of the New Zealand struthian for the Zoological Society's *Transactions*, seemed to suggest a connection between its affirmation of the validity of the Cuvierian method and the containment of the mass meeting at Kennington Common organized by the Chartist Convention on 10 April 1848. Returning the proofs on 13 May, he told Owen:

> Here you have the two last sheets of Dinornis etc., corrected in the midst of reports from Kennington and such cases as a London police court alone brings forth. The paper is real good work and you make your claim for comparative anatomy to take rank as an exact science modestly and forcibly. You will not be all devoured by that insatiable animal the many-headed monster this time. The precautions taken were complete and instead of 100,000 with which we were threatened there were not, according to my information, more than thousands there—*because they knew of the precautions and that every one would do his duty.* By five o'clock the blessed rain came down and they quietly dispersed.[107]

FIGURE 3.4. Henry William Pickersgill, *Richard Owen*. Oil on canvas, 1844. Natural History Museum, London. This painting, commissioned by Prime Minister Robert Peel for his own private collection, shows Owen in his Hunterian robes holding a leg bone of the dinornis. © The Trustees of the Natural History Museum, London.

Owen's paper, which was first delivered at the Zoological Society in January 1848, boldly asserted the "claims of Comparative Anatomy to the rank of an exact science by virtue of the predictive power with which its rules may be applied," insisting that the "prevision of an unseen part, founded upon the laws of the correlation of animal structures, becomes, by virtue of the nature of those grounds, something more than a mere guess."[108] Reading these "sheets of Dinornis etc." while he tried arrested Chartists from the bench of the Thames Police Court, Broderip evidently considered his friend's scientific "real good work" in the same light as the resolute "precautions," which included the mobilization of troops and artillery, that had forestalled the threat of revolution.

Buckland, having by now left Oxford and ascended to the rank of dean of Westminster, responded to the "demonstration on the part of the mob" by "enrolling special constables, &c." and even threatening to "himself . . . stand and knock down everyone" entering Westminster Abbey "with a crowbar."[109] Unlike his increasingly eccentric friend, Broderip, by intertwining the two topics in his missive to Owen, implied that "forcibly" enunciating the law of correlation was the only coercion necessary to repel the craven masses (just as the Chartists themselves were split between advocates of "physical force" and "moral force").[110] After all, Owen had visited Ireland in August 1843 at the height of the so-called monster meetings demanding the repeal of the Act of Union arranged by Peel's bitter adversary Daniel O'Connell. The "villainous repeal agitation," as the Irish naturalist Robert Ball warned Owen, was a "very ugly business," and he was fearful that the "people whom O'Connell has round him may prove to him as unmanageable as Frankenstein's man did to him."[111] Broderip, on the other hand, merely reflected complacently, while Owen was away, that he was "astonishing Paddy with Danger Birds."[112] The dinornis, to which Owen gave the popular soubriquet "danger-bird" while in Cork, would, Broderip was confident, once more afford an effective bulwark against the mainland's own querulous "many-headed monster."[113]

Handmaid to a Trustworthy Natural Theology

Having imbibed the finer points of natural theology at Oxford with Buckland, Broderip's blithe confidence resided, in part, in the manifest evidence of a divinely ordained order in both nature and society afforded by Owen's spectacular feat in identifying the dinornis. The "*providential*" strain of natural theology in which Broderip had been trained proposed, as seen in the previous chapter, that human society was a complex organism in which hierarchy was natural.[114] As Buckland proclaimed in an Easter sermon delivered on 23 April 1848, less than a fortnight after the containment of the Chartist meeting at Kennington Common:

> Notwithstanding the feeble outbreaks of a few unquiet and discordant spirits . . . the God of Nature has determined that moral and physical inequalities shall . . . be co-extensive with his whole creation. He has also given compensations co-ordinate with these inequalities, working together for the conservation of all orders and degrees in that graduated scale of being, which is the great law of God's providence on earth. . . . Equality of mind or body, or of

worldly condition, is as inconsistent with the order of nature as with the moral laws of God.[115]

Far from merely refuting the irritating cavils of Blainville, Owen's accurate induction of the structure of a previously unknown creature from just a single part of its anatomy compellingly corroborated precisely these conservative creeds.

In fact, amid the profound political tensions of early Victorian Britain, as Adrian Desmond has proposed, "Owen's work could only have been welcomed as a powerful defense of God and the divinely instituted social order."[116] It is important to recognize, however, that this was not the reactionary social order of the paternalistic high Tory elite (as Desmond seems to imply), but rather the newer liberal Tory view that the competitive social market was a divinely ordained mechanism that ensured that each individual was at their appropriate rank.[117] It is the "law of Christianity, as it is the law of Nature," Buckland avowed in his 1848 Easter sermon, that "inequalities of worldly condition . . . follow the unequal use of talents and opportunities originally the same."[118] No longer able to ignore the manifest misery and inequity of contemporary society, Buckland's response, according to Boyd Hilton, was to "stand Paley's theory of providence on its head. . . . The machinery of Creation remained perfect, but the demonstration of its efficacy was now its propensity to produce not happiness but justice."[119] Such a theodicy, of course, could be readily accommodated with the laissez-faire economics of early Victorian capitalism. The prediction of the dinornis's past existence by means of Cuvierian correlation indisputably affirmed that only providential wisdom could have produced such a perfectly integrated mechanism as the giant struthian's anatomical structure. Ineluctably, it also implied that the capitalist social order, no matter how iniquitous it might appear at periods of strain such as the 1840s, was a product of divine design that could not be altered by mere mortals, whether through paternalistic melioration or Chartist revolution.

While Owen's daring inference occurred too late to be included in the sustained reassertion of the argument from design in the eight *Bridgewater Treatises* that were published successively from 1833 to 1836, the dinornis was soon incorporated into subsequent statements of natural theology. In the posthumous third edition of Buckland's contribution to the series, *Geology and Mineralogy Considered with Reference to Natural Theology* (1858), Owen himself added a new footnote regarding how he "had previously indicated the former existence of a Struthious Bird . . . in the island of New Zealand." Although never as committed to William Paley's mechanistic conception of

divine contrivance as the book's late author (and instead more commonly ex-
pounding an idealist archetypal understanding of vertebrate design), Owen,
at the behest of Buckland's son Frank, ensured that the dinornis belatedly
received its due in a tome that, as Buckland fils put it, "reappears, again . . . in
furtherance of the wishes of the Earl of Bridgewater."[120] Since the publication
of the initial two editions of Buckland's *Bridgewater Treatise* in the mid-1830s,
though, opposition to the model of divine superintendence of a static nature
that was integral to Paleyite natural theology had become considerably more
vehement and widespread.

It was noted earlier that much of the original copy in *Chambers's Journal*
was contributed by Robert Chambers. In the number for October 1843, he
presented the recent "discovery of the bones of the dinornis" as "one of the
most palpable links which connect the present with the past order of being."
Indeed, Chambers endeavored to co-opt the New Zealand struthian as evidence for the "great natural law which peoples the earth with beings perfectly
adapted to its progressive conditions" that, a year later, he would describe as
the successive "adaptation of all plants and animals to their respective spheres
of existence" in his anonymous best seller *Vestiges of the Natural History of
Creation*.[121] This sweeping epic of evolutionary development propelled by
self-sufficient natural laws, with God demoted to merely a distant First Cause,
enchanted many of the same middle-class audience who purchased *Chambers's*. One such reader, Florence Nightingale, visited the Royal College of
Surgeons in the summer of 1846 and compared the dinornis bones with those
of modern birds, keen to witness, as she told her cousin, "how the species ran
into one another, as *Vestiges* would have it."[122] No mention of the dinornis
was actually made in *Vestiges*, but Nightingale's approach to the Hunterian's
collections after reading the book was precisely the way of interpreting the
bird's bones Chambers had earlier encouraged in his eponymous periodical.

With articles in *Chambers's* often "aimed at preparing a wider audience"
for *Vestiges*, as Secord has suggested, it is perhaps unsurprising that, as seen
earlier, it repackaged Blainville's criticisms of Cuvier's law of correlation
for its popular audience, and was induced to finally acknowledge Owen's
Cuvierian prediction of the dinornis only very belatedly and even then by
means of "scissors and paste" reprinting.[123] After all, Cuvier's argument that
the perfect correlation of the animal frame would require holistically coordinated alterations that were so complex as to render evolutionary change
untenable had, as seen in chapter 1, done much to check the transformist
theories of his own contemporaries in Paris. Across the Channel, the subversive potential of Jean-Baptiste Lamarck's transformism had been quickly
recognized by working-class atheists, and reviews of *Vestiges* now warned of

its own potential to unleash a revolutionary tumult in Britain. The book's apparent rejection of an "overruling Providence" would, as the Cambridge geologist Adam Sedgwick opined, "undermine the whole moral and social fabric, and inevitably . . . bring discord and deadly mischief in its train."[124] Owen, as will be discussed in chapter 9, was at this time himself beginning to formulate a saltational process of evolutionary change by sudden deviations from embryonic types similar to that proposed in *Vestiges*, although the furious response to the book made it impossible for him to publicly acknowledge this.[125] Tellingly, he declined all requests to review *Vestiges*, but, as the debate over the best-selling book became increasingly heated, it was not long before his timely identification of the New Zealand struthian was invoked, in the reviews of others, in opposition to the same developmental natural laws that Chambers had seen it providing evidence for.

In 1845 a scathing notice of *Vestiges* in the *North British Review*, the organ of the evangelical Scottish Free Church, warned, like Sedgwick, that the best-selling book was "prophetic of infidel times . . . with a fair chance of poisoning the fountains of science, and sapping the foundations of religion." In response, the review, penned anonymously by David Brewster, endeavored to "controvert the theory of development" by showing that the "order of succession" in the fossil record was far from progressive. The recent discovery of the "*Dinornis*, a bird one-third larger than the African ostrich . . . resuscitated by Professor Owen," afforded a valuable new instance of an extinct animal that was far larger and better developed than analogous extant forms. Although Brewster also proposed that these "gigantic birds of 'fearful magnitude'" exhibited the "infinite skill and variety of contrivance which distinguish all the works of creation," he did not otherwise allude to the induction that had enabled Owen to intuit the struthian's divine design from just its femur.[126] As with other Edinburgh evangelicals such as Thomas Chalmers and John Fleming, Brewster had little interest in the natural theological implications of Cuvierian correlation. It was precisely those same implications, however, that were later invoked as the most effectual rejoinder to *Vestiges* in the pages of the very same periodical.

In 1858, with *Vestiges* having by then sold out its tenth edition, another anonymous article in the *North British Review* averred that the "incidents connected with the determination of the bones of the Dinornis, though often told, will bear repetition." The article's account of the now famous discovery dismissed Rule as merely a "seafaring man" with no inkling of his bone's avian origins, and resolutely insisted that the subsequent inference from the femur revealed the "exquisite beauty and accuracy of relation in nature!" Notably, it was suggested that such a reiteration of the discovery's details would help to "shelter the Church from the effects of the crude speculations of imaginative, would-be savants" who, as Chambers had in *Vestiges*, promote "wild theories . . . of spontaneous and equivocal generation, and of transmutation of species." In fact, the article's author proclaimed: "We cannot over-estimate the advantage both to natural science and natural theology, in having at such crises men like Owen." The "finely reverent spirit" in which he could "reconstruct the entire animal by getting the key to it in a small fragment" was a veritable "handmaid to a trustworthy natural theology."[127]

The article's author was John Duns, a Scottish Free Church minister who, although close friends with Fleming, had pointedly avoided "giving an opinion" on his fellow evangelical's intemperate attacks, discussed in the previous chapter, on Cuvier's view of the constant relation of teeth and hooves.[128] Duns was also the *North British Review*'s editor, and he sent an advance copy of his adulatory article to Owen, explaining that it was an "attempt to set some of your works in a relation to natural theology, which they have not hitherto held in the estimate of non-scientific readers."[129] Duns soon received a "frank and valued acknowledgement" and within months began sending proofs to Owen for his comments and correction.[130] Like Wills at *Household Words*, Duns also offered Owen a more direct route into the *North British Review*'s pages, proposing: "You have looked at some of my North British articles, and if you would give me even a few lines which I could make public use of I would feel ever obliged."[131] The same friendly relations with periodical editors that had enabled Owen and Broderip to surreptitiously expunge Rule from the story of the dinornis now helped to further recast it as an exemplar of natural theology and scourge of the radical threat of transmutationism.

Conclusion

John Rule had not disappeared altogether, though. In the mid-1850s he returned to the Antipodes for the final time, settling on New Zealand's south island, from where he wrote occasional letters to the local press asserting that it was he who had discovered the colony's now famous giant bird.[132] He seems to have died sometime in the early 1870s. By this time, naturalists resident in New Zealand had become increasingly resentful of Owen's authoritative interpretations of the colony's extinct fauna from faraway London, and began insisting that, with the dinornis especially, it was necessary to view the remains in situ.[133] With Rule now proudly adopted as a New Zealander, contesting Owen's account of the giant struthian's original identification became a further means for the colonial periphery to refuse to passively accept the scientific authority of the metropolitan center and instead assert its own interpretative autonomy.[134] Like Rule, Walter Mantell had also emigrated to the colony, and he evidently retained the indignation his father had felt on first hearing Rule's story at his Belgravia home. Two decades after that shattering encounter, at a meeting of the New Zealand Institute in 1871, the president, William Travers, along with

> the Hon. Mr. Mantell, alluded to the injustice that had been done to the late Mr. Rule, of Nelson, who took the first Moa bone to Professor Owen, and who had been represented in some quarters as being an illiterate seaman, ignorant of such matters, whereas he was an educated medical man, who was perfectly aware that the bone was that of a bird when he took it to England.[135]

While these colonial complainants were seemingly unaware that the condescension with which Rule had been "represented in some quarters" was directly attributable to Owen's informal contacts in the British press, some of their compatriots would soon entertain precisely such suspicions.

Two decades after Travers's and Mantell's remonstrations, the New Zealand Institute's *Transactions* for 1892 published an article by William Colenso, the missionary who had collected the bones that confirmed Owen's original induction back in 1843. Colenso had since asserted that, like Rule, he too recognized their avian characteristics long before hearing of Owen's inference in faraway London. Having only recently been alerted to the observations on the dinornis the "Quarterly Reviewer . . . made forty years before," and consulting the article in the single "copy of the *Quarterly Review* vol. xc" then available in New Zealand, Colenso discovered that a "certain infelicitous animus pervaded it, with regard to Dr. Mantell," who, as seen earlier, had supported his own claim to priority over Owen. He presciently discerned that the "reviewer . . . who wrote the body of the said review, did not write" the sections pertaining to Mantell, as the "tenor, tone, and language are so very different, so discourteous, so largely exceptional, so far from truth!"[136] This, of course, was the collaboratively authored review of Owen's scientific accomplishments in which Owen's extensive, and pettily vindictive, additions to Broderip's prose had perturbed the *Quarterly*'s own editor, and been palpable to Mantell himself. Colenso had already grown to resent Owen's metropolitan hauteur, and he slyly revenged himself with the very first public intimation of Owen's furtive contribution to the *Quarterly*'s effusive review of his own work. Even within Owen's lifetime, which ended in December 1892, his carefully orchestrated attempt to enforce a particular interpretation of the dinornis's discovery was beginning to unravel under the pressure of New Zealand's burgeoning scientific sovereignty.

FIGURE 3.5. Dinornis maximus: *Front View of Skeleton, and of the Author of the Present Work*. Richard Owen, *Memoirs on the Extinct Wingless Birds of New Zealand*, 2 vols. (London: John Van Voorst, 1879), vol. 2, plate 97. In a photograph from the late 1870s, Owen is pictured in the robes of the Hunterian Professor, holding the original broken femur brought to him in 1839 in his right hand while his left hand rests proprietorially on an articulated skeleton of *Dinornis maximus*. Even forty years after his famous prediction of the dinornis's erstwhile existence in New Zealand, Owen still exercised scrupulous control over how the discovery was represented. Reproduced by kind permission of the President and Council of the Royal College of Surgeons of England.

Remarkably, however, such reservations about Owen's version of events remained confined, for a further seventy years, to this far-flung corner of the British Empire. Elsewhere, Owen and his small coterie of loyal supporters were entirely successful at ensuring that the alternative accounts peddled by Rule and Mantell were suppressed, and his own highly partisan rendition of the dinornis's discovery became enshrined as an incontrovertible reality (fig. 3.5). In fact, the most triumphant and potent validation of Cuvier's law of correlation was now generally agreed to have occurred in London, via New Zealand, rather than Paris. It has hitherto been assumed that the industrialized print culture of the mid-nineteenth century, with its mass-circulation magazines and proliferation of literary replication, helped usher in the new developmental theories presented in *Vestiges*, and later Charles Darwin's *On the Origin of Species* (1859), by "forging 'a reading public' for liberal scientific views of progress."[137] Owen's successful control of a variety of new print formats, even of the meanings generated by the process of literary replication, shows that—as will be confirmed in subsequent chapters—large reading audiences could still also be recruited for overtly nonprogressive outlooks such as Cuvier's law of correlation.

Owen and Broderip's covert manipulation of the press's coverage of the dinornis was, as this chapter has proposed, partly a response to the particular political and theological anxieties of early Victorian Britain. Its remarkable effectiveness, though, meant that Cuvierian correlation, and the iconic New Zealand struthian that epitomized its seemingly miraculous powers, remained inescapable features of the nineteenth-century cultural landscape long after the country emerged from economic depression and violent social unrest. This long posterity of the dinornis will be examined in detail in later chapters. More immediately, correlation, as the next chapter will show, became inextricably entwined with an innovative approach to book production that was developing into Victorian Britain's most distinctive and prevalent mode of publication: serialization.

4

Paleontology in Parts

In November 1844 *Hood's Magazine* carried a letter that recounted the super-natural "school-myths" that haunted the author's vivid imagination during a medical apprenticeship in the north of England, inducing a debilitating anxi-ety concerning the cadavers he was charged to dissect. Although this gothic tale's climactic moment exposed its putatively ghostly encounter as simply the result of an "excited imagination," after which "every trace of supernatu-ralism now vanished," it still partook of the familiar narrative pleasures of the ghost story while at the same time resolutely debunking its claims to au-thenticity. This spectral missive appeared in the fifth installment of a series of droll and digressive articles on supernatural phenomena entitled "Recollec-tions and Reflections of Gideon Shaddoe, Esq.," and was signed "Your sym-pathising and admiring reader, Silas Seer." Before signing off, Seer revealed that, despite his subsequent skepticism, he had "mentally vowed while in my mortal agony never again to desecrate the Christian corpse" and thus to quit the profession of medicine. He closed the letter by promising, "I may tell you some day, Mr. Gideon, how this resolution was kept," which would, he sug-gested, involve the telling of "another ghost story." Shaddoe responded by declaring "how happy I shall be to hear from him again, as the conclusion of his letter leads me to hope I may," enticing readers with a suggestively equivocal promise of a further installment that would both offer the same at-tractions as Seer's first letter and also afford something new by resolving the loose ends in his earlier narrative.[1]

One of the founding principles of *Hood's Magazine* had been to "avoid Se-ries or Continuations, against which," the editor remarked, "there has grown up a strong prejudice from the badness of so many of them—the feeling against them has become very general; & makes us anxious for *independent* articles."[2] It was therefore only with "Gideon Shaddoe, Esq.," and especially the letter from Silas Seer, that *Hood's Magazine* began to feature some of the conventions of serialized writing and strategies for sustaining readerly interest that had been developed during the previous decade by novelists such as Charles Dickens and William Harrison Ainsworth. Their innovative style of fiction in novels such as *Nicholas Nickleby* (1838–39) and *Jack Sheppard* (1839–40) offered continuing narratives whose periodic interruptions left the audience in a state of expectant suspense.[3]

When Seer's second letter finally appeared in "Gideon Shaddoe, Esq.," in March 1845, the sense of readerly expectation was again invoked, with Shaddoe exclaiming: "*Here is*, to my own great satisfaction and doubtless to that of our friends who remember his former communication, a second letter from my correspondent Silas, just come in. Silas, thou hast not forgotten thy implied promise, and I thank thee." This combination of anticipation of new installments with the remembrance of former parts was a particular characteristic of the new modes of serialized narration that were emerging in this period. Seer's second letter related a further tale of supposed encounters with gory specters that again were in fact amenable to more prosaic explanations. If the resolution of the question left dangling at the end of the previous letter was disappointingly straightforward, with Seer simply finding that his "anatomical passion soon returned, and all resolves and scruples were forgotten," it nonetheless confirmed that the affects of suspense and expectation were generated as much by the procedure of serialization itself as by the intrinsic qualities of the narrative.[4]

What is particularly notable about this skillful deployment of the controversial new techniques of serialized narration in *Hood's Magazine*, which prompted the exasperated editor to complain, "I think it very bad to have 'continuations'—& do not like 'series'—(we have one already)," are the identities of the pseudonymous Gideon Shaddoe and Silas Seer.[5] Shaddoe was the quaintly sinister sobriquet adopted by the magistrate and naturalist William John Broderip, while well-connected readers of *Hood's Magazine* may already have heard both of the ghost stories contained in Seer's letters told over fashionable dinner tables, in a compelling verbal style, by none other than the foremost expert on comparative anatomy and paleontology in Victorian Britain.[6] Richard Owen's remarkable penchant for gothic ghost stories has never previously been noticed by historians, for whom he generally appears as a narrowly "autocratic and ambitious personality," but it was well known to his contemporaries.[7] Even the Prince of Wales recalled having being told them by Owen when a child.[8] Notably, the royal heir apparent drew a direct connection between his former tutor's propensity for spinning spectral narratives and his paleontological prowess, observing: "Whether he was explaining to you the mysteries of some old fossil bone . . . or whether he was telling one of his vivid ghost stories, one felt that one was under the charm of his presence."[9] When these same vivid supernatural tales subsequently appeared in print, in successive numbers of *Hood's Magazine*, their connection with Owen's renowned ability to infer meaning from isolated osseous parts became still more apparent.

The close "relationship between forms of communication . . . and the ideas they contain" has long been recognized by historians, with Thomas Broman insisting that "genres of writing and scientific theories develop together and become established in particular historical circumstances."[10] Other instances where issues of style and form were significant for the success or otherwise of Georges Cuvier's law of correlation have been noted in previous chapters, but serialization, which became unprecedentedly popular during the 1830s and 1840s, makes this putative connection between form and content strikingly manifest. Indeed, Nick Hopwood, Simon Schaffer, and James Secord have stressed the "significant connections" between "serial modes of organization, production and communication" and the nineteenth century's increasing recognition that the "world was serial in its basic structure."[11] In particular, serial publishing was predicated on the same assumptions about the unerring relation between part and whole that were, of course, integral to the paleontological procedures for which Cuvier, and then Owen, had become famous.

Serialization presented readers with small, disconnected parts from which they had to make inferences about the nature of a work that would often not be completed for several months or even years to come. It was on the basis of these projections that they made commercially crucial decisions whether to continue purchasing a particular serial. Tellingly, paleontology and serial publication shared very similar specialist terminologies of "parts," "fragments," and "fasciculi," with Owen himself regularly employing these overdetermined terms interchangeably, both describing the "fasciculus of bone-tendons" and "petrifiable parts" of the *Dimorphodon*, and also commending the "last published fasciculus of the 'Ostéographie'" (1839–64) of Henri de Blainville, while previously having been no less approving of the same work's "first part" (notwithstanding its persistent criticisms of Cuvier).[12] Paleontologists and serial readers alike had painstakingly to relate each individual part, either in osseous or paper form, to a larger and generally still conjectural whole, whether it was a vast prehistoric creature or a protracted narrative. Both practices, moreover, could induce analogous states of expectant anticipation while awaiting the arrival of new parts.

While Owen and Broderip collaborated regularly on a number of non-fictional publications, their sole foray into quasi-autobiographical ghost stories in "Gideon Shaddoe, Esq.," reveals that they were both acutely aware of the potential uses of suspense and readerly expectation in serialized works. A similar recognition of the affective possibilities of the serial format, as this chapter will argue, is no less evident in the two men's more overtly scientific works appearing in installments. As with Owen's skillful manipulation of the press coverage of the discovery of the dinornis discussed in the last chapter, this appreciation of the potential of serialization for inducing particular ways of reading once more demonstrates his unrivaled mastery of the new publishing formats that began to appear in the 1830s and 1840s. In fact, almost all of the initial accounts of Owen's celebrated reconstructions of prehistoric creatures from just fragmentary parts of their remains were published sequentially in serial form, and they were rendered considerably more remarkable and compelling by the suspense and anticipation involved.

Owen and Broderip were, at the same time, themselves also keen participants of precisely such narrative affects, and they arranged their mutual reading of serial novels so as to intensify their feelings of excited anticipation. Owen was particularly enthralled by the dynamics of serial fiction, and his literary reading practices shed important light not only on his own deployment of serialized publishing formats but also on his Cuvierian paleontological procedures. This connection between correlation and serialization was one that was swiftly recognized by many of the leading serial novelists of the period, including William Makepeace Thackeray and Henry James, who, in conjunction with numerous literary reviewers, adopted metaphors from paleontology to describe their own authorial practices. Significantly, the correspondence between serialized fiction and paleontological reconstruction became an effective means of vindicating the aesthetic credentials of novels published in parts, especially in response to widespread critical disparagement that depicted them as just as monstrously unwieldy as lumbering prehistoric creatures.

Serialization became such a distinctive and pervasive aspect of this period that, as Graham Law and Robert Patten have proposed, only the "concept of 'revolution'" can sufficiently "define the changes taking place in instalment publication during the nineteenth century."[13] Cuvierian correlation, as this chapter will show, was inextricably implicated in this serial revolution that transformed the nineteenth-century publishing industry, ensuring that it retained a highly conspicuous presence in the emerging consumer culture of early and mid-Victorian Britain.

A Favourable Medium of Making Known Successive Discoveries

Only months before his death in May 1832, Cuvier had presented his extensive collection of fossil fish to Louis Agassiz. Despite such prestigious patronage and his own burgeoning reputation, the young Swiss naturalist lacked the
financial resources to publish the results of his ichthyological researches as a
monograph and so, from September 1833, began issuing his *Recherches sur les
poissons fossiles* (1833–43) in *livraisons*. In this arrangement, the profits from
the preceding number helped fund the production of the next one, although
in practice it remained a highly precarious system. Unlike the nascent British
practice of publishing parts at regular intervals, whether weekly, monthly,
or quarterly, the French equivalents, *livraisons*, often came out only intermit-
tently. Agassiz, in a subsequent serialized work, conceded that he was not "able
consistently . . . to confine myself to a regular mode of publication" and so
would have to remain "an author . . . who publishes his labors in Livraisons"
that "appear at irregular intervals."[14] As his colleague Jules Marcou later observed, Agassiz "never knew beforehand what his work would be, even approximately, as to quantity of text and plates."[15]

   This impromptu approach certainly dismayed some Teutonic publishers and readers. The Stuttgart publisher Johann Cotta withdrew from the
project at an early stage, and the geologist Christian Leopold von Buch
informed Agassiz that he considered the "method of issuing your text in
fragments from different volumes, altogether diabolical." Another German
reader, though, was more understanding of the potential advantages of such
a circuitous mode of publication for the classificatory endeavors of the paleontologist. In May 1835 Alexander von Humboldt, to whom *Recherches* was
dedicated, reflected ruefully on the work's serialization that, like Buch, "I also
complain a little . . . but I suppose it to be connected with the difficulty of
concluding any one family, when new materials are daily accumulating on
your hands."[16] Agassiz's initial motivation for issuing his work in parts had
been entirely pecuniary, but, as Humboldt grudgingly acknowledged, the
form was particularly suited to the requirements of scientific researches in
which new specimens that might transform previous categorizations were
constantly coming to light.

   Serialization, however, had a still greater advantage that Humboldt did
not appear to recognize. It was an especially effective means of both disclos-
ing and actually corroborating the successful outcome of the daring inferences made by paleontologists employing the Cuvierian method of correlation. Agassiz's accomplishments in identifying extinct fish from individual

fossilized scales have already been noted in chapter 2, and he later explained to an audience of New Yorkers:

> In the year 1833 I delineated in the first number of my work on "Fossil Fishes," a scale of a fossil fish sent me from England, and from it drew the fish to which I considered it to belong. In the following year, 1834, the whole remains of the fish were collected, and the drawing was given in the third number of my work. I have the satisfaction of saying that the two delineations do not differ in any essential way, even in the details.[17]

It was, as Agassiz clearly realized, the sequential appearance of the parts of *Recherches* that, notwithstanding Buch's diabolic disdain, rendered his predictive powers far more striking, as well as demonstrably verifiable, than if they had been expounded in the more expensive format of a completed monograph as he had originally envisaged. This was precisely what Owen, although without Agassiz's financial constraints, also recognized with the triumphant confirmation of the accuracy of his own paleontological predictions in successive issues of serialized publications. In fact, the British preference for regularly issued parts instead of the intermittent *livraisons* common in France, as well as the sudden burgeoning of the market for suspenseful works of serialized fiction in the late 1830s, enabled Owen, with Broderip's assistance, to utilize the opportunities of serialization even more effectively than Agassiz.

Owen's "conjectural" deduction based on a single fragment of femur bone, in November 1839, of the past existence in New Zealand of a giant struthious bird was first published—as an abstract of his original oral paper—in the *Proceedings of the Zoological Society*. The suggestion that further "similar bones" were likely to be "found buried in the banks of the rivers" of the colony, as well as Owen's dramatically avowed "risk[ing]" of his "reputation" on the accuracy of his inference, left the statement hanging suspensefully in anticipation of a confirmation or refutation that, it might be expected, would come in a future number of the same journal.[18] The "proper attitude of the naturalist," as Owen later remarked, "is the 'expectant' one," and he acknowledged that even he experienced a certain degree of suspense while waiting for corroboration of the "possible big bird of New Zealand." Indeed, he reflected that the "years 1840 and 1841 passed, and I began to doubt, but misgiving went no further than as to locality; of the bird itself I may say I was 'cock-sure.'"[19]

The celebrated proof of Owen's confident conjecture that arrived in January 1843 was soon relayed in the Zoological Society's *Proceedings* with Owen's careful reminder that the intact femur he had just received "proves the specific identity of the present remains with the fragment, upon which I ventured to affirm, three years ago, that a large Struthious Bird 'of a heavier and more sluggish species than the Ostrich' had recently become extinct . . . in New Zealand."[20] Owen's self-quotation explicitly linked the two abstracts published three years apart—the auditors of the original paper at the Zoological Society would have been unlikely to recall the exact wording in the same way as readers of the *Proceedings* who could refer back to the earlier number—in a way that emphasized both the long interval of expectant waiting and the jubilant resolution of the sequence of events effected by the eventual joining together, at least in the minds of readers, of the two periodical accounts.

Of course there were still questions left unresolved at this stage, especially regarding the important matter of the age of the bones, and Owen, at the end of the second installment of his 1843 abstract, "promised for a future Meeting [and thus a future number of the *Proceedings*] . . . details of the analysis of the earthy salts" that would shed light on the "recent character of the bones."[21] The full consideration of this particular avian genus would actually require many further installments in the *Proceedings* over several more years, each one referring back to its predecessors as well as looking forward to new discoveries and possible sources of information. Owen's famous prediction, from the evidence of just a single fragment of bone, of the existence of a struthious bird that he was subsequently able to reconstruct, was first conveyed to the specialist readers of the Zoological Society's *Proceedings* in a linked series of abstracts that involved the same dialectic of memory and anticipation as his own ghost stories in *Hood's Magazine* as well as the burgeoning genre of the serialized novel.

Owen seems to have acknowledged that the serial format of its publication played at least some part in the triumphant confirmation of the accuracy of his original conjecture, later expressing his "deep obligations to the Zoological Society of London for the favourable medium of making known successive discoveries of the extinct Birds of New Zealand in their 'Proceedings.'"[22] One particular advantage that he might have had in mind was that serialization allowed the discreet revision of erroneous inferences no less than the exultant confirmation of those that proved accurate, and Owen himself, when considering plans to "issue in 'Parts'" his *History of British Fossil Reptiles* (1849–84), acknowledged that a "benefit [which] flows from publication" in such installments was "the correction, viz. of errors into which the author may have fallen."[23] Crucially, such corrections could be introduced into later parts without requiring publishers to incur the extra costs involved in producing a revised edition of an expensive monograph. The alleged suspensefulness of Owen's abstracts in the Zoological Society's *Proceedings* might therefore seem, at best, merely an incidental, and entirely unintended, effect of periodical publication, which, after all, was the primary means of scientific communication in the period.

From One Article to Another

The reprinting of the very same material in another serialized work, although one organized in a different way and intended for a different audience, nevertheless makes clear that Owen, as well as Broderip, was well aware of the potential for suspense and expectation in the reporting of paleontological discoveries. This work was the *Penny Cyclopædia* (1833–43), the ambitious reference work published by Charles Knight for the Society for the Diffusion of Useful Knowledge (SDUK) in which, as seen in the previous chapter, Broderip duplicated the very same hyperbolic account of the discovery of the dinornis that he had related in private letters to the Oxford don William Buckland. Serialization was an essential component of the SDUK's endeavor to flood the market with reputable reading material that would be affordable to all, and the *Penny Cyclopædia* was issued from January 1833 in weekly numbers that encouraged the purchasers "who expend their Weekly Penny in this work" to feel that they were "laying it up in a Savings Bank of Knowledge."[24]

This emphasis on the financial incentives of part publication notwithstanding, the patrons of the *Penny Cyclopædia* were less comfortable with other attributes of serialization, and instead proposed that "every Number contain something valuable and interesting," which would best "be effected by avoiding many references from one article to another, and by rendering each, as far as may be, complete in itself."[25] The SDUK, as Adrian Johns has observed, was "rather conflicted in its relation to the steam press," both enthused by its capacity for the mass diffusion of sanctioned forms of knowledge and at the same time fearful of its potential for eliciting less acceptable responses among an undisciplined proletarian audience.[26] The modes of reading induced by serial publication were an important factor in such concerns. In a sermon delivered at Rugby School in 1839, Thomas Arnold lamented the "peculiar mode of publication, of the works of amusement of the present day," which "keep alive so constant an expectation" among their excitable readers.[27] Arnold's jeremiad echoed the editor's anxieties regarding Broderip and Owen's use of series and continuations in *Hood's Magazine*, and the SDUK's General Committee, which had already prohibited the publication of fiction or any other imaginative literature that might hinder its didactic objectives, was similarly wary of the absence of completion and continual anticipation of new elements that serialization made inevitable.[28]

Accordingly, the *Penny Cyclopædia*'s entries were arranged alphabetically, an arbitrary, rational, and efficient mode of organization pioneered in the scientific encyclopedias and dictionaries of the Enlightenment.[29] It was nonetheless rare to combine strict adherence to alphabetical ordering with publication in such short and frequent installments, and, as Knight later observed, the "novelty was not to consist in producing a Cyclopædia under one alphabetical arrangement, but in its issue in weekly sheets."[30] In a serialized format the restrictions of an alphabetical mode of organization would, after all, make it difficult to incorporate new information about subjects, such as anatomy, that had already been treated in early numbers, and leave others, such as zoology, largely unrepresented until the final concluding parts. With a large-scale serialization such as the *Penny Cyclopædia*, these could take years or even decades to appear, and, as the *Mechanics' Magazine*—a persistent thorn in the SDUK's side—complained after the initial annual volume had only reached the end of the first letter of the alphabet, "The young men who have begun taking it in the letter A, will be old men before they arrive at the letter Z."[31]

It was the entries on rapidly advancing areas of science that were most at risk of the obsolescence that was made virtually inevitable by the serial's alphabetical ordering and prolonged period of publication (fig. 4.1). Looking back from the mid-1850s, the *Times*, after observing that "few things have been more remarkable than the progress of natural history during the last 20 years," bemoaned the amount of "intelligent labour . . . required to render the information supplied to the purchaser of the *Penny Cyclopædia* of real use to the unscientific reader of the present day."[32] The article on "Anatomy," for example, had been published in late 1833 and was almost immediately condemned by the *Mechanics' Magazine* for being "'knocked off' in . . . a dozen lines" when it "might be expected to occupy some dozen columns," but it remained unaltered for a further twenty years at a period of enormous changes in the discipline.[33]

Perhaps rather unfairly, this same entry was still being picked out in the late 1840s as a perturbing instance of "how little the true nature of the science of comparative anatomy . . . is comprehended" in the "latest summaries of human knowledge published in this country."[34] The author of this sniping criticism was Owen, who otherwise commended the "excellent 'Penny Cyclopædia,'" and had in fact, as seen in the last chapter, regularly assisted Broderip with his numerous entries on natural history. While Knight later lauded these same entries as "models of scientific exactness and popular attraction," Broderip often blatantly disregarded the *Penny Cyclopædia*'s alphabetical organization and opposition to cross-referencing.[35] In the period immediately prior to his and Owen's experimentations with serialized

FIGURE 4.1. Charles Jameson Grant, *The Penny Trumpeter!* Lithograph, 1832. British Museum. This satirical print shows the SDUK's founder, Henry Brougham, as a newsboy selling the *Penny Magazine*, carrying a sack on his back from which protrudes a paper stating: "Materials for the Penny Cyclopædia to commence in 1833 & to end the Devil knows when." The serialization of the *Penny Cyclopædia* over a long period of time, without a clear end point, presented serious problems for its alphabetical ordering, and especially for entries on rapidly advancing areas of science such as anatomy. © The Trustees of the British Museum.

narrative in *Hood's Magazine,* Broderip instead exploited the *Penny Cyclopæ-dia*'s own serial format both to continually update and revise his entries on his friend's paleontological reconstructions and to endow them with a sense of suspense and anticipation.

Incidents Linked Together by a Chain of Interest

At the conclusion of his entry on "Struthionidæ" published in 1842, Broderip reprinted the final paragraphs of the abstract of Owen's inference of the past existence of a giant wingless bird in New Zealand carried two years earlier in the Zoological Society's *Proceedings.* Broderip made no further comment regarding the bold conjecture contained in the abstract, and instead pointed to the completion of the subject of struthious birds by expressing a hope that "no material omission will be found" in his "abridged account."[36] Owen's abstract, though, still retained the same suspenseful anticipation of a con-firmation or refutation of its central prediction as in its original periodical format, even if the *Penny Cyclopædia*'s alphabetical organization now meant that there was no clear means of affording the required rejoinder.

Clearly, as both Broderip and Owen must have recognized, the outcome of such a striking exemplification of Cuvier's law of correlation could not be stymied by the arbitrariness of the *Penny Cyclopædia*'s alphabetical mode of organization, especially once the initial prediction was almost totally vin-dicated in the following year. In fact, another serialized reference work to which Owen was then contributing, Robert Todd's *Cyclopædia of Anatomy and Physiology* (1835–59), offered just the precedent that they needed, for the editor acknowledged that "it has been found necessary, in a few instances, to depart from the strict alphabetical arrangement, either by placing articles under names not commonly used, or by clubbing together two or more sub-jects." The "necessity for such modifications," as Todd explained, "arose out of contingencies to which all works are liable, when they are published in Parts," which, potentially, allowed Broderip and Owen to employ a similar justification.[37]

Charles Lyell later advised Owen that the "frequent coming out of new parts keeps the public interest alive," and, unlike the three-year intermission between relevant numbers of the *Proceedings of the Zoological Society,* read-ers of the *Penny Cyclopædia* had to wait only a matter of months before they were informed of the outcome of Owen's earlier inference.[38] However, by this time—the spring of 1843—the parts that were being issued had reached the potentially unpromising letter *U.* Undaunted by this awkward predicament, Broderip, halfway through an entry on the "Unau," the arboreal two-toed sloth of South America, made an abrupt, and only tangentially justified, shift by observing that the "close approximation of the Sloths to the Birds in many parts of their organization calls upon us here to notice a discovery which will make the year in which we write (1843) a very remarkable one in the zoological calendar" (see fig. 3.2). The imperative tone of Broderip's rhetorical swerve implied that he was compelled to change subjects almost against his own will, although he then conceded that the presumed connection between sloths and birds was in fact only "superficial" and did not survive "closer examination" of their respective anatomies. Regardless of this superficiality, Broderip urged readers that it was "curious and instructive" to "look back to Professor Owen's description of the fragment of bone" in his earlier entry on "Struthionidæ," before entering into the particulars of the astonishing confirmation of Owen's prediction and reprinting short extracts from the abstract published in the Zoological Society's *Proceedings* earlier that same year.[39] The very different reading audience for the parts of the *Penny Cyclopædia* were invited to experience the dramatic character of Owen's conjecture and its subsequent confirmation in a similar way, again predicated on the sequential nature of serialization, to the specialist readers of the *Proceedings of the Zoological Society.*

What is particularly striking about the entry on the "Unau" is that Broderip was apparently so willing, on the flimsiest of premises, to override the alphabetical ordering on which the *Penny Cyclopædia's* endeavor to diffuse useful knowledge had been based for the last decade. After all, a reader anxious to learn more about the much-discussed dinornis would have been unlikely, especially after 1843 when all the parts could be bound together into twenty-seven volumes, to consult entries under the letter *U* and, in the absence of a general index, would have found it extremely difficult to locate relevant information. Instead, Broderip's entry on the "Unau" was more attuned to readers who treated the *Penny Cyclopædia* in the same way as a periodical or even a serialized novel rather than a static compendium of useful knowledge. They would read each published installment, or at least those elements of it that were of most interest, in turn, making connections between the separate parts, such as those containing the entries on "Struthionidæ" and the "Unau," without particular regard to their overall alphabetical arrangement.

When approached in such a manner, Broderip's *Penny Cyclopædia* articles closely resembled other contemporaneous serialized publications in which "incidents were linked together by a chain of interest strong enough to prevent their appearing unconnected," as Dickens had proposed in the preface to *The Pickwick Papers* (1836–37).[40] The various entries in the *Penny Cyclopædia* were envisaged as precisely such a connected chain by James

Pycroft, who advised followers of his *Course of English Reading* (1844) that, on a number of topics,

> you may consult the "Penny Cyclopædia", which excels all others in the variety of its subjects. You can read each article, more or less attentively, according to the degree of interest which casual notices of those topics in books or conversations have excited. When you have read them all, cast your eye again over the [first] article . . . and you will feel that the several parts of your newly-acquired knowledge have a propensity to "fall in", as the drill sergeants say.[41]

Although some might be attended to more closely than others, the articles on a particular subject, in which interest may have been prompted by casual conversation, needed to be treated as a connected sequence rather than isolated entities. Inevitably, those who read the *Penny Cyclopædia* in such a way rather than as a collection of discrete, unconnected entries faced a sometimes prolonged interval before the next relevant entry on a particular topic became available.

For such serial readers, Owen's Cuvierian feats of reconstruction were revealed sequentially rather than all at once, as would have been the case in an encyclopedia published only in a completed form. What these accounts lost in ease of reference was more than made up for by the suspense and anticipation involved in their serialized disclosure by Broderip. The additional exercise of retrospective rereading of the *Penny Cyclopædia* proposed by Pycroft, moreover, would have had the same effect as Owen's self-quotation regarding the dinornis in the Zoological Society's *Proceedings*, impressing upon readers once again just how bold and courageous the original conjecture regarding the existence of a giant struthian bird in New Zealand, made on the evidence of only a single fragment of bone, had been.

When, ten years later, the material from the *Penny Cyclopædia* was repackaged, with additions and revisions, as the *English Cyclopædia*, published by Knight using stereotype plates bought from the SDUK following its dissolution in 1846, the entry on the "Dinornis" was eminently easy to find within the Natural History division, which was now separated from other subject areas and published in four volumes. In this new edition of Knight's massive publication, as the *Medical Times and Gazette* remarked, a "consistency is observed which we in vain look for in the parent Cyclopædia," while the *Times* observed approvingly that the *Penny Cyclopædia*'s "scattered portions are now all brought together."[42] It was Broderip's extensive contributions to the original, as the *Athenæum* noted, that "still form the basis of the Natural History division of the 'English Cyclopædia.'"[43] Tellingly, however, the "Dinornis" article compiled from the relevant parts of Broderip's earlier entries on

"Struthionidæ" and the "Unau" noted only briefly the circumstances of the bird's initial identification before moving on to newer information regarding the extensive collection of remains unearthed in New Zealand by Walter Mantell. Broderip's two paragraphs of hyperbolic eulogy on the "glimmering light which he [i.e., Owen] extracted from that fragment" and how "every word that he then wrote has come true to the letter" were condensed into simply: "It was not long before an opportunity occurred of testing this very remarkable statement, and of proving the sagacity of the naturalist who had thus staked his reputation upon his conviction of the truth of the general principles of the science of comparative anatomy."[44] Without the suspense and anticipation inherent in the serialized format and sequential installments of the original, the emphasis shifted from the spectacular accomplishments achieved by Cuvierian correlation to the more mundane details of the various locations in which the bones were found.

Read Every Word before Going to Bed

Within hours of its publication on 1 October 1846, the entire stock of the opening number of Dickens's latest serial novel *Dombey and Son* (1846–48) had sold out. Its publishers, Bradbury and Evans, quickly printed another five thousand copies, but these too were soon snapped up by eager readers, and, as the novel's author acknowledged with delight, it was now entirely "out of print."[45] Among the many frustrated readers unable to begin *Dombey and Son* until the following month were Owen and Broderip. On the very first day of November, Owen's wife, Caroline, as she recorded in her diary, "Sent out for Nos. 1 and 2 of 'Dombey and Son.'" Her husband had decided to refrain from reading the novel until his own *Lectures on the Comparative Anatomy and Physiology of the Vertebrate Animals* was published later in the same month, and "Mr. Broderip begged to be allowed to take the two numbers home in his pocket." Owen, Caroline noted, "told him that he might, on condition that he did not look at them to-night—upon which Mr. Broderip said that he should read every word before going to bed."[46] Broderip's jovial insubordination, which contrasted with his customary professional sternness as a magistrate, indicated that the eager anticipation of Dickens's newest serial novel would override the injunction of his friend. It might even prevent Broderip going to bed until the early hours of the morning.

The eager anticipation that both Broderip and Owen evidently felt for the novel's delayed opening number, only intensified by the latter's resolve to defer reading it until even later, suggests strongly that the understanding of the potential uses of suspense and readerly expectation evinced in "Gideon Shaddoe, Esq.," and their entries for the *Penny Cyclopædia* derived, at least in part, from the two naturalists' mutual enjoyment of the nascent format of serial fiction. Broderip's real fondness, though, was for novels of the previous century such as William Beckford's gothic fantasy *Vathek* (1786).[47] It was Owen who was particularly enthralled by the modern serialized novels that, by the 1840s, had become the predominant mode of literary production, and he continued to await the parts of *Dombey and Son* with almost unfaltering enthusiasm for the next seventeen months. While this was only one of many facets of what an obituary in the *Church Quarterly Review* called the "many-sidedness" of this "polished gentleman of varied accomplishments," the extent of Owen's passionate enthusiasm for this particular form of literature is nonetheless striking. As the same obituarist went on to observe: "Mrs Owen kept him well supplied with the novels of the day; and he sat up half the night over . . . the serial stories of Dickens."[48]

This attachment to serial novels appears to have begun with Dickens's pioneering *The Pickwick Papers*, which, after a slow beginning, had become a publishing phenomenon by the end of 1836 and effectively invented the format of fiction appearing in monthly parts. On Christmas Day that year, Caroline recorded that she and Owen enjoyed a "very pleasant evening, with readings from 'Boz' [Dickens's early pseudonym], playing and singing."[49] Her husband's enthusiasm for the innovative fictional format was not confined to such moments of playful relaxation, and soon became closely intertwined with his scientific activities. He regularly referred to the geologist John Brown, an important collector of Pleistocene fossils, as "Mr. Pickwick," informing his sister Eliza that he was the "closest approximation to Boz's famed type" and "like the founder of the Pickwick Club, he solaceth himself with virtuosoizing in antiquities; but, as the immortal Cuvier hath it, 'of a higher order' than those which amuse the F. A. S.'s [i.e., Fellows of the Antiquarian Society in *The Pickwick Papers*]."[50] As this curious juxtaposition of the distinctive styles of the youthful originator of serialized fiction and the late initiator of the law of correlation might suggest, reconstructing Owen's literary reading practices sheds important light on his exactly contemporaneous paleontological researches.

At first sight, it appears that the serialized novels supplied by his wife (fig. 4.2) offered Owen merely an engrossing escape from the pressures and demands of his scientific labors. The regularly published installments of a novel, as Jennifer Hayward has argued, accorded with the increasing separation of public work and domestic leisure in the mid-nineteenth century, with

FIGURE 4.2. William Clift, *Sketch of His Daughter*. Chalk, undated. Caroline Owen Commonplace Book, MS0283, Royal College of Surgeons, London. This informal domestic portrait shows Caroline Clift sometime before her marriage to Owen in July 1835. It was Caroline, of whom this is the only known picture, who kept her husband supplied with the latest serial novels. Reproduced by kind permission of the President and Council of the Royal College of Surgeons of England.

the "ritualization of its consumption . . . help[ing] to mark off work time from leisure time."[51] This certainly seems to be the case in the following entry from Caroline's diary made during the spring of 1844:

> *May 3.*—After a hard day's work, R. deep in "Martin Chuzzlewit". My father [i.e., William Clift] came in before going to the Royal Society, and talked to

R. without mercy; but R., whose thoughts and attention were so entirely given up to Mrs. Gamp and Jonas, could only answer at random. As soon as my father was gone, we laughed over Mrs. Gamp till bedtime.

Owen's almost total absorption in the novel denuded him of the ability to partake in a presumably professional and scientific conversation with his father-in-law and erstwhile mentor. Rather, following an exhausting day amid the abundance of fossilized remains at the Royal College of Surgeons, the particular installment of Dickens's serial novel confined Owen, both physically and mentally, to the gendered space of the domestic sphere and the company of his wife, who, it would seem, was able to take precisely the same pleasure as himself in the bathos of the bilious nurse Sairey Gamp. As the Owens's living quarters at this time were a cramped apartment in the same building as the Royal College of Surgeons, where Caroline sometimes had to "keep all the windows open" and prompt her husband to "smoke cigars all over the house" to counteract the smell of putrid animal cadavers, their mutual enjoyment of the monthly numbers of a novel such as *Martin Chuzzlewit* (1843–44) evidently helped mark off the ambivalent space of the home from the constant intrusions, olfactory as much as bureaucratic, of the adjacent college.[52]

This theme of an absorption in fiction, following the completion of his scientific activities, that was so rapt that it kept him aloof from his intellectual peers is also discernible in entries in Caroline's diary relating to Owen's reading of serialized novels in other environments. On 22 January 1848 she recorded that

> after hearing a lecture of Whewell's, he went on to the [Athenaeum] Club, and took up Thackeray's "Vanity Fair" to read. He became so deeply absorbed . . . that he sat on, oblivious of the fact that everyone else had disappeared one by one. . . . Then, having looked at his watch and found it considerably past 2 A.M., he rushed wildly out of the Club, and, like a scientific Cinderella, left his umbrella and great coat behind.[53]

Rather than engaging in any form of conversation or sociability with his exclusively male fellow club members, or indeed discussing the Friday Evening Discourse that William Whewell had just given at the Royal Institution, Owen was instead isolated by his immersion in the latest installment of Thackeray's novel. He finally departed from the Athenaeum in the ambiguously gendered position of a scientific Cinderella.

Much of the original audience of *Vanity Fair* (1847–48) read the monthly parts of Thackeray's initial foray into serialized fiction alongside those of Dickens's almost contemporaneous *Dombey and Son*, with their experience of the respective serials shaping their responses to both.[54] In the month following Owen's evening of engrossment at the Athenaeum, Caroline's diary recorded: "On February 29 [1848] No. 18 of 'Dombey' appeared and he 'stayed up very late reading it'" (as had Broderip with the initial numbers), and it very much appears that he read the two novels concurrently.[55] By the late 1840s Dickens and Thackeray were established as Owen's favorite serial authors, and their absorbing fictions continued to afford him relaxation and escape from the pressures of his various professional and public commitments. As Owen himself reflected at the Royal Literary Fund:

> Often after the labours of the day, the nervous system, oppressed by the atmosphere of a dissecting room . . . he had rejoined the family circle, too exhausted, perhaps, for the enjoyment of social conversation. And where had he found the best restorative? Sometimes in listening to the genial humour and touches of exquisite pathos which are yielded by the pages of a DICKENS.[56]

Such a predictable binary between a public, masculine world of the scientific and a feminized, domestic realm of the literary, however, is of questionable validity in accounting for Owen's conspicuous predilection for serialized fiction.

Even the intimation that Owen listened passively while his wife read aloud from Dickens's novels does not necessarily segregate this aspect of their domestic leisure from his scientific researches, for Caroline would also assist her husband by similarly reading to him from paleontological works. As her diary for December 1836 recorded, following "readings from 'Boz'" on Christmas Day, three days later she also "read aloud from Cuvier," with Owen an equally rapt auditor on both occasions.[57] As with the Pickwickian fossilist John Brown, moreover, Owen also continued to compare his scientific acquaintances with fictional characters, finding an American "youth in spectacles" who used polygenist arguments to endorse slavery the very "model of Dickens's Jefferson Brick."[58] Reviewers of *Martin Chuzzlewit*, in which the obnoxious Yankee journalist appeared, had already begun to warn that serialized novels, consumed regularly over an extended period of time, might actually distort their readers' perception of reality. Each "new number of Dickens," the *North British Review* advised, is "not a mere healthy recreation like . . . a game of backgammon. It throws us into a state of unreal excitement, a trance, or dream. . . . But now our dreams are mingled with our daily business."[59] While Owen was partial to a "hit of backgammon" when "tired of my work," reviewers suggested that his much greater fondness for serialized fiction would not afford the same respite from his professional labors and, alarmingly, might instead become inextricably confused with them.[60]

It therefore becomes still more problematic to separate Owen's marked enthusiasm for this particular format of fiction from the activities that he undertook during his long working hours at the Royal College of Surgeons, and maybe even from the kind of cognitive processes that he employed there. In the still innovative practice of "reading one instalment, then pausing in that story," as Linda Hughes and Michael Lund have suggested, "the Victorian audience turned to their own world with much the same set of critical faculties they had used to understand the literature."[61] Rather than merely immuring him in a distinctly feminized and domestic realm of literary leisure, it is evident that Owen's experience of reading novels such as *Martin Chuzzlewit* as part of a repeated pattern extending over a period of almost two years not only helped shape his own serialized writing but also exercised a lasting influence on crucial aspects of his paleontological practice.

By a Comparison of Incidents and Dialogue

In *Nicholas Nickleby* the eponymous hero expressed Dickens's indignation with the piratical adaptations of his serial novels that appeared on the London stage before they were even completed by remonstrating with a "literary gentleman . . . who had dramatised in his time two hundred and forty-seven novels as fast as they had come out—some of them faster than they had come out." He and other similarly prolific dramatists, Nicholas heatedly insists, "take the uncompleted books of living authors," "finish unfinished works," and "vamp up ideas not yet worked out by their original projector." It is "by a *comparison* of incidents and dialogue, down to the very last word he may have written a fortnight before," Nicholas goes on, that you "do your utmost to *anticipate* his plot" (italics mine).[62] Notably, the very same terms in which Nicholas portrays the disreputable methods of this piratical literary gentleman, and by implication the less culpable practices of many other readers of serial novels who likewise attempted to imagine how a story would unfold, were used by Owen to describe his own paleontological practices. The "interpretation of . . . fossil remains," he observed, "requires a *comparison* of them with the corresponding parts of animals now living, or of previously determined extinct species."[63] In Cuvier's analysis of extinct elephants, Owen proposed, these comparative procedures had meant that a "rapid glance . . . over other fossil bones, made him *anticipate* all that he afterwards proved" (italics mine).[64] The audience for serial fiction made careful comparisons between the available parts of a novel—as well as with a broader taxonomic repertoire of plots and characters drawn from their reading of other novels—in order to anticipate the resolution of its plot in a way that closely resembled the methods and language used by paleontologists. In fact, the above passage from *Nicholas Nickleby*, first published in the fifteenth number in May 1839, offers a striking parallel with what Owen would do only five months later when anticipating the structure of an unknown flightless bird from New Zealand by what he termed an "exhaustive comparison" of its femur with "similar-sized portions of . . . skeletons" in the Royal College of Surgeons.[65]

Even more strikingly, it is clear that Owen was actually reading the monthly installments of *Nicholas Nickleby* in the period immediately prior to his receiving this enigmatic bone. On 8 September 1839 he told his mother-in-law: "I read the last number of 'Nicholas Nickleby' in bed the other night."[66] Having stayed up late to finish the penultimate part of Dickens's novel, it is likely that its concluding double number, which was published three weeks later at the start of October, would have occasioned no less excitement and been read with a similar alacrity. Dickens's usual practice was for the initial parts of his next novel to appear simultaneously with the final installments of the previous one (as had happened with *Oliver Twist* and *Nicholas Nickleby*), but the as yet untitled "NEW WORK BY 'BOZ'" advertised in the wrappers of the last number of *Nicholas Nickleby* was not due to commence until the following March. The "Author of these pages" acknowledged of his readers, as Dickens noted archly in the novel's final part, "that on the first of next month they may miss his company at the accustomed time as something which used to be expected with pleasure."[67] By the middle of October, Owen, for the first time since his enthusiastic response to *The Pickwick Papers* in the winter of 1836, was deprived of his regular fix of serialized fiction and its attendant readerly activities of comparison and anticipation. Only days later, on 18 October, he was sent a letter "offer[ing] for sale a portion or fragment of a bone" brought from New Zealand, with the mysterious "part of the femor" affording a timely surrogate with which he could engage in similar procedures.[68]

That Owen's reading of serialized fiction involved precisely such anticipations based on comparisons of the evidence available from previously published installments is made clear by his sense of grievance that his apparently rational predictions about the fate of Mr. Carker when reading the serial parts of *Dombey and Son* were stymied by his abrupt and sensationalist death beneath a train in the eighteenth number. The "character of Carker as drawn throughout the book," Owen observed disgruntledly, "makes it evident to me that he was not the man either to act or to be acted upon in such a way."[69] Nor was Owen alone in this view, and it is important to recognize that mid-nineteenth-century readers held shared conceptions of appropriate literary form and design (going back to Aristotle's *Poetics*) that actually meant that serial novels were expected not to depart from established tenets of storytelling and thus, in certain ways, to be predictable. The *English Review*, for instance, complained of the "exaggerated . . . portraiture" of *Dombey and Son*, contrasting it with Thackeray's "thoroughly self-consistent 'Vanity Fair,'" in which "nothing is forced, nothing artificial."[70] Owen's own strictures regarding the inconsistent characterization of Carker may likewise have been conditioned by his simultaneous reading of the parts of *Vanity Fair*. In fact, during the spring of 1848 he seems to have spent his evenings comparing the individual parts of the two novels, in their respective green and yellow wrappers, before predicting the likely fates that awaited characters such as Carker. During the day, meanwhile, he was employing analogous comparative procedures to arrange a miscellaneous collection of dinornis bones recently arrived from New Zealand. His earlier engrossment in the parts of *Martin Chuzzlewit* similarly coincided with his being painted by Henry Pickersgill "in the act of lecturing, holding the dinornis bone" (see fig. 3.4).[71] Owen's reading of serialized novels consistently overlapped with his professional activities as a paleontologist.

The connection is made even clearer by Owen's account of the sequential temporality of paleontological study and how in waiting for new fossil remains that might confirm initial predictions about the structure of prehistoric creatures "one's interest is revived and roused year by year as bit by bit of the petrified portions of the skeleton come to hand."[72] Indeed, few things, as Owen noted, could "equal the excitement of that in which, bit by bit, and year after year, one captures the elements for reconstructing the entire creature of which a single tooth or fragment of bone may have initiated the quest."[73] These emotive descriptions, with their emphasis on eager expectancy and the periodic revival of interest in a quest narrative that began with merely a single component, closely resembled Owen's grandson's portrayal of his own attitude toward serialized novels: "He watched for the monthly numbers of Dickens's works with great eagerness, and read them with much enjoyment as they came out."[74] And even if, as was evidently the case with the eighteenth number of *Dombey and Son*, this eagerness soon gave way to disappointment, this was, again, not so very different from Owen's experience of paleontological frustrations such as a long-awaited "moa's head (so called)" that, as his wife recorded in July 1844, had "just arrived" at the Royal College of Surgeons. After "so much excitement," Caroline observed, "it was perhaps a little trying to find that this enormous head proved to be nothing more than the skull of a seal. A bit of dinornis skull was thrown in."[75] From the late 1830s, Owen was perpetually waiting, with apparently equal anticipation and excitement, for fossilized remains coming bit by bit and novels arriving part by part, in the expectation, albeit tempered by occasional disappointment, that both would verify earlier predictions made largely on the basis of comparison.

A Serial Story Is a Monstrosity

Notwithstanding Owen's fervid enthusiasm, serialization remained a controversial issue for many Victorian reviewers. Whereas a critic could judge the overall proportion and harmony of a work of fiction that was published as a whole in bound volumes, the successive numbers of a serial novel, liveried in gaudy paper wrappers, precluded such aesthetic evaluations, and instead focused attention only on the partial vignettes of each installment. The "serial tale," according to the *Prospective Review*, was "probably the lowest artistic form yet invented; that, namely, which affords the greatest excuse for unlimited departures from dignity, propriety, consistency, completeness, and proportion."[76] On the other side of the Atlantic, William Gilmore Simms, writing in the *Southern Quarterly Review*, similarly complained: "There is something decidedly unfriendly to art, in the present popular mode of writing for *serial* publication. . . . The author soon becomes indifferent to all general proportions in his work,—to all symmetry of outline."[77] Far from inducing their readers to relate each individual part to a larger whole, the numbers of a serialized novel were, in the view of many critics, isolated fragments of only transitory appeal that could never afford the harmony, proportion, and sense of design required of a genuine work of art.

Notably, many of these same antagonistic reviewers also depicted serial novels in the very same terms used to describe lumbering, ungainly prehistoric creatures. The American journalist Champion Bissell insisted that "no one can gainsay that a serial story . . . is a monstrosity," and Simms derided such works as "incongruous, halting and indefinite."[78] Back in Britain, *Sharpe's London Magazine* likewise warned of the "monstrous joining together of . . . anomalous members" to be found even in Dickens's latest serial productions.[79] These descriptions closely paralleled the language used to describe the megatherium by men of science such as Charles Darwin, who corresponded with his sisters in the 1830s about this so-called "Monster M" and subsequently reflected that "their ponderous forms . . . seem so little adapted for locomotion."[80] Periodicals such as the *Church of England Magazine* similarly commented on the creature's "monstrosity of outward form," and observed that it was "ill-proportioned, with its clumsy and incongruous members."[81]

This particular vein of novelistic criticism also recalled an earlier tradition of depicting convoluted nonfictional works such as the *Penny Cyclopædia* as, in the words of the *Mechanics' Magazine*, "an ill-shaped monster, with its head too big for its body, and its body too big for its legs."[82] The original meaning of the term *monster* was a creature amalgamating incongruous parts, with connotations of great size only subsequently being added to its primary meaning. By the early nineteenth century it was frequently used more broadly to refer to almost all prehistoric animals, and the threat of producing a composite and distorted monstrosity haunted paleontologists working with osseous fragments that were often widely dispersed and mixed up with the remains of other creatures.[83] It was the huge and ungainly megatherium, apparently composed of elements of armadillo, anteater, sloth, and rhinoceros, that, in the wake of Cuvier's own distaste for the creature discussed in chapter 2, became an exemplar of this kind of awkwardly monstrous—and even potentially fictive—amalgam.

The Novelist Puts This and That Together

William Makepeace Thackeray had made sardonic play with the "Megatheria of history" since at least the mid-1840s, when he adopted the Greek name Cuvier had given to the colossal edentate mammal for the metropolitan club to which the holidaying narrator of his *Punch* sketches "Brighton in 1847" longs to return.[84] While this club, "our beloved Megatherium," would continue to appear regularly in Thackeray's writing, including many of his best-known novels, the comic potential of the huge and purportedly awkward creature was also used in a number of other contexts.[85] In fact, Thackeray's very first reference to this prehistoric giant had come in his Christmas book for 1846, *Mrs. Perkins's Ball*, and, significantly, he initially perceived its immense and cumbersome frame as analogous not to a large metropolitan club but rather to a ruinously colossal literary publication. The characters introduced in *Mrs. Perkins's Ball* include "Poseidon Hicks, the great poet," who is the author of, among other works, "The Megatheria." Although this epic poem is "'a magnificent contribution to our pre-adamite literature', according to the . . . reviews," Thackeray's more skeptical narrator, Michael Angelo Titmarsh, reflects: "I know that poor Jingle, the publisher, always attributed his insolvency to the latter epic, which was magnificently printed in elephant folio."[86] This was only the first of many such disparaging references to what the American poet James Russell Lowell soon after designated the "megatheria of literature."[87] Within eight years, however, Thackeray would exhibit a very different, and considerably more subtle and sympathetic, understanding of the putative relation between the megatherium and amply proportioned works of literature, and, notably, it was in precisely this same period that he both began publishing serialized novels and became acquainted with the eminent paleontologist who was himself such an aficionado of the format.

From the early 1840s Thackeray and Owen regularly encountered each other in the familiar purlieus of the metropolitan literary and intellectual elite. Following Owen's rapt absorption in the monthly numbers of *Vanity Fair*, the same novel that finally brought Thackeray both fame and fortune, the two struck up a friendship, with Owen recording that at a dinner at the Royal Academy in 1850, he saw "Thackeray, who sent to me across the table to take a glass of wine." They remained close for more than a decade until the novelist's sudden death in 1863, with Owen telling a correspondent, "Poor Thackeray's departure was a sorrowful shock to me." Their friendship would undoubtedly have brought Thackeray into contact with aspects of Owen's scientific thought. He attended his friend's anatomical lectures, even if only to accompany his children, and after Owen commenced his Fullerian Lectures at the Royal Institution in the spring of 1859, he told his sister Eliza, "I have capital audiences. . . . Thackeray told me the other day that 'two young ladies (I suppose his daughters) were among my great admirers.'"[88] This friendship with Owen, who in the late 1830s departed from both Cuvier and William Buckland in showing the perfectly integrated design of the seemingly ill-proportioned megatherium, occurred at exactly the same time as a conspicuous shift in Thackeray's treatment of the sloth-like quadruped as a model for the formal structure of literary works.

Whereas in *Mrs. Perkins's Ball* Titmarsh's ironic narration indicated that "The Megatheria" was merely a verbose and grandiloquent epic whose unwieldy size had bankrupted its publisher, the narrator of *The Newcomes* (1853–55), Arthur Pendennis, actually likens his own novelistic effusions to the same enormous creature. He reflects:

> As Professor Owen or Professor Agassiz takes a fragment of a bone, and builds an enormous forgotten monster out of it, wallowing in primaeval quagmires . . . so the novelist puts this and that together: from the footprint finds the foot; from the foot, the brute who trod on it; from the brute, the plant he browsed on, the marsh in which he swam—and thus in his humble way a physiologist too, depicts the habits, size, appearance of the beings whereof he has to treat;—traces this slimy reptile through the mud, and describes his habits filthy and rapacious . . . points out the singular structure of yonder more important animal, the megatherium of his history.[89]

The passage seems, at first, merely to reflect Thackeray's characteristic cynicism about narratorial omniscience, with Pendennis appearing to suggest that many of the details of the history of the Newcome family are based on questionable inferences and dubious hypothetical reconstructions. This is how the passage has usually been interpreted by critics of *The Newcomes*, with George Levine contending that the Cuvierian method is invoked "half-mockingly."[90] Pendennis has, after all, already advised that "in the present volumes . . . the public must once for all be warned that the author's individual fancy very likely supplies much of the narrative." Still worse, he then likens the doubtful status of his narrative to another scientific exemplar, the "descriptions in 'Cook's Voyages'" that were notoriously fabricated in the eighteenth century.[91] The subsequent comparison of Pendennis's avowedly deficient style of narration to Owen's and Agassiz's much-heralded ability to infer the existence and life habits of prehistoric creatures from just small fragments of their remains might, then, imply that this too was predicated merely on capricious and predominantly fallacious guesses.

As the narrative of *The Newcomes* slowly accrues, though, the fragmentary nature of almost all evidence, as well as the instinctive necessity of filling in the gaps and correcting that which is defective, becomes increasingly evident. There are, moreover, dependable scholarly methods for taking such suppositions beyond the realm of the merely speculative, and Pendennis gives the "example, when you read such words as QVE ROMANVS on a battered Roman stone, your profound Antiquarian knowledge enables you to assert that SENATVS POPVLVS was also inscribed there." The possibility of errors cannot be removed entirely, but at least, in line with such antiquarian methods, a narrator should "tell your tales as you can, and state the facts as you think they must have been."[92] Similar to the scholarly knowledge of the ancient world gleaned from the artifacts of Herculaneum, a perception of past events that is at least relatively reliable can, Pendennis suggests, still be attained. When he moves beyond his customary aesthetic and antiquarian frame of reference, Pendennis offers a scientific, and even more reliable, procedure for extrapolating from mutilated fragments. Tellingly, this allusion to the contemporary practices of Owen and Agassiz is considerably longer and more detailed than any of the previous references to more conventional scholastic methods, while, still more notably, Richard Altick has suggested that it is in fact the "most extended topical" analogy in the entire corpus of Victorian fiction.[93]

It is likely that Thackeray gained a firsthand—if slightly misremembered —knowledge of such paleontological procedures not just from Owen himself but also from Agassiz. Now based at Harvard, the émigré ichthyologist met Thackeray during the novelist's successful lecture tour of America in the early 1850s. Thackeray remarked, in March 1853, that "Professor Agassiz [is] a delightful *bonhommious* person as frank and unpretending as he is learned and illustrious in his own branch."[94] Thackeray actually began writing *The Newcomes* only three months after his return across the Atlantic and, later in the same novel, made reference to the naturalist he had recently got to know and whose character and abilities he apparently regarded so highly.

This reinforces the sense that the inclusion in the same passage of Owen, with whom Thackeray had a still closer friendship, similarly derived from their personal contact (he could, after all, have instead used Cuvier) and that the invocation of the correlative methods for which both Agassiz and Owen were well known was not predominantly cynical or sardonic.

Significantly, Pendennis's portrayal of the "novelist [who] puts this and that together" as, in the same vein as Owen and Agassiz, "a physiologist too" becomes still more pertinent in relation to the process of serialization in which Thackeray was engaged when writing *The Newcomes*. Edgar Harden has revealed that, despite the conventional image of him as a careless and perennially unpunctual writer, Thackeray was both proficient and highly conscientious in his composition of serial parts. The exacting demands of the format nevertheless ensured that, as Harden observes, "he had to work out various details of his conception during the processes of composition."[95] As such, Thackeray, like other serial novelists who regularly wrote the next installment only after the previous one was already published, had to painstakingly relate each individual part to a larger and often still conjectural narrative whole in order to build up both character and plot. This arduous authorial practice, as Thackeray seems to have recognized while composing the parts of *The Newcomes*, bore an uncanny resemblance to the paleontological procedures employed by his recent acquaintances Owen and Agassiz, who likewise began with just a small fragment and gradually constructed an "enormous . . . monster out of it."

But if the serialized novel assumes the "singular structure" of that "important animal, the megatherium of his [i.e., the novelist's] history," it is not simply on account of its prohibitive dimensions, as with Poseidon Hicks's gargantuan epic poem, or the cumbersome and incongruous arrangement disparaged by critics such as Simms and Bissell. Rather, Owen, as the allusion to him in *The Newcomes* suggested, had been able, starting from just a single extra tooth in a fragmentary cranium, to explain the relation between all the apparently anomalous elements of the megatherium's anatomy. He showed that their harmonious relation to one another allowed a mode of feeding that, while seemingly ungainly, was closely suited to the particular environment in which the gigantic creature had lived (fig. 4.3). As Owen proposed:

> In the remains of the Megatherium we have evidence of the frame-work of a quadruped equal to the task of undermining and hawling down the largest members of a tropical forest . . . which gives the explanation of the anomalous development of the pelvis, tail, and hinder extremities [which allowed it to remain upright]. No wonder . . . that their type of structure is so peculiar; for

FIGURE 4.3. Benjamin Waterhouse Hawkins, *Diagrams of the Extinct Animals Prepared for the Department of Science and Art; Vertebrate Animals; Sheet 5; Class: Mammalia—Edentata*. Lithograph, circa 1856. Natural History Museum, London. This pedagogic diagram, part of a series of six, shows a pair of megatheriums, with huge posteriors and diminutive heads, uprooting trees. Their seemingly ungainly structures were, as Owen showed, in fact perfectly adapted to this particular mode of feeding. © The Trustees of the Natural History Museum, London.

where shall we now find quadrupeds equal, like them, to the habitual task of uprooting trees for food?[96]

This account politely amended Buckland's interpretation of the creature's fossorial feeding habits in his own earlier explanation, examined in chapter 2, of how the megatherium's "egregious apparent monstrosity" and the seeming "incongruities of all its parts" were "in reality systems of wise and well contrived adaptation."[97]

In his *Description of the Skeleton on an Extinct Gigantic Sloth* (1842), Owen observed that the principle manifested in the "admirable adaptation" of the multifaceted "fore-foot of the extinct Megatheroid quadrupeds" was "beautifully set forth by the poet" in the following lines:

> In human works, though labour'd on with pain,
> A thousand movements scarce one purpose gain:
> In God's, one single can its end produce;
> Yet serve to second too some other use.[98]

When properly understood, the apparent monstrosity and incongruity of the megatheroid structure, Owen suggested, was in fact best represented by the neoclassical formal coherence of Alexander Pope's *Essay on Man* (1734). Pope's epigrammatic poem was an exemplar of Enlightenment natural theology, articulating the conviction that "Whatever IS, is RIGHT" and:

> All nature is but Art, unknown to thee;
> All chance, Direction, which thou canst not see,
> All Discord, Harmony not understood;
> All partial Evil, universal Good.[99]

Owen's use of the same poem to depict the exquisite functional adaptations of a creature in which, as he later observed, the "fertility of the Creative resources is well displayed" implies a connection between aesthetic and divine design.[100] The serialized novel, as a species of literary megatherium, might too reveal an underlying design behind its seemingly ill-proportioned parts that would render it as aesthetically unassailable as the most revered literary works of the previous century, and could even suggest a parallel between the initially enigmatic but nonetheless perfectly integrated designs of the serial novelist and those of the omnipresent author of the natural world.

Precisely the kinds of structural deficiencies for which critics condemned Thackeray's writing at the time of the commencement of his friendship with Owen in 1850, such as the *Spectator's* reproach, in December of the same year, that "Mr. Thackeray cannot or will not frame a coherent story, of which all the incidents flow naturally from one another, and are so necessarily connected with each other to form a whole, whose completeness would be marred equally by taking away or adding to it," were those that Owen had been able to show operating perfectly in the ostensibly just as incoherent skeletal structure of the megatherium.[101] In this colossal creature, Owen observed, "all the characteristics which co-exist in the skeleton . . . conduce and concur to the production of the forces required for uprooting and prostrating trees; of which characteristics, if any one were wanting, the effect could not be produced."[102] Early in *The Newcomes* Pendennis had conceded exactly the point raised by the *Spectator*, acknowledging: "In such a history events follow each other without necessarily having a connexion with one another."[103] Following his comparison of the serialized novel to a megatherium later in the book, however, Pendennis shifts his position and insists that in this "story . . . which for three-and-twenty months the reader has been pleased to follow," there were developments, central to the ongoing plot, that the "acute reader of novels has, no doubt, long *foreseen*" (italics mine).[104] Owen would doubtless have considered himself just such an acute reader of serials. By employing reading practices analogous to his own paleontological techniques that involved moving from part to whole and predicting how the events of each number would fit in, or correlate, with a harmoniously proportioned overall framework, the original audience of *The Newcomes* could, as Pendennis avows, confidently anticipate—or more precisely *foresee*—its unfolding plot and structure in exactly the same way that the no less necessarily connected skeletal structures of prehistoric creatures such as the megatherium could, as Owen insisted, "be *foreseen* through the . . . physiological principle of correlation of forms" (italics mine).[105]

Thackeray had long been intrigued by the comic potential of this gargantuan and purportedly awkward creature. However, in equating the "singular structure" of *The Newcomes* with the megatherium rather than any other comparably bulky antediluvian monsters (such as the "forty feet long . . . Megalosaurus" invoked by Dickens two years earlier in *Bleak House* [1852–53]), Pendennis instead postulates a means of appreciating novelistic design and structure that, drawing explicitly on Owen's elaboration of the megatherium's perfectly integrated anatomy, vindicated the aesthetic credentials of serialized fiction on precisely the grounds on which it had been most vociferously condemned.[106]

Large Loose Baggy Monsters

Prehistoric monsters nevertheless continued to afford reviewers a convenient analogue for a variety of prodigious and overblown works of literature. Even during the serialization of *The Newcomes*, the *New Monthly Magazine* cautioned its readers that a new collection of essays was "too capacious and bulky" and that this "Megatherium 'at large' will be too heavy for their shelves."[107] In fact, far from ceasing following the publication of Thackeray's novel, this particular strain of criticism was taken up again, at the end of the nineteenth century, by a new generation of writers frustrated with the stylistic conventions of the mid-Victorian novel. Most notably, the American novelist Henry James launched a hugely influential critique of what his compatriot and friend James Russell Lowell had continued to deride as "those megatheria of letters."[108] Despite his notoriously fastidious tastes, James himself was not indifferent to the curious inhabitants of the prehistoric world, and, just as with Thackeray, images of extinct megafauna sometimes intruded into his fiction. In the early story "Gabrielle De Bergerac" (1869), the protagonist's old-fashioned father is compared to "those big-boned, pre-historic monsters discovered by M. Cuvier," while in *The Bostonians* (1886), James evoked the shabby Upper East Side of New York, in which appeared the "fantastic skeleton of the Elevated Railway, overhanging the transverse longitudinal street, which it darkened and smothered with the immeasurable spinal column and myriad clutching paws of an antediluvian monster."[109] This awkwardly gargantuan imposition of urban modernity looms over and smothers the long street with its clutching paws in exactly the same way that Owen had famously shown the megatherium to clasp the stems of tall trees in its clawed forefeet. Such a peculiar mode of feeding was facilitated by a skeletal structure that closely resembled cognate industrial mechanisms, and the frequent identification of the megatherium with railways will be examined further in the next chapter. For James, however, it was literature that was most likely to assume the form of a monstrous prehistoric creature.

An aesthetically accomplished novel, as James adumbrated in the preface to the New York edition of *The Tragic Muse* (1908), required a "deep-breathing economy and an organic form" that precluded any superfluous "waste." This flawlessly integrated natural structure could be achieved only by the "premeditated" design of the "artist, the divine explanatory genius," which would ensure the "complete pictorial fusion" of all the different elements in a "long story." In a novel lacking such deliberate composition, on the other hand, the "joining together of these interests, originally seen as separate, might, all disgracefully, betray the seam," thereby rendering the work, whatever its other merits, a composite amalgam of incongruous parts.[110]

While the shoddily stitched seam, as well as the hyperbolic "mortal horror" James had for it, might suggest the gothic fiend that Victor Frankenstein composed from anonymous cadavers, it was another prevalent nineteenth-century form of monstrosity that James seemingly had in mind. He observed:

> A picture without composition slights its most precious chance for beauty. . . .
> There may in its absence be life, incontestably, as "The Newcomes" has life, as
> "Les Trois Mousquetaires," as Tolstoi's "Peace and War," have it; but what do
> such large loose baggy monsters, with their queer elements of the accidental
> and the arbitrary, artistically *mean*?

Monster, as was noted before, was the customary terminology for almost all prehistoric creatures, as James had recognized with his allusions to antediluvian "monsters" in both "Gabrielle De Bergerac" and *The Bostonians*. Without any coherent design, the "large loose baggy monsters" of his preface to *The Tragic Muse* were pieced together from any accidental and arbitrary elements that might come to hand in the same way as the allegedly apocryphal reconstructions of early paleontologists working with haphazard osseous fragments. Accordingly, their "organic . . . structural centre" was, James observed, rarely in the "proper position," and such an awkward and unbalanced novelistic monster was generally "condemned to the disgrace of legs too short, ever so much too short, for its body."[111] The *Mechanics' Magazine*, as seen earlier, had similarly reproached the *Penny Cyclopædia* as "an ill-shaped monster, with its . . . body too big for its legs," and James's legendary epigram—perhaps the most famous and influential of all descriptions of the nineteenth-century novel—becomes immediately explicable as a variety of prehistoric megafauna when located in the long-standing tradition of literary criticism drawing on paleontological analogies.[112]

Although literary critics have never previously made this connection, James's much discussed portrayal of the monstrous amorphousness of the mid-Victorian novel was actually very closely related to contemporaneous reconstructions of prehistoric creatures. In fact, its principal terms exactly replicated the language employed in the 1870s by the American paleontologist Edward Drinker Cope to describe the "rulers of the waters of ancient America, viz, the *Pythonomorphs.*" This "*monster* of the ancient sea," as Cope proposed on the basis of the Cuvierian "mode of reconstruction of extinct animals from slight materials," would likely have been endowed with a "peculiar shape" to facilitate its "habit of swallowing large bodies," and "hence the throat in the *Pythonomorpha* must have been loose and almost as baggy as a pelican's" (fig. 4.4).[113] Cope's distinctive description was revived and corroborated in a paleontological dispute of the early twentieth century when Charles Hazelius Sternberg, in 1903, rejected more recent interpretations and instead avowed: "I am led to think Cope was right when he suggested that this animal had the power . . . to force food down a loose, baggy throat."[114] Significantly, Sternberg's affirmation of the loose, baggy structure of this aquatic monster was published, and subsequently reprinted in more popular forms, only three years before the appearance of James's famous New York preface to *The Tragic Muse*.

If the vocabulary of loose bagginess potentially derived from Cope's and Sternberg's accounts of the fearsome marine reptiles of the Pythonomorpha order, the initial novel that James identified as an exemplar of the large, loose, baggy monster suggests that he might also have had a more familiar prehistoric mammal in mind. In *The Newcomes*, of course, Pendennis compared the novel's singular structure to that of the megatherium elaborated by Owen, and Thackeray's playful penchant for the colossal quadruped seems to have struck a chord with James, who mournfully reflected to a friend, "I sit alone in the great dim solemn library of this Club (Thackeray's Megatherium or whatever)."[115] James, as R. D. McMaster has shown, was an especially astute and attentive reader of *The Newcomes*, and by specifically invoking it, rather than any other equally gargantuan Victorian novel, in the preface to *The Tragic*

FIGURE 4.4. Sydney Prentice, *Kansas Cretaceous Sea*. Erasmus Haworth, *The University Geological Survey of Kansas*, 9 vols. (Topeka, KS: W. Y. Morgan, 1896–1908), vol. 6, plate 23. The mosasaur basking in the foreground displays the loose and baggy throat that was a defining characteristic of the Pythonomorpha order of aquatic reptiles. © The British Library Board, Ac.2692.i/2.

Muse, he was seemingly identifying the large, loose, baggy monster with the megatherium that Thackeray's novel avowedly resembled.[116] The Modernist poet Ezra Pound certainly considered that James's attitude to the defective, shapeless novels of the previous century was closely linked to this particular creature, inventively interpreting a passage from Edmond De Goncourt thus: "If ever one man's career was foreshadowed in a few sentences of another, Henry James's is to be found in this paragraph. It is very much as if he said: I will not be a megatherium botcher like Balzac."[117] As Pound's conjecture intimates, for James the literary megatherium represented an ungainly and ill-adapted monster, and his insistence on the disordered loose bagginess of the nineteenth-century novel was a conscious repudiation of the model of perfect functional adaptation that Thackeray, at least in *The Newcomes* following the commencement of his friendship with Owen, had discerned in this prehistoric creature's complex skeletal structure.

In a later memoir James recalled the "prolonged 'coming-out' of The Newcomes, yellow number by number" when he had first read the novel, and it is notable that the two other works he designated as "large loose baggy monsters," *War and Peace* (1865–67) and Alexandre Dumas's *Les trois mousquetaires* (1844), were themselves both initially published in serial form.[118] Much of James's own fiction, including *The Tragic Muse* itself, was also serialized in magazines and reviews, although he never disguised his abhorrence at being one of those "anxious novelists condemned to the economy of serialisation."[119] Like Thackeray in *The Newcomes*, in the preface to *The Tragic Muse* James depicted the serial novelist as in the same position as a paleontologist, piecing together small fragments to assemble a larger structure. However, with a very different conception of the artistic significance of fiction, James contended that the authors of such quasi-paleontological serials could merely invest their monstrous creations with a clumsy vitality while the more important attributes of form, harmony, and premeditated design remained conspicuously absent.

This altered understanding of the creative possibilities of the parallel between paleontology and the novel had already been noted by reviewers of James's fiction even before the publication of his preface to *The Tragic Muse*. Whereas Thackeray had compared the composition of serialized fiction to the process by which Owen or Agassiz built enormous monsters from just a fragment of bone, and that other megatherium botcher Honoré de Balzac eulogized Cuvier, as seen in chapter 1, as the nineteenth century's greatest poet, James's narrow insistence on the primacy of artistic "material *as* material," according to William Crary Brownell, rendered him incapable of actually moving beyond the mere contemplation of a single osseous fragment. As Brownell observed in the *Atlantic Monthly*: "Cuvier lecturing on a single bone and reconstructing the entire skeleton from it is naturally impressive, but Mr. James often presents the spectacle of a Cuvier absorbed in the positive fascinations of the single bone itself,—yet plainly preserving the effect of a Cuvier the while."[120] The serial novelist remained an artistic equivalent of the Cuvierian paleontologist, but the fundamental distinction between James's fastidious formalism and the more spontaneous fiction of his mid-Victorian forebears came down to a different attitude to what the likes of Cuvier, Owen, and Agassiz could conjure from a solitary fossilized bone.

Conclusion

James's celebrated critical epigram in the preface to *The Tragic Muse* was, of course, a retrospective summation of the defective novels of an earlier generation, made amid the radical cultural changes occurring in the opening years of the twentieth century. But in perpetuating the debates of Thackeray and various hostile reviewers from the middle of the previous century by giving *The Newcomes* and other such amorphous serial novels the paleontological designation "large loose baggy monsters," James revealed how pervasive and enduring the connection between serialization and Cuvierian correlation had been. Although it was most fully developed by novelists such as Thackeray or in Owen's own enthusiastic reading of serialized fiction, the resemblance was actually integral to the successful marketing of a vast range of works, nonfiction as much as novels, published in parts. An early notice in the *Medico-Chirurgical Review* of Todd's serialized *Cyclopædia of Anatomy and Physiology* stated that "we prophecied . . . when the first part of this work came out, that . . . the issue would be successful," and now, following the appearance of the next two numbers, the "complete success of the work may be foretold, from the excellence which characterizes the portion of it already published."[121] Such prophetic inferences from serial installments, which formed the basis of long-standing financial commitments, assumed an unerring relation between part and whole that was identical to the understanding of anatomical structure that enabled paleontologists to wield what Owen called the "prophetic power" of Cuvier's "principle of palæonotological research" in their own inferences from fragmentary bones (and Owen's description of this prophetic power itself appeared in a serialized work).[122]

Although serialization was a technique that had been developed by writers and publishers long before the nineteenth century, it became unprecedentedly pervasive from the 1830s onward, when the erstwhile practice of issuing cheap reprints in installments was supplanted by a trend toward original material that was specially commissioned for publication in parts. With the revolutionary impetus afforded by the phenomenally successful model of Dickens's fiction of the 1830s, serialization quickly emerged as the prevalent format across almost every sector of the mid-Victorian publishing industry. Crucially, it allowed both publishers and purchasers to spread their costs over a manageable period of time, thus dramatically increasing output by making new books less risky and capital-intensive for publishing houses and more affordable to a broader and more diverse range of consumers. Additionally, as Robert Patten has noted, "*buying* is the key to serial fiction," and the habit of repeat purchasing induced by the temporal dynamics of serialization, with its attendant intervals and suspenseful anticipation of new parts, proved hugely remunerative for a new generation of market-savvy publishers such as Bradbury and Evans.[123]

Serialization's intimate connection with correlation, as witnessed by Broderip and Owen's skillful deployment of the pioneering technique in their own works, the latter's absorption in the serial novels of Dickens and Thack-

eray, and the novelists' own discernment of incisive paleontological analogies, ensured that the famous Cuvierian principle remained at the forefront
of the entrepreneurial innovation and progressive modernity that were becoming enshrined as the emblematic values of mid-Victorian Britain. As
was noted in the last chapter, the liberal Tory view that the competitive social market, notwithstanding its massive inequities, was a divinely ordained
mechanism meant there was no conflict between laissez-faire capitalism and
certain versions of the natural theological argument from design. Indeed,
with Thackeray utilizing Owen's elaboration of the ungainly megatherium's
perfect adaptation to its particular antediluvian environment to reveal a similarly harmonious design in his purportedly no less ill-proportioned serial
fiction, the natural theological connotations of correlation could be readily
accommodated with the most innovative forms of modern publishing. The
next chapter will continue this focus on the law of correlation's imbrication
with mid-Victorian modernity, exploring its close relation to the central symbol of this new age of entrepreneurship, industry, and consumerism, and
switching from paper serials to brick-and-mortar models.

5

Correlation at the Crystal Palace

When the Canadian geologist William Logan traveled to Britain in January 1851, he left behind a "slab of sandstone, showing foot-prints," which, because of its weight, he "could not conveniently carry . . . across land to the sea-board at Boston." Instead, Logan brought a lighter "plaster cast" that, on arriving in London, he showed to Richard Owen.[1] The paleontologist soon proposed that the "shape of the body and the nature of the limbs indicated by the foot-prints accord best with those of the Chelonian reptiles," and he subsequently "inferred that the species was a fresh-water or estuary tortoise."[2] The original slab had been excavated from Potsdam sandstone at the base of the Lower Silurian system, and Owen's confident supposition prompted Charles Lyell to add an addendum to his annual address as president of the Geological Society excitedly proclaiming: "Assuming the Chelonian origin of these foot-prints, they constitute the earliest indication of reptile-life yet known."[3] Although, as Logan acknowledged, it "would have been more satisfactory to have exhibited the original than the cast," his plaster replica of this "curious fossil track" nonetheless facilitated what appeared another of the remarkable triumphs of the paleontological methods that had enabled Georges Cuvier to surmise the dentition of creatures that had left cloven footprints with more certainty even than Voltaire's fictional Zadig, and that Owen now utilized with such skill and precision to recreate the extinct denizens of the ancient past.[4]

Logan's main purpose in crossing the Atlantic at the beginning of 1851 was to arrange the collection of Canadian economic minerals that were to be displayed, alongside other raw materials from across the British Empire, at the Great Exhibition that summer. The mineralogical exhibits arrayed in Hyde Park's spectacular Crystal Palace included "beds of silicious conglomerate . . .

called the Potsdam sandstone" that, as well as yielding putatively chelonian footprints, was extensively "quarried for building purposes."[5] The plaster cast of the slab containing the fossilized footprints was itself also displayed under the shimmering iron-and-glass edifice, allowing Owen's triumphant inference to be included among the exemplars of British inventive genius that the Crystal Palace, and the vast exhibition within it, were intended to embody.[6]

The thirteen thousand exhibits on display at the Great Exhibition attracted more than six million visitors, and compelled those who attempted to describe the unprecedented experience to resort to hyperbole. Samuel Warren's *The Lily and the Bee* (1851) combined prose and poetry to "describe that astounding spectacle" in which "objects of every form and colour imaginable, far as the eye could reach, were dazzlingly intermingled."[7] While this effusive tone was also evident in the reports of newspapers such as the *Illustrated London News*, *The Lily and the Bee* had a particular focus for its excitement.[8] In a reverie under the Crystal Palace's "transparent fabric," Warren's breathless narrator suddenly declaims: "There is OWEN, profoundly pondering a shapeless slab of stone, neglected, and perhaps unseen, by millions; yet may he read in it an immense significance." Overlooked by the hordes of visitors, this "plain grey stone" from "cold Canada" is in fact the "True Philosopher's Stone . . . mystically writ upon . . . Telling of Life, and Air sustaining it." With it "now, here, Within our Palace!" the narrator implores that the stone slab ought to be recognized as no less remarkable than the famous Koh-i-noor diamond, which was the Crystal Palace's most visited exhibit. The narrator then encounters a mysterious "deep philosopher" who relates an imaginative vision of the creature who made the footprints. Initially, he observes that

> A reptile crawled, slowly, painfully:
> Now moving on: then resting for a while,
> Tired, or, perchance, looking for food.

The "thought-worn" sage then reveals, more specifically, that it was a "Tortoise he these prints that made," while the evidence of tidal currents marked in the sandstone even shows that "zephyrs swept his horny back." In an explanatory endnote appended to this volubly prophetic verse, Warren observed that the "method of coming at these results . . . appears to the author . . . profoundly interesting and instructive," and he recounted at length the "delicately exact . . . observation!" by which "our far-famed zoologist, Owen," was able to discern all that was "deducible solely from these faint footprints."[9]

When Logan again made the journey from Canada in January 1852, however, he brought with him originals of the sandstone slabs bearing footprints that, as Owen soon acknowledged, revealed a "character . . . that was not so distinctly recognizable in the casts" and that, crucially, was "irreconcilable with their having been formed by the . . . foot of a Chelonian." Owen, with chastened hesitancy, instead suggested as the "most probable hypothesis, that the creatures which have left these tracks . . . belonged to an articulate and probably crustaceous genus."[10] The only certainty, as Logan reflected, was that the "creature will be no tortoise," thereby refuting Lyell's eager avowal that it was the earliest reptile yet discovered.[11]

Two years later, in 1854, a second edition of *The Lily and the Bee* was published to coincide with the reopening of the glass "Temple of Wonder and Worship" on the southern outskirts of London. Warren assured readers that the section dealing with Owen's lauded inferences from the "Philosopher's Stone" could still "be depended on, as being in conformity with the existing state of knowledge on the subject." This was achieved by simply amending the sage's imaginative vision, so that his assertion "A reptile crawled" now became "Crawled a mailed reptile," with an accompanying footnote strategically conflating taxonomic differences by suggesting that this was the "*Crustacean*, of modern naturalists." The erstwhile "Tortoise" was transmuted into a less specific "mail-clad creature," while the "zephyrs" that had "swept his horny back" now instead "swept his mailed back." These tactical modifications notwithstanding, *The Lily and the Bee* retained its hyperbolic eulogies for the "wonderful . . . Stone" and the still more impressive "instructed eye" that could decipher its "import . . . sublime."[12]

Many critics had been aghast at Warren's "gaudy abomination" in which he discoursed on paleontology with "all the connoisseurship of an Antediluvian," and *The Lily and the Bee*'s publisher, William Blackwood, came to regard the book as a mistake.[13] It was still inexpensive enough to issue a second edition in a literary marketplace where continual revision was the norm, and new editions of books operated similarly to the serialized publications considered in the previous chapter.[14] Such serials, as was seen, enabled paleontological predictions that were subsequently relinquished to be revised and updated in later parts without requiring publishers to incur any extra costs. Owen's erroneous supposition that the footprints in Logan's plaster cast had been made by an unprecedentedly ancient tortoise could be similarly corrected when a sufficient market for a new edition of *The Lily and the Bee* was created by the reopening of the Crystal Palace as part of a vast entertainment complex at Sydenham (fig. 5.1). The no less hypothetical re-creations of prehistoric creatures that adorned the grounds of this new Crystal Palace, however, were not amenable to the same cost-effective methods of correction. Rather, the form and dimensions of these huge, and colossally expen-

FIGURE 5.1. George Baxter, *The Crystal Palace and Gardens*. Color print, 1854. This view shows the newly completed grounds of the reconstructed Crystal Palace in Sydenham as seen from the nearby railway, with the models of extinct animals in the foreground already attracting the attention of early visitors. Wellcome Library, London.

sive, three-dimensional brick-and-mortar models were, as James Secord has put it, literally "set in stone."[15]

The creation of these life-size models of extinct animals, as well as the reconstruction of the Great Exhibition's celebrated iron-and-glass structure, was financed by the Crystal Palace Company, a commercial syndicate of railway owners who sought to turn a profit by selling tickets, and train fares, to the new attraction. With a floatation of one hundred thousand shares priced at five pounds, the company raised capital of half a million pounds, of which nearly fourteen thousand pounds was reportedly spent on the prehistoric models alone.[16] The financial stakes could hardly have been any higher, and it was imperative that every ticket available be sold, at prices ranging from one shilling for daily entry to two guineas for a season ticket. The patronage of the paying public was, in part, dependent on the perceived accuracy of the re-creations of extinct animals, which might otherwise be dismissed as no less fanciful than Warren's "printed abuse of public confidence" in *The Lily and the Bee*.[17] This gave the financial speculators of the Crystal Palace Company a vested interest in upholding the unerring accuracy of the paleontological methods on which the models were based, especially as it would be impossible to amend or update them. And with Owen collaborating—though not unproblematically—with the artist Benjamin Waterhouse Hawkins on the design and construction of the models, it was inevitably Cuvier's famous law of correlation that, as this chapter will show, became entwined with entrepreneurial capitalism in the grounds of the rebuilt Crystal Palace.

The demands of mid-nineteenth-century commerce gave a new impetus to the endorsement of Cuvierian correlation, ensuring that it remained central to the new forms of print culture, and modes of visual education and entertainment, that were emerging in the 1850s. With railways and other forms of steam-driven technology integral to the enterprise of the Crystal Palace Company, and regularly compared to the integrated anatomical mechanisms apparent in the design of Owen and Hawkins's prehistoric models, even the natural theological connotations of correlation that harked back to William Paley could be readily accommodated with the cutting edge of Victorian modernity. In the new age of trade and technology inaugurated by the Great Exhibition of 1851, the fortunes of Cuvier's law of correlation reached their apogee.

A Spectacle Apparently Passing out of the Public Mind

Despite the necessity of the awkward revisions to the poem's transmuted tortoise, Warren dedicated the second edition of *The Lily and the Bee* to Owen's "TRUE PHILOSOPHIC SPIRIT" and "PROFOUND RESEARCHES" as a memorial of their "CORDIAL FRIENDSHIP."[18] It was the two men's close friendship that had actually precipitated Warren's composition of the poem, for he told Owen when it was completed: "Shall I ever forget our visit to the Exhibition? Never; ever may you."[19] *The Lily and the Bee* was a self-proclaimed "Apologue of the Crystal Palace," an ancient genre of allegory that employs heightened imagery to convey a particular moral, although Warren evidently intended to imbue it with an intense Romantic sensibility too.[20] Reporting how an early reader "became manifestly moved very strangely, as if taken by storm; but bit his lips, & recovered himself for he is a very proud man," he immodestly defied Owen "to read out, a great deal of this work, without being profoundly affected," asking him as a "special favour:—not to read it rapidly, *but slowly*; and I know what will be then the result."[21] The transcendent triumphs of Owen's paleontological methods, when rendered in Warren's emotive verse, might even elicit manly tears of veneration.

Owen's advice had clearly informed *The Lily and the Bee*'s effusive passages concerning the Philosopher's Stone and its putatively chelonian footprints, although in assuring Warren of the absolute veracity of his inferences, he seems to have kept some vital details from his friend. Even before publication, as Warren confided, the "*Stone* has excited such wonder among the few that have read the poem that I *sweat* at this idea of a plaster of Paris cast," and he reminded Owen: "I had it from you yourself that the original which you had seen, was at this moment in London."[22] Within days of *The Lily and the Bee*'s publication, Warren urgently contacted Owen regarding

> a business about which I have been very nervous from the beginning—*the Philosopher's Stone*. We are already found out! Yesterday a friend of consummate critical judgement . . . dined with me. . . . Guess my horror, when he coolly charged me, good naturedly, with having drawn on my imagination about *The Stone*:—for the instant he had read the "dazzling picture of geology", away he went—and . . . saw only "a big hideous plaster of Paris cast"—& on enquiring (as I instructed him) was told that it was all they had ever had! . . . He was *reluctantly* somewhat pacified; but insists on my putting in a foot note to the 2nd Edition saying that . . . in the Exhibition [is] only a faithful plaster of Paris cast.

Warren proposed having the original "Stone placed *over* the plaster of Paris cast"; otherwise, a "splendid opportunity for giving a vast impetus to the study of geology or zoology maybe lost!!!" although, as Owen well knew, no such original was yet in London.[23]

Owen, it would appear, was willing to hoodwink even a close friend to ensure that his correlative paleontological methods were enshrined among the glories of British genius commemorated at the Great Exhibition. In so doing, those same methods would be prominently associated with the progressive values of liberal free trade, internationalism, and inclusive social unity promoted by what Owen himself lauded as the "Great and happily-conceived Exhibition; from which every one, no matter what his social or intellectual grade, must have derived . . . lasting instruction and benefit."[24] While *The Lily and the Bee* was "never written for the million," it too was, as Warren assured Owen, aimed at "*THOSE WHO INFLUENCE* THE MILLIONS."[25] And with Warren apparently "nervous from the beginning" about "*the Philosopher's Stone*," it may actually have been Owen himself who, as with his similar collaborations with William John Broderip, was the driving force behind the poem's embellished celebration of his powers. Owen, in this instance, was certainly no less unscrupulous or even disingenuous than historians have generally considered him to be, but these dubious attributes were not exercised merely on behalf of an entrenched and antiquated Anglican establishment.[26] Instead, Owen's undoubted guile and cunning were used to associate his science with the progressive modernity of the Great Exhibition.

Within two years, Warren was concerned that the "Great Exhibition of

1851" had become a "spectacle . . . apparently passing out of the public mind without having had its true significance adequately appreciated." In his next book, *The Intellectual and Moral Development of the Present Age* (1853), he loyally overlooked his friend's fabrications with the fossil footprints and assured readers that Owen had "carried comparative anatomy much beyond the point at which it had been left by his illustrious predecessor Cuvier," before recounting the celebrated story of the discovery of the New Zealand dinornis.[27] The same need to keep the successful yoking of the purported marvels of correlation with the wondrous spectacle of the Great Exhibition in the public mind may have suggested to Owen the "idea of exhibiting restorations of certain extinct animals, as in life, in the geological part of the grounds of the Crystal Palace at Sydenham."[28] After all, Owen assured the Society of Arts in 1855 that "restorations of associated extinct animals" were a "long cherished scheme and aspiration of mine for public instruction," and the particular plan to have them at Sydenham, as he later insisted in an unpublished letter to the editor of *Nature*, "originated with myself. I communicated the idea in 1853 to . . . the Directors of the proposed New Crystal Palace, and I was requested to take the requisite steps toward its realization."[29]

Porcupine-Like Jealousy

Or at least this idea might have been suggested by Owen's collaboration on *The Lily and the Bee* if his retrospective claim to priority had been any more truthful than his specious assurances to Warren that he had seen more than just a plaster cast of the sandstone footprints. In reality, the Crystal Palace Company had already determined by August 1852, as the minutes of a meeting of its directors state, "that a Geological Court be constructed containing a collection of full sized models of the animals & plants of certain geological periods."[30] The Crystal Palace Company's directors had to juggle several discordant aspirations, wishing, according to Henry Atmore, "to profit themselves and their shareholders, and to edify, instruct and entertain the crowds expected to flock to Sydenham."[31] In accordance with this hybrid business model, the mooted "full sized models" would have to be crowd-pleasing spectacles as much as the more sober sources of "public instruction" that Owen subsequently claimed was his intention. Indeed, far from Owen having any involvement with the scheme at this stage, it was resolved that "Dr. Mantell be requested to superintend the formation of that collection."[32] Owen was regarded with an "intense feeling of hatred . . . by the majority of his contemporaries," as Thomas Henry Huxley claimed at the end of 1851, but it was Gideon Mantell who, according to Huxley, was his "arch-hater."[33]

In the 1820s Mantell had famously discovered the colossal iguanodon after comparing some "teeth and bones of a fossil herbivorous reptile" with those of "recent lacertæ," including a modern iguana, and inferring that the extinct creature "bore the same relative proportions" and must therefore have been "upwards of sixty feet long."[34] Since then, he had penned a series of popular treatises such as *The Wonders of Geology* (1838) that, through successive editions, deployed elaborate rhetorical tropes to "embody these inductions in a more impressive form," and show how paleontology could "satisfy the most enthusiastic lover of the marvellous." With Mantell regularly entreating his readers to "imagine the fossil bones to be clothed with appropriate muscles and integuments," he must have seemed the perfect candidate to superintend the Crystal Palace Company's proposed three-dimensional restorations.[35]

Initially, Mantell responded favorably to the proposition, attending the erection of the new Crystal Palace's first column and offering recommendations about the purchase of coal sections for its geological displays.[36] Despite his snobbish disdain for the "ignorant mobs" of "vulgar . . . country people" who attended the Great Exhibition, Mantell was not necessarily averse to the nakedly commercial aspirations of the Crystal Palace Company.[37] Writing to his son Walter on 11 August 1852, he exclaimed: "The Crystal Palace is demolished, but a private company has purchased it. . . . Half a million of money has been subscribed towards . . . a splendid building of glass & iron . . . to be open to the public at 1s cash." The prospect even moved him to Warrenesque effusions:

> I think there is every reason to conclude that it will succeed. It will be a glorious thing for this enormous Babylon, to have a temple where the great mass of the population may breathe on fresh air, and be surrounded by objects to amuse & yet elevate the multitude.[38]

Yet when, only nine days later, a representative of the Crystal Palace Company called on Mantell to apprise him that the "plan intended to be carried out as to Geology, was merely to have models of extinct animals," his attitude changed and he at once "declined the superintendence of such a scheme."[39]

This abrupt change of heart has been noted before by historians, who have attributed it to Mantell's "not shar[ing] the projectors' optimism for the scheme" and, more decisively, his distaste for the "idea of life-sized models—as opposed to more scientific displays."[40] While Mantell's fulsome missive to his son suggests that he was no less sanguine about the project than its financial backers, he was similarly far from disliking all life-size models or regarding them as necessarily unscientific. A year earlier, in *Petrifactions and Their Teachings* (1851), he had noted that, in the British Museum's "noble

Gallery of Organic Remains," the "model of the gigantic *Megatherium* arrests the attention of the visitor." This model, he observed, was "constructed with great care from the original bones," and although the

> attitude given to the skeleton, with the right arm clasping a tree, is of course hypothetical; and the position of the hinder toes and feet does not appear to be natural; altogether . . . the construction is highly satisfactory, and a better idea of the colossal proportions of the original is conveyed by this model, than could otherwise have been obtained.

The British Museum's model was in fact only a reassembled composite skeleton that used replica ribs and vertebrae to complete missing sections and enable the grasping megatherium to stand upright in an allegedly lifelike pose. But *Petrifactions and Their Teachings* went further in offering its readers a woodcut of what a model of a full restoration of the creature, complete with musculature and skin, might look like (notably, Mantell's restoration was a bulky quadruped in opposition to the actual model's bipedal configuration). Significantly, this "Restored Outline" resembled, albeit with less artistic detailing, the images of the three-dimensional models of extinct animals that subsequently appeared, without Mantell's proposed input, in the official guidebook to the Crystal Palace's geological displays (fig. 5.2).[41]

The British Museum's model megatherium, however, was atypical in being based on an almost complete set of bones, whereas what the Crystal Palace Company was proposing was the construction of life-size models of creatures that, in many cases, were known only from incomplete and fragmentary remains, and whose restoration would require a considerable degree of hypothetical inference. Mantell, since the very beginning of his paleontological researches three decades before, had trusted the application of what he called "Cuvierian sagacity" to fossilized fragments, relying on the French savant's techniques in his celebrated discovery of the iguanodon.[42] As recently as February 1848 he had eulogized Owen's famous identification of the dinornis from a single "fragment of bone" as the "*experimentum crucis* of the Cuvierian philosophy."[43]

But Mantell, despite his hyperbolic tone, had already begun harboring doubts that Cuvier's principles might only be able to yield results that, as he conceded in 1846, were "presumptive and not conclusive," and could even lure paleontologists into making pronouncements that were "premature and unphilosophical."[44] In considering the vertebra of an unknown saurian, he acknowledged that it would normally be "expected that some estimate should be given of the probable magnitude of the reptile," only to insist, in direct contrast to his Cuvierian procedure with the iguanodon, that "calculations

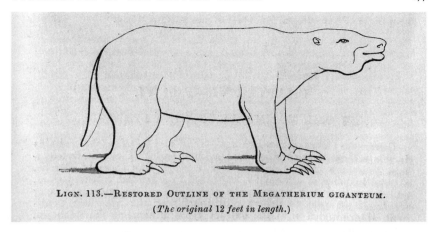

LIGN. 113.—RESTORED OUTLINE OF THE MEGATHERIUM GIGANTEUM.
(The original 12 feet in length.)

No. 7. Megalosaurus.

FIGURE 5.2. *A, Restored Outline of the* Megatherium giganteum. Gideon Mantell, *Petrifactions and Their Teachings* (London: Henry G. Bohn, 1851), 479. *B,* Megalosaurus. Richard Owen, *Geology and Inhabitants of the Ancient World* (London: Bradbury and Evans, 1854), 20. The similarity of this woodcut showing a model of a full restoration of the megatherium (*A*) to the images of restored extinct animals such as the megalosaurus in the official guidebook to the Crystal Palace's geological displays (*B*) suggests that Mantell was not entirely averse to the idea of three-dimensional models as proposed by the Crystal Palace Company. *A,* Reproduced by permission of the University of Leicester. *B,* © The Trustees of the Natural History Museum, London.

of the length and proportions of the original animal taken from a single bone . . . can afford but vague and unsatisfactory results."[45] When, as seen in chapter 3, John Rule visited him in August 1848 alleging that he had alerted Owen to the ornithic character of the famous fragment from which he had purportedly inferred the dinornis's existence, neither Mantell's already faltering trust in Cuvier's paleontological methods nor his relationship with Owen ever recovered.

Mantell's precipitous loss of faith in Cuvierian correlation was even more dramatic than the volte-face over his involvement with the new Crystal Palace, and the two may well have been directly connected. By 1850 Mantell was entirely "disclaiming the idea of arriving at any certain conclusions from a single bone," lambasting such attempts as mere "guesses at truth."[46] Two years later, at the very time that he was considering the Crystal Palace Company's proposal, Mantell, as he told his son on 11 August 1852, was "hard at work on a new edition of the Medals of Creation."[47] In its first edition, from 1844, this geological primer had related the famous anecdote of how, when considering some small mammalian fossil bones in Paris in 1804:

> From the character of the jaws and teeth, Baron Cuvier pronounced that the animal was related to the Opossum, and confidently predicted, that the two peculiar bones which support the pouch in these animals, would be found attached. . . . Accordingly he chiselled away the stone, and disclosed the marsupial bones; thus proving the truth of those laws of correlation of structure, which he was the first to enunciate and establish.[48]

While the revised second edition of *The Medals of Creation* retained this passage, it now added a pointed caveat:

> There are true marsupials in which the *ossa marsupialia* are merely rudimentary, for example, in the . . . *Thylacinus* . . . . Thus the fossil pelvis of the Thylacinus . . . would not have afforded the certain evidence of its marsupial character to which Cuvier triumphantly appealed in demonstration of the Didelphys . . . yet the Thylacinus would not therefore have been less essentially a marsupial animal.[49]

Mantell was rescinding his earlier faith in even the most renowned instances of Cuvier's predictive powers at precisely the moment the Crystal Palace Company was requesting him to undertake hypothetical three-dimensional restorations whose huge expense and need to attract paying spectators required them to be, or at least seem, infallibly accurate. No longer confident of Cuvierian correlation, it is hardly surprising that, even with his optimism for the scheme's commercial prospects and advocacy of models when based on sufficient evidence, Mantell turned down the chance to superintend the creation of the monstrous creatures that would soon emerge in the grounds of the new Crystal Palace.

Instead, it was Owen, the philosophic hero of Warren's ornate account of the Great Exhibition, who was contracted to oversee the construction of the models of prehistoric megafauna that had already been commissioned from the artist Benjamin Waterhouse Hawkins.[50] When, only three months after rejecting the Crystal Palace Company's offer, Mantell succumbed to various long-standing illnesses, his replacement on the Sydenham scheme penned an anonymous obituary that spitefully noted his "intrinsic want of exact scientific, and especially anatomical, knowledge," which had engendered in him an "extreme susceptibility of any doubt expressed of the accuracy . . . of that which he advanced."[51] A further, and still worse, affront came two years later, during a celebratory banquet held at Sydenham in the mold from which the nearly completed model of the iguanodon had been cast. At the end of the meal, Owen, hosting this elaborate publicity stunt for the soon-to-open Crystal Palace, proposed a toast to the "discoverer of the beast in the model of which the company had just dined." However, the "memory of Mantell," he informed the assorted grandees, would, notwithstanding his many achievements, "ever be associated with . . . that porcupine-like jealousy which he always displayed lest any person should sacrilegiously dare to cut off an inch of the tail of the monster which he had constructed from a single fossil tooth."[52] Amid the very models that he himself had initially been asked to superintend, Mantell's bitter rival not only alluded to the resentment that was allegedly occasioned by his scientific failings, but also, in a final indignity, posthumously reconverted him to Cuvierian correlation.

Bran-New Resuscitations

Unlike either the Great Exhibition or the grand museums in central London, the new Crystal Palace on the capital's outskirts only rarely displayed original artifacts, and instead mostly offered casts and replicas of the greatest works of human artistry and ingenuity from around the world. Now twice the size of the original structure in Hyde Park, the Sydenham Crystal Palace, as *Fraser's Magazine* marveled, would enable the "artisan who will . . . pay his shilling and pace these splendid galleries" to "see more of the fine arts of Europe than any nobleman who goes 'the grand tour' at the cost of thousands."[53] But it was not merely a cheaper and more convenient equivalent of their social superiors' travels across Europe that such plebeian visitors experienced at the Crystal Palace. Its unashamed emphasis on ersatz reproductions also facilitated a distinctive form of temporal tourism.

Whereas collections of classical art on the Continent, as well as at the British Museum, offered only crumbling, faded remnants of the glories of ancient Greece and Rome, the Crystal Palace aspired to recreate these past civilizations as they were actually experienced by their original inhabitants. As S. J. Hales has proposed, "Throughout the Crystal Palace, nothing was original. . . . The Palace board clearly thought that the ubiquitous use of casts, reconstructions, and reproductions was one of their great strengths . . . staging a prefect pretence rather than the crumbling, piecemeal reality of ruins."[54] Reflecting the Crystal Palace Company's fusion of commercial and pedagogic aspirations, such re-creations would offer both a vivid spectacle, and, in accordance with the theories of the Swiss educationalist Johann Heinrich Pestalozzi, convey knowledge of the past directly through the sensual experience of objects—and visual perception especially—without the need for any textual mediation.[55] The Pompeian Court, for instance, was a meticulously detailed reconstruction of a domestic house in the ancient city before it was engulfed in volcanic ash, enabling visitors to put themselves directly in the sandals of the Romans and experience the same sensations they had (fig. 5.3).

The pallid antiquities of Athens, including the Parthenon frieze, were, in the same vein, painted in the polychromatic hues that, purportedly, the "Greek eye would have demanded." Anxious to place "before the public . . . a result that might have existed" even when "there is so little to guide," Owen Jones, who superintended the Crystal Palace's artistic displays, acknowledged resorting "to analogy for the proposed restoration of the several colours." Although Jones's enthusiasm for analogical polychromatic reproductions had already provoked "hard words" from "those critics who stand on the ground

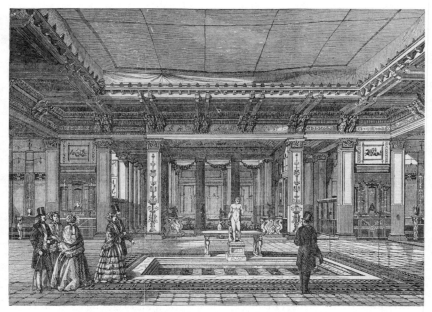

FIGURE 5.3. *The Pompeian Court, at the Crystal Palace, Illustrated London News* 26 (1855): 64. The Crystal Palace staged vivid re-creations of past civilizations that enabled Victorian visitors to experience, among other things, a domestic interior in the Roman city of Pompeii. Reproduced by permission of the University of Leicester.

of traditional opinion," his insistence that the Crystal Palace's restorations should be rendered in their presumed original colors even extended into its grounds.[56] Reporting how "antediluvian monsters have settled in strange and terrible groups" amid the "mountains of loose earth which now abound . . . in the park," *John Bull* noted, in April 1854, that

> Mr. Owen Jones, whose theories of colour would thus appear to transcend the regions of art-decoration, declares that analogy would afford rules for the decoration of these monsters. His proposition is that those of them who were preyed upon by others ought to be a dull colour . . . while those which were accustomed to chase their interesting dinosaurian brethren ought to be of a gay or lively hue.[57]

If the completed models of extinct animals that greeted the paying public when the gates of the park finally opened two months later were not bedecked in the gaudy multihued tints proposed by Jones, their more drably painted forms had nonetheless been conceived as part of the same agenda of ersatz restoration that had guided the so-called "Colour King" in fashioning the Crystal Palace's iridescent interior.[58] Akin to the re-creation of the visual experience of the inhabitants of ancient Greece, the prehistoric models would offer something equivalent to viewing living animals at the Zoological Gardens in Regent's Park.[59]

In fact, with the polychromatic adornments mired in controversy and Jones compelled to issue a defensive *Apology for the Colouring of the Greek Court in the Crystal Palace* (1854), the models of extinct megafauna he also sought to color were heralded as exemplars of the educational value of the policy of re-creation. Like many other cultivated visitors to the Crystal Palace, the art critic William Michael Rossetti bemoaned the "influence of individual views developed conjecturally, as in restorations and improvised polychromatism," preferring to "leave the thing as it was found—defective possibly, but not false." Rossetti, however, also acknowledged that the "Crystal Palace has its own real usefulness and influence," acclaiming its capacity to make the "discursiveness of its lesson so palatable" as "without exaggeration unique." And it was "Mr. Hawkins's and Professor Owen's bran-new resuscitations" of "præadamite" creatures that, notwithstanding his somewhat facetious tone, particularly impressed Rossetti. An "uninformed person," he observed,

> will immediately and without labour gain from the models in the Sydenham garden-lake as tolerable an idea of the appearance, habits, and affinities, of the antediluvians, as the first science of the day can put into shape . . . such as no quantity of . . . helpless inspection of authentic débris would have supplied him with.

For a "person already learned in the question," the "study of actual fragments" inevitably remained much preferable to such "inadequate material symbol[s]" of paleontological inductions, but the prehistoric models
were, even for an aesthetic connoisseur such as Rossetti, superior embodi-
ments of the Crystal Palace's pedagogic aspirations than any of its artistic
reconstructions.[60]

Another artistically inclined visitor, the auctioneer Samuel Leigh Sotheby,
was similarly dismayed at the "theory adopted by Mr. Owen Jones in respect
to" the "most exquisite works of the early Greeks" being "barbarously be-
daubed." Like Rossetti, he had much more faith in the scientific method
that "enabled the master-minds of . . . Owen and Waterhouse Hawkins, to
represent accurate proportions of those [extinct] animals, were it only by
the discovery of a *single* bone." This was, Sotheby opined, "rather different to the *ideal* representations of Palaces, founded upon a few remains of
'polychromatic decorations' on the stuccoed walls."[61] In fact, Sotheby actually consulted with Hawkins on what the latter agreed were the "important
subjects of miscoloration," and the modeler supported his "convincing arguments against the painting of the . . . Assyrian Court . . . &c. &c."[62] The
models created by Hawkins and Owen were, for both Rossetti and Sotheby,
the standard-bearers of the historical authenticity of the Crystal Palace's vast
array of other, more controversial reconstructed artifacts, even though their
monstrous forms and dimensions were generally surmised from much less
evidence than the re-creations of classical antiquities.

Sotheby, though, had considerably more at stake than Rossetti. He was
also, as he boldly announced himself, "one of the Shareholders of 'THE
CRYSTAL PALACE COMPANY.'"[63] It was the "multiplication of small stakes"
and "property, in the shape of hard money, every shilling earned by some
individual," which, according to Harriet Martineau, had "built this palace
and laid out these grounds," and shareholders such as Sotheby, who spent
£2,710 on 542 shares, demanded a voice in how the business was managed.[64]
Controversies such as that over the accuracy of Jones's polychromatic reconstructions threatened the value of their investments. The reputed precision
of the paleontological method by which the prehistoric restorations had been
inferred, on the other hand, helped underpin not only the models' own acceptance by the public but the Crystal Palace Company's wider policy of reconstruction and, with it, its shareholders' entire financial speculation.

Unsurprisingly, the investors and businessmen who funded the ambitious
and potentially hugely remunerative undertaking at Sydenham were, without
exception, staunch advocates of Cuvierian correlation. Sotheby had actually
been warned in private, by the zoologist John Edward Gray, that the "restored

fossil animals . . . are a gross delusion, Cuvier himself never attempted to give the external surface of the fossil animal, he merely gave an indistinct outline of what might be its general form."[65] In his public *A Few Words by Way of a Letter Addressed to the Directors of the Crystal Palace Company* (1855), however, Sotheby only briefly alluded to any potential variance of "opinion as regards the correctness in the restoration of the exterior coatings of these terrestrial monsters," before acclaiming the absolute accuracy of Owen's masterful inductions from a "*single* bone."[66] Barely a week after receiving Gray's letter, Sotheby clearly resolved to ignore its authoritative admonitions, respectfully sending Owen "a copy of my Letter to the Directors of the Crystal Palace Company" along with some photographic "views of the monster animals" taken by his wife, who, Sotheby recalled to Owen, "had the pleasure of meeting you when she was last at the workshop of Mr. Waterhouse Hawkins."[67]

With Sotheby disgruntled at the "present depressed state" of his shares and what he perceived as the mismanagement of Crystal Palace Company's directors, Owen's paleontological prowess was perhaps the only thing on which he agreed with the company's chairman, the railway magnate Samuel Laing.[68] At the Sydenham site's opening ceremony, Laing proudly declared:

> The restoration from a single fossil fragment of complete skeletons of creatures long since extinct, first effected by the genius of Cuvier, has always been considered one of the most striking achievements of modern science. Our British Cuvier, Professor Owen, has lent us his assistance in carrying these scientific triumphs a step further and in bringing them down to popular apprehension.[69]

Hawkins himself acknowledged that "Owen's sanction and approval" had rendered the "restorations truthful and trustworthy," and it was the modeler's "sincere wish . . . that, thus sanctioned, they may . . . prove one of many sources of profit to the shareholders of the Crystal Palace Company."[70] Owen's paleontological authority, and his unrivaled mastery of Cuvier's law of correlation, evidently had significant pecuniary value on the mid-Victorian stock market.

Acres of Swamp and Mud

But, despite the tributes of Sotheby, Laing, and Hawkins, the vaunted British Cuvier had little involvement in the actual creation of these precious prehistoric commodities. For one thing, with the reconstruction of the Crystal Palace hampered by a "continuous succession of rain and unfavourable weather" and Hawkins's onsite workshop "almost inaccessible for deep ruts and acres of swamp and mud," Owen, with the exception of the banquet in the iguanodon mold on New Year's Eve 1853, never visited the Crystal Palace's uncongenial site before its official inauguration the following June.[71] Hawkins, with "mud-boots and woollens . . . his only protection against the winter wet and blast," labored on alone, assisted only by a "small, but valiant staff" (fig. 5.4).[72] Even once the Crystal Palace had opened, Owen's appearances remained extremely infrequent, with his encounter with Sotheby's wife in the winter of 1854 seemingly a rare exception. In March of the following year, Hawkins, mindful of the mud, pleaded that the "weather having so favourably changed leads me to hope that I may have the pleasure of seeing you here."[73] Two months later, he was still forlornly hoping "you will bestow a leisure day upon me here."[74] Earning what Sotheby complained were "little more than journeyman's wages," Hawkins lacked the social status to protest more vociferously.[75]

Owen's prolonged absences, which left Hawkins having to "refrain from moulding" the models while he waited "to obtain the benefit" of the paleontologist's "valuable criticism," did not go unnoticed at higher levels of the Crystal Palace Company.[76] Joseph Paxton, the designer of the original iron-and-glass structure and now the principal manager of its new incarnation, gently remonstrated with Owen less than a week before the Crystal Palace's opening: "I wish you would come down here soon, again, and see what we are about."[77] In his official *Guide to the Crystal Palace and Park* (1854), Samuel Phillips claimed that "Mr. B. W. Hawkins" had been "busily employed, under the eye of Professor Owen."[78] In reality, however, Owen must have first seen the finished prehistoric restorations he had nominally cocreated only days ahead of artisanal visitors paying their shilling entrance fees.

More significantly, Owen's own conception of the Crystal Palace's geological displays, notwithstanding his later claims to have originally formulated the plan for having restored prehistoric models, seems to have differed markedly from those of its financial backers. Conducting their collaboration almost exclusively through correspondence, Owen apprised Hawkins, in March 1854, of the imperative "scientific value" of a "cabinet collection of invertebrate fossils . . . now offered for purchase to the Crystal Palace Company." While Hawkins dutifully promised to "lay this offer before the Directors," he had to acknowledge that the "acquisition of such valuable collections does not come within the present views of the Directors of the C. P. Com."[79] A preview, in *Hogg's Instructor*, of the "marvellous illustrations of the science of geology" soon to be revealed at the "People's Palace at Sydenham" reported that it was "intended" to combine "actual specimens of fossil remains, and life-sized restorations of the monsters of the pre-Adamite world," although

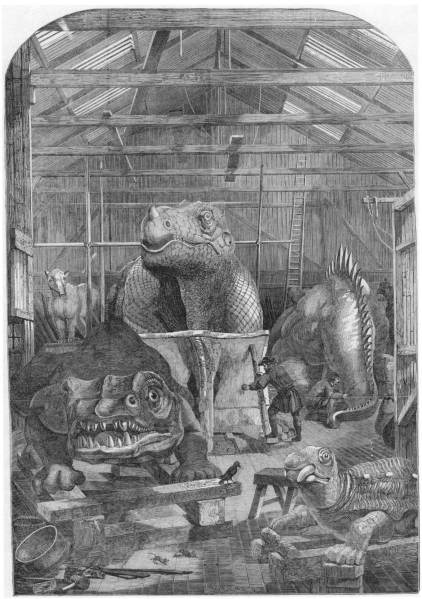

FIGURE 5.4. *"The Extinct Animals" Model-Room, at the Crystal Palace, Sydenham, Illustrated London News* 23 (1853): 600. With the exception of the famous banquet held in the iguanodon mold on New Year's Eve 1853, Owen never actually visited Hawkins's crowded and often muddy workshop before the opening of the Crystal Palace. © Illustrated London News Ltd. / Mary Evans.

any visitors subsequently seeking these "actual specimens" would have been disappointed.[80]

Owen's desire to supplement the three-dimensional models with genuine fossils of course contravened the Crystal Palace's pervading ethos of ersatz historical reconstruction. It actually had more in common with Mantell's proposal that the British Museum should, like his own Mantellian Museum in Brighton, provide "restored figures of the animals whose remains are in the cabinets . . . painted or suspended on the walls." While such "pictorial illustrations" might be "pleasing to the eye, and instructive to the mind," they would, in Mantell's plan, remain secondary to the "surpassing interest" of the authentic fossil specimens, serving merely to "render them intelligible to the uninstructed observer."[81] Owen too, despite their bitter rivalry, seems to have shared Mantell's distaste for the Crystal Palace Company's scheme "merely to have models of extinct animals," and entreated its directors to purchase fossil collections that would have transformed the geological displays into something resembling Mantell's vision of the British Museum.[82]

The Crystal Palace's commercial and curatorial agenda, however, was very different to that of the national collections that had recently been moved to their own palatial government-funded premises in Bloomsbury. That "there is a parallel collection recently purchased by the Trustees of the British Museum" was, as Hawkins noted, a further reason why Owen's proffered invertebrate fossils would not appeal to the Crystal Palace Company, which endeavored to offer a wholly distinct, and more vivid, readily accessible, and, above all, marketable experience.[83] While courteously tactful when responding to Owen, Hawkins himself agreed that the "British Museum, though containing some of the finest fossils that have been collected . . . offers little more than objects of wonder, literally only dry bones or oddly-shaped stones to the majority who see them."[84] Even a connoisseur such as Rossetti recognized that the "British Museum is a better mistress . . . than the Sydenham Museum, but she takes a longer time in giving her curriculum."[85] In any case, some visitors to the Crystal Palace evidently could not distinguish between the synthetic reconstructions and actual fossil remains, with Martineau observing disdainfully: "The extinct animals have already been roughly handled by strangers—whole rows of their teeth having been pulled out. . . . The culprits suppose they are carrying away the real teeth of preserved animals."[86] Rather than abruptly withdrawing from the scheme like Mantell, Owen accommodated his aspirations with the more demotic plans of the Crystal Palace Company, even if his perpetual absence from the site made his indifference glaringly obvious.

Nor did Owen fare much better with the design and construction of the three-dimensional models that were gradually emerging from Hawkins's muddy workshop. Their collaboration, according to Owen, involved him furnishing the artist with

> a drawing of the skeleton of one of the extinct animals, which I thought might be restored, as living, of the natural size. . . . Hawkins prepared a small model from the drawing: I suggested requisite corrections, and gave the dimensions, estimated from the largest known fossil of the species, for a corresponding enlarged model.[87]

Hawkins, though, was a much less compliant collaborator than Warren had been with *The Lily and the Bee*. He was equally adamant that, having been "enabled to interpret the fossils that I examined and compared" by studying the "elaborate descriptions of Baron Cuvier" and the "learned writings of our British Cuvier," it was his role to "make the preliminary drawings" on which the initial scale models were based. Only then were these "sketch models . . . submitted . . . to the criticism of Professor Owen," whose "profound learning," Hawkins conceded, was "brought to bear upon my exertions to realise the truth."[88] Without this "valuable criticism," as Hawkins told Owen in private, "I never think my works endowed with vitality or endurance," although this again implied that Owen's contribution came only in the final stages.[89] In fact, Hawkins later told Gray that Owen had "afforded no assistance" at all during the "time when the models were being prepared."[90]

Whatever the truth of Owen's degree of involvement in the creation of the models that were, at some stage, actually submitted to his scrutiny, there were, he later complained, "other 'restorations' which Mr. B. W. Hawkins added, without reference to or consultation with me," and these, he added sniffily, "I could not recommend to be exposed to view."[91] Owen was sufficiently piqued to make his reservations public in his official contribution to the Crystal Palace Library, *Geology and Inhabitants of the Ancient World* (1854), which notified visitors that the iguanodon's nasal horn was "more than doubtful," and warned of the frog-like labyrinthodon that it should "be understood . . . that with the exception of the head, the form of the animal is more or less conjectural."[92] Owen even repeated his warning against the speculative "basis of the restoration of the *Labyrinthodon salamandroïdes*, at the Crystal Palace" in his textbook *Palæontology* (1860), where, notably, he refrained from acknowledging his own involvement in the creation of any of the models.[93] Only once in *Geology and Inhabitants of the Ancient World* did Owen indulge in his customarily effusive rhetoric concerning the anatomical forms that "Baron Cuvier's prophetic glance saw buried in the womb of time, and . . . which verified his conjecture[s]."[94] The necessity of dissociating himself from Hawkins's speculative inventions meant that Owen showed an uncharacteristic reluctance for boldly affirming the accuracy of hypothetical paleontological inferences, whether the insubordinate artist's or even the great French savant's.

This same guidebook also inadvertently revealed Owen's continued failure to actually visit the Crystal Palace, with him advising readers that, in the case of the aquatic mosasaurus, "almost the entire skull has been discovered, but not sufficient of the rest of the skeleton to guide the complete restoration of the animal. The head only, therefore, is shown."[95] Although it was largely obscured from view by a waterfall, Hawkins had, unbeknownst to Owen, modeled most of the creature's long scaly body.[96] Similarly, the guidebook concluded, without any explanatory letterpress, with an isolated image of the dinornis, even though Hawkins's original proposal to have a model of the extinct struthious bird among his prehistoric restorations had not been— and would never be—realized. Owen, in addition, apparently delegated checking the "proofsheet[s]" of *Geology and Inhabitants of the Ancient World* to Hawkins, and his shoddy guidebook, which, at just forty pages for three-pence, was the shortest and the cheapest of all the Crystal Palace Library's official publications, makes his disenchantment with the prehistoric models painfully manifest.[97]

The Great Forge-Bellows of Puffery at Work

Ironically, the Crystal Palace, with its Pestalozzian principle of educating through the eye without the need for textual mediation, generated an unprecedentedly vast array of guidebooks, catalogs, journalistic reportage, topically themed novels, and even newspapers dedicated solely to events at the Sydenham site. As one such newspaper, the *Crystal Palace Herald*, reported: "No palace or public building that has been erected since the invention of printing ever received such universal attention from the press as the Great Glass Museum."[98] With Owen's mastery of the predictive powers of paleontology so integral to the financial value of the Crystal Palace Company's entire scheme of ersatz historical reconstruction, and yet with him only rarely physically present on-site and seemingly disillusioned with the project, this eruption of print afforded a vital opportunity to fashion a more marketable version of his putative contribution.[99] This was especially needful as problematic rumors had begun to circulate. As Gray confided to Sotheby in December 1854, while "the directors profess to shelter themselves under the authority of Professor Owen . . . I have been repeatedly assured by the modeller [i.e., Hawkins] he has never been there (except to the Dinner)."[100] For the Crystal Palace

Company, it was imperative to rebut such damaging insinuations and instead ensure that press coverage reflected its own corporate agenda. In particular, it was Owen's alleged deployment of Cuvier's unerringly accurate law of correlation that most suited the commercial interests of the Crystal Palace's owners. Conveniently, this was also the case for many publishers that, in the fiercely competitive mid-Victorian literary marketplace, also sought to profit from what the *Illustrated Crystal Palace Gazette*, another of the dedicated Sydenham newspapers, heralded as a "new epoch in the pregnant history of popular amusement."[101]

As the self-proclaimed "conductor" of the best-selling weekly magazine *Household Words*, Charles Dickens was acutely aware of what he called the "great Forge-bellows of puffery at work" on behalf of the "CP," considering even its official "guide-books . . . a sufficiently flatulent botheration." Dickens recognized that this noxious gust of publicity artfully concealed less palatable realities (including Laing's deliberate downplaying of fatal accidents to maintain his railway company's share price), and he resolved to "avoid the bellows."[102] By this time, though, *Household Words* had already run an ebullient preview commending the Crystal Palace as a veritable "Fairyland" in which the "King of Animals, Professor Owen" had, "according to some subtle theory," brought "back those antediluvian days when there were giants in the land."[103] Indeed, Dickens, as seen in chapter 3, had been helping to puff Owen's paleontological accomplishments in *Household Words* since the summer of 1852. And other editors were willing to be much less discriminating than him.

The *Illustrated London News*'s extensive and enthusiastic coverage of the Great Exhibition had more than doubled its circulation during the summer of 1851, and the innovative weekly paper was soon induced to similarly promote the Crystal Palace's new incarnation.[104] Helpfully, its proprietor, Herbert Ingram, was a close personal friend of Paxton, and retained a strong influence over the editor Charles Mackay, whom he had appointed personally in 1852.[105] Ingram was among the "one-and-twenty guests" invited, on New Year's Eve 1853, to dine alongside Owen and the "Directors of this truly national undertaking" in the mold of the iguanodon. A week later, the *Illustrated London News* devoted almost half a page to a wood engraving of this highly newsworthy example of the "modern hospitality of the Iguanodon" (fig. 5.5). The accompanying report concluded by noting how, as host, Owen

commented upon the course of reasoning by which Cuvier, and other comparative anatomists, were enabled to build up the various animals of which but small remains were at first presented to their anxious study; but which, when afterwards increased, served to develop and confirm their confident

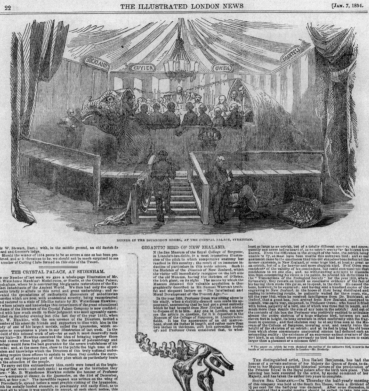

FIGURE 5.5. *Dinner in the Iguanodon Model, at the Crystal Palace, Sydenham.* "The Crystal Palace at Sydenham" and "Gigantic Bird of New Zealand," *Illustrated London News* 24 (1854): 22. This broadsheet page from the *Illustrated London News*'s weekly number for 7 January 1854 juxtaposes a large wood engraving and report of the celebratory banquet in the iguanodon model held on New Year's Eve with an ostensibly separate news article on the dinornis, with its own accompanying wood engraving, that strategically augments the emphasis of Owen's speech at the banquet on the accomplishments of the law of correlation. Reproduced by permission of the University of Leicester.

conceptions—instancing the Megalosaurus, the Iguanodon, and Dinornis as
striking examples.[106]

If Owen was unusually coy about identifying his own involvement in the last
of these "striking examples," the following column of the same broadsheet
page carried an apparently separate article entitled "Gigantic Bird of New
Zealand," which, apropos of no recent events whatsoever, reported his "immense induction of particulars" with the dinornis fifteen years earlier. Tellingly, this same article then reprinted the jubilant passage in which the struthious bird's discovery was "graphically described in Mr. Samuel Warren's
truthful and eloquent lecture on 'The Intellectual and Moral Development of
the Present Age.'"[107] Warren's highly partial attempt to maintain his friend's
profile as the British Cuvier in the wake of the Great Exhibition's closure was
now being recycled, as an ostensible item of current news, in the run-up to
the Crystal Palace's reopening as a private concern. Rather than simply reporting on the celebratory banquet in the iguanodon mold, the *Illustrated
London News* was complicit in stage-managing the event.

While Hawkins was adamant that his "geological restorations" would
boldly "reverse that order of teaching which is described as the names and
not the things" and instead embody the policy of "direct teaching through
the eye" associated with the "name of Pestalozzi," his paymasters in the Crystal Palace Company were evidently less sanguine about the possibility of for-
saking all textual means of guiding how visitors might interpret the prehistoric models.[108] In fact, the very same tactic employed in Ingram's *Illustrated
London News* a week after his attendance at the promotional banquet in the
iguanodon mold—complementing coverage of the restored extinct animals
with information about the Cuvierian law of correlation, and especially the
story of Owen's renowned use of it to predict the existence of the dinornis—
was soon replicated in a range of other publications aimed at very different
reading audiences. Even without the same largesse granted to Ingram, the
parallels with the *Illustrated London News*'s treatment of the same subject, resembling the calculated "literary replication" examined in chapter 3, suggest
strongly that the Crystal Palace Company's promotional agenda was similarly
at work.

The upmarket monthly magazine the *Ladies' Cabinet*, having already
"looked on the treasures of art within the wondrous Crystal Palace" that
were more suited to its usual emphasis on fashion and now turning to the
less seemly "monuments . . . of pre-Adamite ages," considered that a "brief
sketch of the method pursued" in their construction "may not be unacceptable to those who incline to accompany us through this seemingly phantas-

mal scene." Those female readers who were disposed to follow were informed of "certain laws . . . of comparative anatomy" that meant that a "glance at . . . a tooth" could afford a "general idea of . . . the habits and organization of the creature." This process was exemplified by Owen's inference from a "portion of some gigantic wingless fowl" and the subsequent appearance of more substantial remains "incontestably proving he was right." Owen's famous discovery of the dinornis, it was urged, should "convince our readers that the restoration[s] of the extinct animals . . . are not so entirely conjectural as many might suppose." It could even persuade them that "these remarkable restorations" were "perhaps the most bold and original part of the whole wonderful design of this great national school of instruction."[109]

If his contretemps with Hawkins had rendered Owen's authorized handbook for the Crystal Palace Library surprisingly diffident about the accuracy of hypothetical paleontological inferences, the "road to the Palace," as *Punch* reported, was clogged with "vendors of spurious 'Guide Books,'" and these unauthorized publications were, paradoxically, much more willing to toe the Crystal Palace Company's official line.[110] The entrepreneurial publisher George Routledge issued his own *Routledge's Guide to the Crystal Palace and Park at Sydenham* (1854), which, like the *Ladies' Cabinet*, acknowledged that "spectators who for the first time gaze upon these uncouth forms, will be disposed to seek for the authority upon which these antediluvian creatures have been constructed." Accordingly, it advised that the "visitor, before making acquaintance with the various antediluvian monsters . . . will be pleased to learn the interesting circumstances" of the "re-construction, from a portion of fossil bone, of a 'gigantic, wingless bird'" made "by that most able and ardent votary of the science, Professor Owen."[111] Once again, textual mediation was considered necessary before visitors could encounter the models, in the approved Pestalozzian manner, with their own eyes. With *Routledge's Guide* costing just a shilling, its predominantly plebeian purchasers were considered needful of precisely the same prior instruction on correlation and the discovery of the dinornis as the fashionable women who read the *Ladies' Cabinet*. Only then would either group be enabled to experience the actual prehistoric models in the appropriate way.

The same preparatory excursus on how "Cuvier, from a single bone, could describe a creature he had never seen" and the accompanying story of the "great wingless bird of New Zealand" featured in another effusive account of the Crystal Palace in the *London Quarterly Review*. Assessing the various guidebooks available, the article proposed: "By degrees . . . Owen has trained you, till it is now as easy to think of a naturalist building up a skeleton from one bone, as it is to think of a cartwright building a wheel if you give him a felly."[112] The repeated emphasis on correlation and the iconic dinornis in the torrent of guidebooks and journalistic instruction that accompanied the Crystal Palace's opening in the summer of 1854 had, according to the *London Quarterly Review*'s correspondent, successfully trained their readers to regard the paleontological methods by which the prehistoric models had been constructed as no less reliable and accurate than the habitual working practices of even the most artisanal visitor. This was precisely what the Crystal Palace Company, alert to the need to attract paying customers from all social classes, had intended its promotional campaign to achieve.

Its Methodist roots notwithstanding, the *London Quarterly Review* endeavored to maintain a strict impartiality, especially on scientific matters. Only two issues after it had published the eulogy to Cuvierian correlation in its endorsement of the new Crystal Palace, the same journal's July 1855 number carried an article on "Animal Organization" by the paleobotanist William Crawford Williamson. Although making no direct reference to the Crystal Palace, Williamson averred that "we must . . . protest against some of the pretensions set forth on behalf of comparative anatomists." Specifying the "Stonesfield pterodactyls," the "giant Iguanodon," and the "megatherium, which combines the sloths, armadillos, and ant-eaters," all of which had been reconstructed at Sydenham, Williamson insisted: "That these animals *have* been restored by the genius of men such as . . . Owen . . . is unquestionably true; but it has been from no single bones." Rather, their forms had been surmised more prosaically "by the slow and careful comparison of bone with bone, and collection with collection," and even then "gross mistakes have not always been avoided." For Williamson, such "mistakes involve no discredit, because they are inevitable: they only become disgraceful, when associated with the claim of something like infallibility." Anxious that Owen was actually "in danger of receiving more injury in the hands of friends than foes," Williamson resolutely asserted the "impossibility . . . of anatomists accomplishing all that florid writers have attributed to them," instead urging the "necessity of a becoming modesty" on paleontologists.[113]

With Mantell now dead and Gray confiding his concerns that Cuvier's supposed abilities had been grossly exaggerated only in private letters, Williamson's anonymous contribution to the *London Quarterly Review* was a rare expression of scientific skepticism amid the "great Forge-bellows of puffery" that Dickens found so perturbing. Williamson's particular reservations about the inevitability of mistakes were still potentially very awkward for the Crystal Palace Company, especially as more pragmatic shareholders were complaining at the "wild sum . . . that the 'ologies' in and about the Crystal Palace have together cost" and calculating that "'Geology', *i.e.* those detestable, absurd 'extinct animals', cost the incredible sum of 13,729*l*. 0*s*. 6*d*."[114] These exorbitantly expensive brick-and-mortar models, moreover, could not be corrected, in the same way as textual restorations such as the fallacious tortoise in Warren's *Lily and the Bee*, if—or more likely when—any errors came to light. With Gray already privately protesting that the accuracy of the models rested "often on very slender and sometimes on what is now known to be erroneous grounds," the cavils articulated in the *London Quarterly Review* might easily precipitate a financially hazardous controversy like that which had engulfed Jones's polychromatic restorations of classical antiquities.[115] It was thus imperative for the Crystal Palace Company, and its obliging advocates in publishing and the print media, to drown out any opposition to Cuvierian correlation, and enforce its own strict interpretation of the paleontological method's apparent infallibility.

Learn Something of What They Call "the Development Theory"

Inevitably, with the Crystal Palace attracting such a diverse array of paying customers, from the aristocrats and affluent bourgeoisie who arrived in their own carriages and invested two guineas in a season ticket that afforded unlimited access, to the laborers who came only on specified "shilling days" and cut the cost by traveling third class on the railway, not all visitors could be trained to respond in the same mandated manner. Many of the proletarian visitors arriving on cheap excursion trains would, in any case, have been illiterate and thus largely beyond the reaches of even the most effective promotional campaign in the press. Nor, for that matter, could every journalist be induced to adhere to all aspects of the Crystal Palace Company's authorized agenda. Even among the divergent and sometimes contradictory meanings that were engendered by the prehistoric models, however, support for the law of correlation generally remained a given.

Commending the Crystal Palace's reconstruction of the frog-like labyrinthodon, *Hogg's Instructor* observed that the "ascertaining of its true character was one of the triumphs of Professor Owen," before adding: "Unfortunately for the adherents of the 'development' hypothesis, it combined with the general form and contour of the lowest, many of the most important characters of the highest members of its class."[116] The life-size model of the taxonomically hybrid—and therefore demonstrably nonprogressive—labyrinthodon, despite Owen's own reservations about its conjectural nature, afforded a powerful, and highly conspicuous, bulwark against the encroachments of transmutationism. This, according to Adrian Desmond, was also the explicit intention of Owen's stout-limbed dinosaurian reconstructions at the Crystal

Palace, although, of course, such a claim is complicated by Owen's manifest lack of involvement in the models' designs.[117] In any case, such an antitrans-mutationist interpretation of them was by no means inevitable.

For the *Illustrated Crystal Palace Gazette*, the lesson of the restorations, which were presented in a chronological sequence, was the exact opposite.[118] In an account of the famous banquet in the iguanodon mold, it addressed potential visitors who would "rather not have your mind unsettled about any old traditions," entreating them: "You must move on . . . and you must go to school at the Crystal Palace, and learn something of what they call 'the development theory.'" But far from the transmutationist theory "backed by the mysterious author of the 'Vestiges'" negating the correlative principles of paleontology (as Cuvier had insisted it must), the *Illustrated Crystal Palace Gazette*'s correspondent flippantly related how "after there's no telling how many thousands of years of development, there live animals with brains that enable them to prophesy backwards" and "tell you all the strange forms that have been living . . . from a single tooth."[119] The remarkable paleontological powers that had facilitated the construction of the Crystal Palace's prehistoric models might themselves be the result of progressive transmutation.

Not everyone, though, would have noticed even such potentially contro-versial connotations. While the indifference of visitors for the Crystal Palace Company's vaunted educational aspirations became a source of scornful satire in *Punch*, some visitors, by contrast, were overly enthusiastic in their responses to the prehistoric models.[120] Few visitors responded with such visceral curios-ity as the light-fingered "artisans and tradespeople" whose "vulgar mischief" and downright "theft" had, as Martineau lamented, rendered some of the "strange creatures . . . nearly toothless."[121] Nancy Rose Marshall has proposed that "Owen and Hawkins designed almost all their models with open mouths so as to display their teeth. This pose was . . . educational, since teeth greatly preoccupied early paleontologists who often used them to classify and iden-tify fossils."[122] As such, the plebeian purloiners whom Martineau found so dis-concerting may actually have been availing themselves of pedagogic souvenirs directly connected to the particular paleontological methods they might have read about—or had read to them—in cheap guidebooks and newspapers.

An equivalent sense of enthusiasm was also attributed to visitors of a very different social class and educational background. Several commentators suggested that, far from merely dispelling the "fogs . . . from the swamp of ignorance in which the multitude have been lying," as Martineau haughtily put it, the Crystal Palace's prehistoric restorations might in fact materially as-sist the researches of specialist men of science.[123] Although it was itself aimed at more humble visitors, *Routledge's Guide* proposed that scientific men who had devoted a long life to the accumulation and study of fossil remains, who had put together the skeletons of these gigantic monsters, and seen them in imagination roaming over the pathless forests of our island, had never yet beheld the entire animal reproduced before them;—geologists were for the first time to gaze upon the fruit of their industry, and the results of their science.[124]

The life-size reconstructions at Sydenham, it was claimed, represented the culmination of generations of geological scholarship, and would enable the science to now progress in new ways. As well as advancing a heretically trans-mutationist reading of the prehistoric models, the *Illustrated Crystal Palace Gazette* rather brazenly suggested that the "British Association for the Advancement of Science has many points in common with the Crystal Palace," and it might seem that such proposals that the geological restorations could genuinely augment scientific research were no less wayward than yoking them with the scandalous development hypothesis.[125] Mantell and Gray had, after all, already expressed varying degrees of skepticism, and even Owen himself appeared largely indifferent. This high regard for the Crystal Palace's scientific credentials, however, was not confined to shilling guidebooks and partisan promotional newspapers.

In its "Scientific" column recording the activities of the metropolitan scientific societies, the prestigious weekly the *Athenæum* reported that

> to the great majority of the public these restorations will present all the novelty of a first acquaintance, and even many students of geology have hitherto been unable to realise the true forms and size of those extinct animals, with the names of which they might however be perfectly familiar.[126]

The Crystal Palace's three-dimensional models, it was implied, effected an innovative process of visualization by which geologists could, for the very first time, comprehend the true nature of the objects of their scientific scrutiny. Models have customarily been marginalized in favor of two-dimensional texts and images in understanding the production of scientific knowledge, but, as historians have recently begun to recognize, they often played a crucial role in guiding scientific research.[127] In fact, with even journalists such as the *Leisure Hour*'s correspondent baffled by the "monsters which we cannot pretend to name, and which Adam never named at all," scientific practitioners who were already acquainted with the recondite Graeco-Latin names may have benefited most from the Crystal Palace's Pestalozzian policy of educational visualization, which was increasingly recognized as being inadequate for other visitors who required labels.[128]

Certainly, not all men of science shared the distaste felt by Gray, who complained that the garish touting of Owen's supposed Cuvierian accomplishments was "out Barnuming Barnum himself" and reducing the geological restorations to a vulgar spectacle.[129] In June 1855 Charles Darwin preferred to miss a Royal Society council meeting to instead "go to the Crystal Palace to meet the Horners, Lyells, and a party." Although Darwin's desire to "conceal the scandalous fact" suggests that neither he, Lyell, nor Leonard Horner was engaging in any serious scientific study, the "scenery & foliage" of the Crystal Palace's grounds clearly afforded a congenial environment that was not vulgar or unscientific where the three eminent geologists, accompanied by their families, could meet and talk.[130] In fact, Darwin, having purchased a "pair of Season-tickets," thought the "C. P. . . . thoroughly enjoyable," while Horner regularly "went in a carriage to the Crystal Palace" and "walked in the garden and looked at the beautiful objects inside."[131] Frustratingly, neither of them recorded their opinions of the prehistoric reconstructions. However, no less an authority on aesthetic taste than John Ruskin considered that the "model of the Iguanodon, now the guardian of the Hesperian Gardens of the Crystal Palace," was an "exact counterpart of the . . . saurian of Turner's" 1806 painting *The Goddess of Discord in the Gardens of the Hesperides* (fig. 5.6).[132] Ruskin remained anxious that Paxton's iconic Crystal Palace had reduced the

FIGURE 5.6. Joseph Mallord William Turner, *The Goddess of Discord in the Gardens of the Hesperides.* Oil on canvas, 1806. Tate Britain. The saurian Ruskin compared to the Crystal Palace's model of the iguanodon sits atop the rock formation in the center right of the painting. © Tate, London, 2014.

"first principles of architectural sublimity" to merely the "lustre of a few rows of panes of glass," but the adjacent grounds featuring elegantly Turneresque behemoths were evidently an appropriate space for refined gentlemen of taste and understanding.[133]

Ironically, Darwin, living nearby in rural Kent, seems to have spent considerably more time at the Crystal Palace than the perennially absent Owen. But if Owen was rarely present, his involvement, no matter how nominal, lent the geological restorations the imprimatur of the foremost expert in the field. This was reinforced by his official guidebook, which, although somewhat underwhelming, acknowledged its author's "F.R.S." status on the title page, and provided several footnoted references to Owen's exhaustive "Report on British Fossil Reptiles" for the British Association. As with the other handbooks in the Crystal Palace Library, *Geology and Inhabitants of the Ancient World* was far from assuming that all visitors lacked prior knowledge of the subject and required merely vulgar simplification. This was certainly not the case with those who belonged to the Geologists' Association, which, even into the late 1870s, arranged "excursion[s]" to the Crystal Palace to "afford Members an opportunity of inspecting those objects of geological interest," including the "life-size models . . . which are well executed."[134] The Crystal Palace, as Sadiah Qureshi has argued of its ethnological displays, served multiple functions, providing, among other things, a crucial site of scientific instruction for both lay and learned audiences.[135] Such scientific training, moreover, was not consumed passively by any of its audiences, and, as with the contrasting attitudes in *Hogg's Instructor* and the *Illustrated Crystal Palace Gazette* on the vexed issue of transmutationism, the prehistoric models were interpreted in different ways and used for a diverse range of purposes.

Like a Megatherium Smoking a Cigar

Even more than the putative accuracy of its ersatz reconstructions, the financial speculation of the Crystal Palace Company's shareholders depended on the efficiency and capacity of the railways that had begun to spread across the country only in the previous two decades. As the novelist William Makepeace Thackeray observed skeptically after attending the erection of the first column in August 1852:

> They have bought a most charming park at Sydenham where the New Palace is to be: but who will go to it?—not hundreds of thousands of people I doubt very much, for you must go to London Bridge first and then 7 miles by railway so that the "palace of the people" will be a failure I think.[136]

Within two years, Thackeray had changed his mind, becoming resentful that the publisher he shared with the Crystal Palace Library, Bradbury and Evans, "were too busy with the Crystal Palace to attend to his affairs."[137] He even hoped to cash in on the initial success of the grand undertaking in faraway Sydenham, proposing to "B & E" a humorous work that might be "sold at the Crystal Palace," where the company had opened a new office.[138]

Thackeray's ongoing serial novel *The Newcomes* (1853–55), the conclusion of which alluded whimsically to a "Crystal Palace Exhibition in fable-land," carried regular advertisements for the real Crystal Palace in the wrappers of its monthly parts (fig. 5.7). In fact, the novel's famous comparison, discussed in the previous chapter, between the method by which "Professor Owen . . . takes a fragment of a bone, and builds an enormous forgotten monster out of it" and how the novelist creates the "megatherium of his history" was first published in December 1854 and would therefore have been read alongside the exactly contemporaneous, and similarly adulatory, journalistic accounts of Owen's paleontological powers in relation to the Crystal Palace's prehistoric reconstructions.[139] Certainly, early installments of *The Newcomes* were extracted, in close proximity to such accounts, in the *Crystal Palace Herald*.[140] Even if it were not read in relation to these journalistic reports, the novel's unparalleled density of allusions to exotic commodities, classical sculpture, and other ancient artifacts, in addition to antediluvian megafauna, recreated textually the experience of visiting the crowded exhibition spaces at the Crystal Palace.[141] And with reviews of Thackeray's serialized fiction suggesting that a "level number" would "amuse for an hour . . . the traveller by railway," at least some readers of *The Newcomes* would, despite its author's initial skepticism, have read it on the excursion train from London Bridge to Sydenham.[142]

Reading, as publishers such as George Routledge quickly recognized, was the chief form of recreation available to travelers by train, and new formats of print and innovative methods of retailing them emerged alongside the rapidly expanding railway network.[143] Much of the print generated by the Crystal Palace was explicitly designed to be read on trains and was sold at the start of the journey at bookstalls on London Bridge station. Phillips's authorized *Guide to the Crystal Palace* began: "We will presume that the visitor has taken his railway ticket. . . . Before he alights, and whilst his mind is still unoccupied by the wonders that are to meet his eye, we take the opportunity to relate, as briefly as we can, the History of the Crystal Palace." Once on the train, readers of Phillips's *Guide* were informed of the awaiting prehistoric reconstructions such as the "*Palœtherium*," which, they learned, was "justly called the first triumph of comparative anatomy, as, from a few detached pieces of bone Cuvier was enabled to construct the entire animal."[144] Such curious

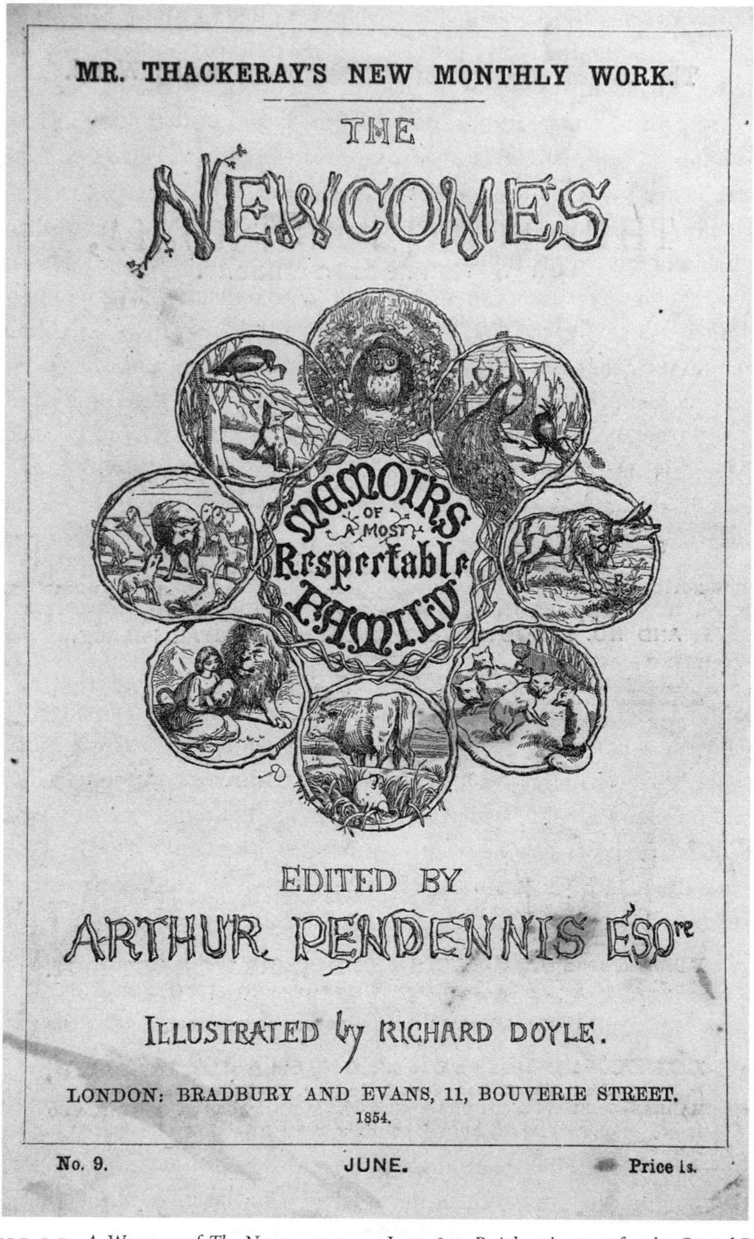

MR. THACKERAY'S NEW MONTHLY WORK.

THE

NEWCOMES

MEMOIRS OF A MOST Respectable FAMILY

EDITED BY

ARTHUR PENDENNIS Esqre

ILLUSTRATED by RICHARD DOYLE.

LONDON: BRADBURY AND EVANS, 11, BOUVERIE STREET.
1854.

No. 9.   JUNE.   Price 1s.

FIGURE 5.7. *A*, Wrapper of *The Newcomes*, no. 9, June 1854. *B*, Advertisement for the Crystal Palace in "The Newcomes Advertiser," p. 2. The monthly numbers of Thackeray's serial novel regularly carried advertisements for the Crystal Palace, especially, as here, in the run-up to its much-anticipated opening. *A* and *B*, Author's collection.

creatures, and the remarkable methods by which they had been discovered, were already a mainstay of the novel publishing formats intended for train travel, with an anthology of *Railway Readings* (1847) relating an anecdote of the "strange induction" by which "from the tooth of the megatherium . . . Mr. Owen inferred that . . . it was a species of leaf-eating sloth, but too large to climb trees."[145] The self-styled "fast" magazine the *Train*, meanwhile, carried a droll dialogue in which a maid struggles to obtain a copy of "Owen on the Megatherium" from "Mr. Moody's [*sic*] cart" because of her defective "memory and pronunciation."[146] While undertaking the epitome of modern industrialized experience, the thoughts of many railway travelers were firmly in the distant past.

When, after "flying over the tops of miserable houses, and skimming the fearful squalor of Bermondsey," the excursion train finally arrived at Sydenham, the very first sight of the Crystal Palace's grounds visible from the carriage windows was the shapes of the prehistoric restorations (see fig. 5.1).[147] As a reporter from *John Bull* noted, the "approach by the Crystal Palace Railway discloses to the astonished and terrified view those strange and monstrous forms."[148] A more flippant correspondent from the *Illustrated Crystal Palace Gazette* proposed that, although he had never actually visited the "abode of the antediluvian beings" on foot, "nevertheless, from the Palace railway we have enjoyed sufficiently close and repeated views to form an opinion."[149] More adventurous visitors who did traipse across the Crystal Palace's extensive grounds and then looked back from the brow of the hill where its iron-and-glass centerpiece stood might have noticed a closer relation between the railway and the models of extinct creatures than mere proximity.

As the *London Quarterly Review* proclaimed of this panoramic view:

> At your feet is the park, on the furthest edge of which the geological monsters stand, while immediately beyond them comes, in full career, as if mocking their impotence, the *megatherium* of the nineteenth century, blowing like a whale, snorting like a wild horse, and making the welkin resound with the thunder of his train.[150]

Since the first passenger railways opened in the 1830s, these slightly fearful marvels of technological innovation had been frequently compared to the no less disconcerting antediluvian megafauna that was just then being reconstructed by paleontologists, often from fossils unearthed during the cuttings and tunneling necessary to construct railway lines. The poet Alfred Tennyson, "standing by a railway at night" in the mid-1840s, fancied the "engine must be like some great Ichthyosaurus."[151] In particular, it was the megatherium that, as in the vista described in the *London Quarterly Review* as well as the allusion in Henry James's *The Bostonians* (1886) discussed in the last chapter, was considered most analogous to the railways. *Fraser's Magazine*, for instance, proposed in 1846 that a railway locomotive's bulky engine and upwardly protruding funnel were "like . . . a megatherium smoking a cigar."[152] While the louche masculine pleasures of smoking were not among the creature's habits inferred by paleontologists, William Buckland, in the mid-1830s, evidently had earlier modes of transport in mind when proposing that the "body of the Megatherium . . . must in some degree have resembled a tilted wagon" (like Cuvier, he assumed that its huge frame was covered in protective dermal armor).[153] And it was an analogy that could be readily updated for the new railway age.

But if railway engines were the megatheriums of the nineteenth century, as the *London Quarterly Review* alleged, it was not just on account of their shared bulk and power. Rather, they were also both instances of complex yet perfectly integrated mechanisms that, as in William Paley's classic instance of the watch, necessarily indicated the presence of a designer, whether human or divine. Even before the advent of the railways, Paley himself, in *Natural Theology* (1802), had pointed to "steam-engines" which "deriv[e] their curious structures from the thought and design of their inventors" as a further analogy of providential design.[154] The ultimate implication, to avoid raising the inventor to a hubristic godlike status, was that the design of the steam engine was directed by divine providence no less than that of an organic structure. Similarly, Buckland's *Geology and Mineralogy Considered with Reference to Natural Theology* (1836) compared the anatomy of the megatherium to the "hammer and anvil of an anchorsmith," which, "though massive," was "neither clumsy nor imperfect," while Owen's later account of the megatherium's structure and habits continued to make similar references to the "efficiency of [its] masticating machinery."[155] Such Paleyite contrivances were no longer used to show the manifest perfection and happiness of creation, but instead to demonstrate, as noted in chapter 2, that the inequalities and apparent cruelty of the competitive social market were in fact divinely ordained mechanisms of justice. As part of this overtly capitalist theodicy, the same perfect integration of component parts that enabled Cuvierian paleontologists to reconstruct prehistoric creatures from single bones, thereby affirming the harmonious designs that natural theology discerned in organic structures, was equally integral to the engines and machinery that powered the mid-Victorian economic boom that the Crystal Palace symbolized.

Alongside its Greek and Roman Courts, the interior of the Crystal Palace also featured Birmingham and Sheffield Courts displaying the latest manufactures of these famously ingenious industrial cities. Such mechanical inno-

vations were no less evident outside, where a colossal hydraulic pump powered spectacular ornamental waterworks, as well as the artificial tidal lake in
which many of the prehistoric models wallowed.[156] In such a mechanized
environment, the extinct creatures could themselves have the appearance of
elaborate contrivances, with *Routledge's Guide* dubbing the megalosaurus a
"truly 'infernal machine,'" and observing of the ichthyosaurus: "Here is a
pre-Adamite monster, who may probably be five hundred thousand years
old, and who carries a tail which is the exact form of the most improved
screw-propeller of the present day." As well as showcasing the "ingenious discoveries of modern science," the Crystal Palace was also a pertinent reminder
that "there is nothing new under the sun."[157]

Such mechanical resemblances were far from unintended. In the revised
and updated third edition of his father's *Geology and Mineralogy* (1858), Frank
Buckland commended the achievements of "my friend, Mr. Waterhouse
Hawkins, who, after a continuous mental and bodily labour of more than
three years, has presented . . . in the gardens of the Crystal Palace at Sydenham, restorations of no less than thirty-three extinct animals." Significantly,
Buckland noted that "Hawkins tells me that in the process of modelling his
restorations, he has received the greatest assistance from Dr. Buckland's plates
in this book."[158] In the original edition of his *Bridgewater Treatise*, Buckland
père had repeatedly demonstrated that the structure of even the most ungainly extinct creature was in fact a "well-contrived and delicate mechanical
instrument," with the engraved plates, in a separate volume, helping display
these "peculiar mechanical contrivances" more clearly (see fig. 2.3).[159]

Hawkins, as an attentive student of Buckland's plates, was no less insistent
on the mechanistic principles that underlay animal anatomy. He proposed
that the artist who wished to represent the

> living mechanism . . . can never fully succeed without having carefully studied
> the parts of its machinery. . . . Neglecting this, he will be in the position of
> the engineer who tries to understand or represent some complex machine,
> of whose structure and uses he knows no more than an outside glance has
> told him.[160]

Once an artist had scrutinized this organic mechanism, they would recognize
just how close the analogy between anatomy and industrial design was. In an
earlier discussion on this topic, Hawkins, as a friend afterward recalled,

> immediately drew a sketch of [Isambard Kingdom] Brunel's bridge across
> the Tamar . . . and pointed out that in the body of an elephant Nature had
> previously adopted the same means of sustaining a heavy weight. Mr. Hawkins

had only to add an elephant's head and tail to his drawing, and the resemblance was rendered very apparent.[161]

With the megalosaurus recently described by Dickens, in *Bleak House* (1852–53), as "waddling like an elephantine lizard," Hawkins's models of that and other similarly enormous extinct megafauna might themselves, as some commentators seemed to intuit, easily be converted to resemble complex feats of engineering.[162]

Buckland's *Bridgewater Treatise* had, of course, adumbrated the classic Paleyite argument that "from the combinations of perfect mechanism . . . we infer the perfection of the wisdom by which all this mechanism was designed."[163] In the updated third edition, his son Frank enlisted Hawkins to supplement the original's plates with a "sketch, from his own pencil, of his marvellous models of ancient marine Saurians, the originals of which are now at Sydenham" (fig. 5.8). In his explanatory notes, Buckland provided an appropriately natural theological gloss:

> In the centre of the sketch . . . Mr. W. Hawkins gives a representation of "Ichthyosaurus communis". . . . Its whole appearance proclaims it to be a

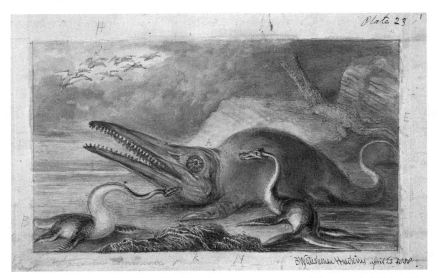

FIGURE 5.8. Benjamin Waterhouse Hawkins, Ichthyosaurus communis. Watercolor, 1858. Natural History Museum, London. This sketch, which was subsequently engraved and appeared as plate 23 in the third edition of Buckland's *Geology and Mineralogy Considered with Reference to Natural Theology* (1858), was based on Hawkins's models of the ichthyosaur and plesiosaur at the Crystal Palace. © The Trustees of the Natural History Museum, London.

sea-monster, whose structure (especially its array of powerful teeth) is admirably adapted to devour and keep in check the numerous hard-scaled fish . . .
which abounded in the depths of the sea.[164]

The functional adaptation of the ichthyosaurus exemplified in Hawkins's
model was so perfect, according to Buckland, that it even ensured that the
population of other species remained regulated in harmony with the overall
plan of Creation. Hawkins's own support for such an interpretation of the
Crystal Palace's prehistoric reconstructions was made clear in his manual *A
Comparative View of the Human and Animal Frame* (1860), where he eulogized the "harmonious fitness of all animals for that place in Creation, which
they were originally designed to fill," and, in the wake of Darwin's recently
published *On the Origin of Species* (1859), exhorted the "artistic student"
to resist the "grotesque *theory of development.*"[165] When, at the end of the
1860s, Hawkins traveled to America, he supported himself with lecture tours
in which his adherence to the argument from design and vehement anti-
Darwinism tallied with the conservative religious values of his audiences.[166]

Despite any personal misgivings about his perennially absent collaborator, Hawkins affirmed the ability of "our British Cuvier, Professor Owen . . .
to call up from the abyss of time . . . those vast forms and gigantic beasts
which the Almighty Creator designed with fitness to inhabit . . . this part of
the earth."[167] For him, as it was for both generations of Bucklands, the "possibility of restoring these animals by induction, and by reasoning upon the
fragments that have reached us" was the ultimate confirmation of the perfect
cohesion of the component parts of organic structures, and thus of their divine design.[168] The prehistoric models created by this same paleontological
approach, although funded by innovative methods of shareholder capitalism,
publicized by modern marketing techniques, and themselves often resembling cutting-edge steam technologies, embodied the long-standing argument from design. Like the third edition of his father's *Bridgewater Treatise*
that Frank Buckland published in 1858, which "strengthened . . . expunged or
modified" the "original fabric" while leaving the "actual argument . . . in its
original purity," Hawkins's restorations at the Crystal Palace refitted the old
Paleyite creed for a new age of visual spectacle.[169]

As well as provoking evangelical fervor against its materialism and sinful
pride, the original Great Exhibition, as Geoffrey Cantor has argued, afforded
a rich resource for reflecting on divine providence and natural theological
design, especially in its vast array of raw materials.[170] The Crystal Palace's new
incarnation in Sydenham prompted similar religious objections, not least
over the contentious decision to open on the Sabbath. The much-heralded

deployment of Cuvierian correlation in the construction of its models of extinct creatures, bolstered by the modeler's own faith in the argument from design, nevertheless ensured that it too had a prominent emphasis on natural theology.

Conclusion

The vested commercial interests of the Crystal Palace Company and its shareholders ensured that, during the intensive marketing of the geological restorations between late 1853 and the opening of the Sydenham complex in the summer of 1854, the law of correlation received more extensive, and more adulatory, press coverage than at any previous time. Few visitors could have arrived on the excursion trains from London Bridge, or in their own private carriages, without having encountered at least some details of either Cuvier's famous exploits in Paris or Owen's celebrated discovery of the dinornis. This was testament to the proactive approach to the print media pioneered by the Crystal Palace Company, at whose dedicated "News Room," a promotional flyer boasted, the "very Latest News, by special Electric Telegraphic Despatch, is exhibited immediately on its receipt."[171] With Hawkins's prehistoric models also exhibiting aspects of the same mechanical innovation proudly displayed elsewhere in Paxton's iron-and-glass temple of industry, the correlative method of paleontology, even while retaining the natural theological connotations appended to it in the 1820s and 1830s, reached its apogee in the midst of the modern entrepreneurial Britain that the Crystal Palace, both with the Great Exhibition and then at Sydenham, came to symbolize. This remarkable synthesis of correlation, capitalism, and cutting-edge technology, however, would not last for long.

Thackeray's initial skepticism had actually been correct, and, notwithstanding its systematic marketing campaign, the Crystal Palace Company was unable to entice sufficient numbers of visitors to take the train to Sydenham to balance the books. In September 1855, Hawkins, who had continued adding new creatures to his prehistoric menagerie of models, was told to stop. For the "sake of a few hundred pounds," as Sotheby complained, "that gentleman's services, without the smallest notice, [were] most unceremoniously dispensed with."[172] Ironically, it was the restoration of the dinornis, whose reconstruction from a fragmentary bone had been so integral to the Crystal Palace Company's promotional campaign, that, along with an already half-completed mammoth, was so abruptly halted (fig. 5.9). At the end of the 1850s Hawkins continued to "hope that at some future time the restoration of the *Dinornis* will make its appearance on the island already prepared for it at

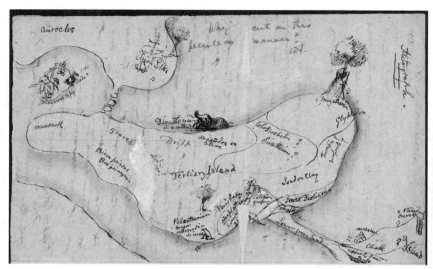

FIGURE 5.9. Benjamin Waterhouse Hawkins, *Sketch of Planned Additions to the Tertiary Island, with Small Nearby Island with Dinornis.* Pen and ink, 1855. Richard Owen Papers, 14:534, Natural History Museum, London. This sketch, showing—in the top left corner—a small island with three figures of the dinornis, giant eggs, and a possible dodo that Hawkins wished to add to his other models of extinct creatures, was made in a letter Hawkins wrote to Owen on 24 October 1855. Hawkins had actually been dismissed by the Crystal Palace Company a month earlier, and none of the additions was ever realized. © The Trustees of the Natural History Museum, London.

the Crystal Palace," but this icon of the almost miraculous predictive powers of Cuvierian paleontology was never realized in brick and mortar.[173] Instead, the Crystal Palace Company shifted to a more remunerative emphasis on plebeian entertainment rather than Pestalozzian education, with cold winters enabling "skating on the lakes in the park," during which, as the *Crystal Palace Herald* reported, the "skaters enjoyed themselves in a manner which no doubt astonished Mr. Hawkins's antediluvians, who have been nearly buried in coats of snow."[174]

It was not just the models themselves that were soon interred and marginalized. The very "Forge-bellows of puffery" that Dickens bemoaned in July 1854 also provoked an impromptu attack on Cuvier and his law of correlation from a young and little-known naturalist that, within two years, would develop into the most effectual assault on the Cuvierian method of paleontology of the entire nineteenth century. The youthful assailant of correlation was Thomas Henry Huxley, and his endeavor to dethrone Cuvier, beginning in the pivotal summer of 1854, will be considered over the three chapters of the following part.

PART III

Overthrow, 1854–62: Scientific Naturalists, Popularizers, and Cannibals

Correlation under Siege

The centenary of the Society for the Encouragement of Arts, Manufactures and Commerce was commemorated in early July 1854 with a dinner held under the "crystal dome" of the vast edifice opened in Sydenham less than a month earlier. The banquet brought together many of the leading representatives of the society's varied activities, who sat at "13 parallel tables" placed within a "large semicircular upper table" seating the aristocratic guests. After the dinner, this "mixed assemblage" was requested to raise their glasses and "drink 'Success to the Crystal Palace Company,'" following which they were told by the company's chairman, Samuel Laing, that the newly opened Crystal Palace "must be associated with the education and advancement of the people of England." Indeed, Laing was "not disposed to underrate the educational advances which may be made by the masses of the population in such scenes, where they may walk peaceably and quietly, surrounded by the most instructive sights, both as regards nature and art."[1] Prominent among these instructive sights of nature, as was examined in the previous chapter, were the giant prehistoric creatures modeled by Benjamin Waterhouse Hawkins. These life-size models were produced, putatively at least, under the guidance of Richard Owen, whose astonishing "ability to reproduce accurate models of the entire structure and correct proportions of extinct animals from the discovery of a single bone" had been extolled at an Ordinary Meeting of the Society of Arts less than two months before its centenary dinner.[2]

Owen's continual absence from the Crystal Palace had already become notorious, and he did not attend this dinner. Instead, as the *Journal of the Society of Arts* recorded: "At the Science table, which was that immediately to the right of the Council table, w[as] seated, amongst others . . . Mr. Thomas

Huxley, F.R.S."[3] Huxley, yet to reach thirty and having made the transition from impecunious journalist to lecturer in natural history at the Government School of Mines just a few weeks before, had little experience of such imposing public events. He must nevertheless have bridled at the lavish post-prandial praise bestowed on the educational capacity of the models so associated with the paleontological methods of Owen, whose once supportive relationship with Huxley had grown increasingly rancorous over the previous eighteen months. Huxley had earlier lauded the "great Crystal Palace," when it had initially housed the Great Exhibition during the summer of 1851, as the foremost "Temple of England" for those who shared his own "state of practical unbelief," and it must have been galling that the same building's new incarnation in Sydenham was now being connected with a method of paleontology that had long been used as an exemplar of natural theology.[4] In fact, just nineteen days after sitting through the toasts and encomiums of the Society of Arts' centenary dinner on 3 July, Huxley used a lecture to the same society's Educational Exhibition in central London to question the degree of skill and expertise required to perform the inferential reconstructions on which Hawkins's models of extinct megafauna were based.

In "On the Relation of Physiological Science to Other Branches of Knowledge," delivered at St. Martin's Hall on 22 July 1854, Huxley observed that "Science is, I believe, nothing but *trained and organized common sense*," before offering his audience, which consisted largely of "scholars, tutors, governesses, schoolmistresses, and others engaged in Education," one particular illustration of this axiom.[5] "So," Huxley proclaimed,

> the vast results obtained by Science are won by no mystical faculties, by no mental processes, other than those which are practised by every one of us, in the humblest and meanest affairs of life. A detective policeman discovers a burglar from the marks made by his shoe, by a mental process identical with that by which Cuvier restored the extinct animals of Montmartre from fragments of their bones. . . . The man of science, in fact, simply uses with scrupulous exactness the methods which we all, habitually and at every moment, use carelessly.[6]

Hawkins had lauded the exceptional abilities of "our British Cuvier" to the Society of Arts on 17 May, insisting that, with the labyrinthodon, it was not "until the mighty genius of Professor Owen placed the teeth and head before us . . . and thus, by induction, the whole animal was presented" that he could commence the project for "revivifying of the ancient world" that was integral to the educational objectives of the Crystal Palace.[7] Only two months later, Huxley told the same society that precisely those inductive methods were simply the exercise of entirely quotidian mental processes that had nothing to do with genius or any other purportedly preternatural capabilities.

The reconstructive feats of Cuvier, and by implication those of his British successor, were equated by Huxley with the procedures of the detective policemen who, since the beginning of the 1850s, had become increasingly visible through the fiction and journalism of Charles Dickens.[8] While supportive, Dickens's depictions of police detectives such as Mr. Bucket in *Bleak House* (1852–53) and his real-life counterpart Inspector Field (who "recognis[es] the Ichthyosaurus as a familiar acquaintance" when patrolling the British Museum) nonetheless presented them as hardworking and shrewd rather than inspired or outstandingly gifted.[9] Huxley was a keen supporter of the Society of Arts' endeavors to improve standards of scientific and technical education, but he insisted, in one of his very first public addresses, that it needed to adopt a less authoritative and exclusive model of science that was notably at odds with the way that, as seen in the last chapter, Owen's role in the construction of the Crystal Palace's antediluvian models had been presented.[10]

That Huxley's ire was focused particularly on Owen's celebrated deployment of Cuvierian methods of paleontology, perhaps sharpened by his recent experience of the Society of Arts' effusive centenary dinner, is made clear by his subsequent more sustained and technical attacks on the very same methods in the following years. Twice in 1856, initially at the Royal Institution and then in the pages of the *Annals and Magazine of Natural History*, Huxley impugned the inductive abilities of Cuvier and Owen, asserting that the ostensibly infallible law of physiological correlation, the unerring veracity of which afforded crucial evidence of the harmonious design of organic structures, was in reality based on prosaic empirical observations of customary correspondences. Assumptions about the absolute necessity of such correspondences, Huxley avowed, were made on the basis of authority, both scientific and religious, and not reason or logic.

At this time, Huxley was vigorously assailing almost every aspect of Owen's scientific reputation, and in July 1856 he accomplished a remarkable pincer movement by simultaneously publishing damning accounts of both the separate functionalist and transcendental strands of Owen's anatomical researches. An unfortunate tendency for his overbearing "authority" to be "employed to the positive detriment of science, in perpetuating error and retarding the progress of truth," Huxley maintained, was no less a feature of Owen's work on the archetypal plan of living invertebrates than it was of his Cuvierian reconstructions of extinct vertebrates.[11] Huxley, a headstrong tyro with a taste for controversy, actually had little practical experience of paleontology, having entered the field only after assuming responsibility for the fossil collection at the Government School of Mines, and it was not long before his impudent accusations prompted a reaction. After an initial riposte from Hugh Falconer, an expert on proboscidean fossils recently returned from India, Owen himself responded imperiously to Huxley's provocations in the autumn of 1856, and the vexed issue of paleontological method became a far more persistent and significant component of the two men's famously vituperative encounters than has previously been recognized.

Despite his mauling by Owen, Huxley received invaluable support from other young men similarly seeking to make a remunerative career in science, including the botanist Joseph Dalton Hooker and the philosopher Herbert Spencer, as well as, most important, from a reclusive contemporary of Owen's who had long harbored a clandestine theory of species change that was no less naturalistic and contrary to conventional conceptions of divine design than Huxley's attitude to paleontological method. The spring of 1856 was precisely the moment at which Charles Darwin began revealing elements of his theory of natural selection to a carefully chosen group of potentially sympathetic naturalists, and Huxley's ferocious and exactly contemporaneous paleontological dispute with Falconer and then Owen helped forge a crucial sense of solidarity among those who would soon emerge as the principal advocates of *On the Origin of Species* (1859).

Other naturalists, most notably Gideon Mantell, had also begun to question the dependability of the law of correlation in the late 1840s and early 1850s, but it was in 1856 that this vein of criticism reached a tipping point. This, of course, was only two years after the apogee of Cuvierian paleontological method represented by the opening of the Crystal Palace, and this chapter will focus on how large parts of the scientific community, including former supporters such as Falconer, subsequently lost faith in the very methods by which Hawkins had, purportedly, been able to create the accurately proportioned models of prehistoric creatures whose educational value was so celebrated, in Huxley's brooding presence, by the Society of Arts.

The Very Foundations of Palæontology Are Assailed

His health irreparably damaged by long exposure to the tropical climate of northern India, Hugh Falconer retired from his post as superintendent of the Royal Botanic Gardens at Calcutta in the spring of 1855 and journeyed back to Britain by the overland route (fig. 6.1). During the long voyage he passed through the Crimean Peninsula just as the yearlong Siege of Sebastopol reached its critical juncture, with British, French, and Turkish troops

FIGURE 6.1. Joseph Dinkel, *Hugh Falconer.* Lithograph, circa 1865. Charles Murchison, ed., *Palæontological Memoirs and Notes of the Late Hugh Falconer,* 2 vols. (London: Robert Hardwicke, 1868), vol. 1, plate 1. This lithograph was drawn from a photograph taken for the series "Portraits of Men of Eminence in Literature, Science and Art" (1865), although Falconer, who claimed to dislike controversies and whose health remained delicate after his return from India, appears more diffident than dominant. © The British Library Board, 7204.df.20.

launching their last and most sustained bombardment of the Russian for-tifications.[12] A year earlier the *Spectator* had observed that the rationale of the initial attack was that "if we would overthrow the power [of the Russian empire], we must sweep away its foundations," while in October, a month after the siege's bloody conclusion, the *Edinburgh Review* proclaimed that "to attack Sebastopol was . . . to assail the stronghold of Russia in the East."[13] Back in London at the end of 1855, where newspaper coverage of the war was still transfixing public attention, Falconer soon encountered a no less "remarkable" verbal onslaught in which, like the besieged ramparts of Se-bastopol, the "very foundations of palæontology, as they have hitherto been understood, are assailed." This ferocious "attack on Cuvier and his followers"

was launched in February 1856, the same month when peace was being negotiated at the Congress of Paris, by a "man of science, of recognized standing, [who] assails generally admitted principles and established reputations"
and whose bellicose manner was, Falconer considered, such "as to require
some notice."[14] Although regularly bedridden and with a "constitutional
aversion from the hispid walks of controversy," the veteran of Sebastopol felt
compelled to single-handedly defend the besieged foundations of Cuvierian
paleontology.[15]

Despite its timing, the onslaught that prompted Falconer's concern in fact
only rarely resorted to the pugnacious militaristic rhetoric that was still dominating the press, though leavened by criticism of the bungling aristocratic
high command, in the early months of 1856.[16] The triumphant assertion that
"Cuvier himself . . . surrenders his own principle" of "physiological correlation," echoing reports in the *Times* of the "surrender of . . . part of her territory . . . about to be imposed on Russia," was the most conspicuous exception
to its generally unmartial tone.[17] The author of this carefully worded attack,
the printed form of which was signed and "authenticated with his initials" as
Falconer noted, had nevertheless displayed a zealous enthusiasm for military
matters only months earlier when writing for the anonymous *Westminster
Review*.[18] There he expressed his "taste for ordnance of all kinds, sea fights,
Minié rifles, and Crimea expeditions," and also revealed a detailed knowledge
of the tactics by which the "allied armies [in] their siege operations [at] the
south side of Sebastopol" could have secured "a *coup-de-main* . . . without serious loss to the assailants."[19] This aficionado of siege warfare even identified
himself, again in the unsigned pages of the *Westminster*, with the "youthful
vigour" of the ongoing Islamic insurgency against the imperialistic "Greco-
Russian Czar-worship, misnamed Christianity" in the Caucasus.[20]

The anonymous author was Huxley, who, as an assistant surgeon in the
Royal Navy at the start of the war, only narrowly avoided active service in
the Crimea, all the while insisting that "nobody can accuse me of an objection to facing the Rooshians."[21] The image of naturalistic science as engaged
in perpetual warfare against a corrupt orthodox theology that he continued
to hone throughout his long career—the famous military metaphor—was
one that had its origins in the real battlefields of the mid-1850s.[22] The metaphorical conjunction of science and militarism, however, was certainly not
exclusive to Huxley, and was, in any case, a decidedly reciprocal interchange.
The very paleontological principles on which Huxley turned the ersatz guns
of his military metaphor had themselves been invoked metaphorically by the
Crimean conflict's new breed of war correspondents reporting from the front

line, with William Howard Russell of the *Times* proposing after the end of the Sebastopol siege:

> If Cuvier or Owen could reconstitute the whole structure of some antediluvian animal from the mere glance at some joint or fragment of bone, it is sufficient for us to examine the emaciated body and empty havresac of any one of the wretched Russian soldiers. . . . From this we can infer fairly enough the condition to which the empire has been reduced.[23]

Back on the home front, moreover, it was Falconer, with the artillery fire of Sebastopol still ringing in his ears, who interpreted Huxley's circumspect lecture at the Royal Institution in martial terms, even comparing its anti-Cuvierian argument to the classic siege tactic of having "dykes breached for a fresh submergence."[24] It is evident that Huxley's opponents were equally willing to apply the language of warfare to scientific controversies during this period, and, unlike the self-proclaimed "Prophet-Warrior," to do it without the cover of anonymity.[25]

This was, of course, a more defensive version of the metaphorical juxtaposition of science and war. However, while discussions of the military metaphor in nineteenth-century science have tended to emphasize its relation only to hostile offensive tactics (with Adrian Desmond, for instance, depicting Huxley "shouldering his .45 to shoot over the ranks of obstructive Anglicans"), when it first began to be employed in the mid-1850s, there was considerable interest in, as well as great admiration for, the defensive strategies adopted by both sides in the Crimea.[26] As James Reid has observed, the "Crimean War was mostly a defensive war in which aggressive forward movements had limited aims. This defensive-minded strategy . . . originated in the post-Napoleonic perception of balance of powers that sought to avoid conquests and far-ranging military campaigns."[27] In September 1855 the Crystal Palace opened a Crimean Court with a scale model of Sebastopol showing the impregnable earthen ramparts built by the Russians according to the defensive principles outlined in *A Proposed New System of Fortification* (1849) by James Fergusson.[28] "Mr. Fergusson's model," as the official guidebook explained, demonstrated the "great principle of his system" in "arming the ramparts and . . . flanking defences," and was something that at "the present moment the visitors will be particularly interested with."[29] The upholders of the paleontological principles also celebrated in the life-size brick-and-mortar models on the same site now felt the need—utilizing a defensive form of the military metaphor—to publicly render the foundations of their Cuvierian methods of reconstruction no less unassailable from concerted attacks.

Huxley's lecture, delivered on 15 February 1856 only hours after dispatches reporting the final destruction of the Sebastopol docks were received by the War Department, was one of the Royal Institution's weekly Friday Evening Discourses, which attracted large, fashionable audiences made up of the institution's members and their guests.[30] As a subject calculated to interest men of science while at the same time not perturbing the fashionable ladies among his auditors, Huxley was simply to "set forth . . . an estimate of the science of Natural History" as a means of instilling "*knowledge*," "*power*," and, most important, "*discipline.*" The initial discussion of natural history as a form of knowledge, however, soon afforded Huxley a reason to adopt a more controversial tone, especially when he turned to the "works of Paley and the natural theologians," who had discerned evidence of a "utilitarian adaptation to benevolent purpose" throughout the natural world. In his *Moral and Political Philosophy* (1785), William Paley had reconciled Christian ethics with those of utilitarianism, and Huxley now attributed a similar emphasis on pragmatic expediency and end-directed teleology to his argument from design in *Natural Theology* (1802). Despite the evident teachings of the rigorous and sustained investigation of the natural world, this "principle of adaptation of means to ends," Huxley complained, continued to be trumpeted, "not only in popular works, but in the writings of men of deservedly high authority," as the "great instrument of research in natural history," and was even "enunciated as an axiom" whose truth was supposedly self-evident.[31]

Huxley then traced this overhyped "doctrine to its fountain head," finding that it was "primarily put forth by Cuvier" in certain passages from his "famous 'Discours sur les Révolutions de la Surface du Globe,'" a separate edition of the "Discours préliminaire" from *Recherches sur les ossemens fossiles* (1812) that was published in 1822. The only viable means of explaining this, Huxley went on, was to conclude that this "prince of modern naturalists . . . did not himself understand the methods by which he arrived at his great results," and that his "master-mind misconceived its own processes." This might, as the thirty-year-old who was still yet to publish any paleontological papers of his own acknowledged to the scientific grandees of the Royal Institution, seem "not a little presumptuous." But if all the "arguments be justly reasoned out" without regard to celebrity or patronage, then it would be recognized, Huxley insisted, that "it is correct." Huxley risked the justifiable charge of impudence in front of such an august audience because it was on the basis of the hyperbolic doctrine of adaptation to purpose that it had been inculcated "that all Cuvier's restorations of extinct animals were effected by means of the principle of the physiological correlation of organs."[32]

This was the same paleontological tenet whose intellectual requirements

Huxley had impugned at the Society of Art's Educational Exhibition less than two years earlier. In front of a still more influential central London audience, he now took the opportunity to again maintain that Cuvier's celebrated procedure was nothing more than "a method used as much in the common affairs of life as in philosophy."[33] Assumptions about the absolute necessity of what were actually prosaic empirical observations of customary correspondences were made on the basis of conceited authority and not common-sense rationality. Although Huxley did not mention it directly, his audience would have been aware that the same unquestioning obedience to rank and authority was, at that very moment, being widely blamed for the dreadful administrative failures of the patrician and nepotistic civil service during the Crimean conflict, most notably by Dickens in his satirical serial novel *Little Dorrit* (1855–57).

Huxley concluded, "Whatever Cuvier himself may say, or others may repeat, it seems quite clear that the principle of his restorations was not that of the physiological correlation or coadaptation of organs." Rather than disputing the actual accuracy of any of Cuvier's famous restorations (as Henri de Blainville had done regularly over the previous two decades), Huxley instead contended that they had been accomplished by methods other than those adumbrated by the great anatomist himself and then unquestioningly regurgitated by his many influential disciples. It was intimated that Cuvier "implicitly" acknowledged the redundancy of his own avowed principles, and, as seen in chapter 1, he had tempered the hyperbolic rhetoric of the "Discours préliminaire" by emphasizing the crucial role of empirical observation, but Huxley held back from a direct accusation of duplicity.[34] What was clear was that the empirical procedures were much simpler and more straightforward than the putative method of necessary correlation, which, supposedly, could be employed accurately only by the extraordinary genius of a select few.

Hooker and One or Two Others

Huxley's vehement attack on Cuvier immediately posed a delicate problem for Charles Darwin and Joseph Dalton Hooker, who, in May 1856, abandoned an attempt to get their headstrong new friend elected to the Athenaeum because of concerns that Owen would sway the club's committee against him.[35] Darwin fearfully imagined Owen with "a red face, dreadful smile & slow & gentle voice" asking "what Mr Huxley has done, deserving this honour; I only know that he differs from, & disputes the authority of Cuvier . . . as of no weight at all." Rather than Owen, however, it was Falconer whom Darwin, as he told Hooker, had "found . . . very indignant at the manner in which

Huxley treated Cuvier in his R. In^n Lecture," adding, "& I have gently told
Huxley so." Significantly, for Darwin and Hooker, Falconer had been "a mutual friend so dear to us both" since the mid-1840s, almost a decade before
either had even met Huxley.[36]

Darwin was evidently anxious to renew the acquaintance when Falconer
came back from Calcutta—via Sebastopol—in late 1855, inviting him, in
April of the following year, to stay at his home at Downe along with "Hooker
and one or two others."[37] Ironically, one of those other guests at the proposed
weekend party was to be Huxley. Even without knowing the identity of the
rest of the company, Falconer declined, and when Darwin finally did meet his
erstwhile friend again a month later, in May, he must have soon realized that
the planned introduction of Falconer to Huxley would have been unexpectedly incendiary. Falconer was incensed at Huxley's impertinent conduct, and
was already penning a stern rejoinder to the younger naturalist's defiance of
the established Cuvierian approach to paleontological reconstruction.

Falconer's final task before leaving India had been to catalog the paleontological collections in the museum of the Asiatic Society of Bengal. Working
"under the pressure of an approaching departure for Europe," he had employed Cuvier's customary methods to quickly bring taxonomic order to the
"appalling confusion, disorder and dilapidation" of the miscellaneous fossils
that were "huddled together in heaps . . . without a label or mark of any kind
whatsoever to indicate whence they came!" even if the lack of time meant that
some identifications had to remain "simply conjectural."[38] A year earlier he
had written to his niece from Calcutta, explaining the procedures employed
in his paleontological work:

> A tooth or the end of a joint . . . is as conclusive evidence of the former ex-
> istence of an animal as if all the structure—skin, flesh and blood, and living
> limbs—were before us. . . . The evidence is fragmentary and inductive, but . . .
> clear and conclusive. . . . For the Almighty has so ordained it that reason can
> safely reproduce all that has been lost, and restore to the tooth all that was
> correlative to it in life.[39]

Now back in London, Falconer was loath to have precisely these same methods repudiated by a brash novice whom he had probably never heard of before reading his lecture in the *Proceedings of the Royal Institution.* Huxley,
after all, was in Australia as a lowly assistant surgeon on HMS *Rattlesnake*
when Falconer had gone to India in 1848.

Falconer's response appeared in the June 1856 number of the *Annals and
Magazine of Natural History,* and his tone was unmistakably terse, impatient,
and condescending. Huxley's charges against Cuvier were "remarkable" and

"startling," and the necessity of "making practical refutations" of them led Falconer to exclaim testily: "All this is familiar knowledge; the only marvel is, that one should have to adduce the facts at the present day in such an argument."[40] The Cuvierian axiom of "necessary correlation" that Huxley had found so untenable was simply "*necessary* in the sense of being demonstrable in such a way that the contrary involves an absurdity and is inconceivable," so that "Mr. Huxley, with the skeleton of a hawk before him, might as well say that, for any physiological necessity to the contrary, that creature might have its jaws with teeth, and its internal organs arranged, like those of a tiger."[41] When applied to actual paleontological reconstructions, Huxley's unwillingness to acknowledge that certain associations in anatomical structures were necessary might permit the creation of absurd, incongruous monstrosities.

The end "result" of all this belligerence, Falconer concluded, was "that after the encounter the law of correlation stands exactly as Cuvier found and left it . . . wholly uninjured by its latest assailant."[42] As noted earlier, Falconer's language was strikingly redolent of the reporting of the siege warfare he had himself witnessed on his journey back from Calcutta. Like Fergusson's famously indestructible earthen ramparts now displayed in a scale model at the Crystal Palace (which, as the *Methodist Quarterly Review* reported, had enabled the "Russians in Sebastopol, behind their impregnable ramparts, [to] laugh at Minnie rifles and Lancaster guns"), Falconer's defense of the besieged foundations of Cuvierian paleontology had decisively repelled the great anatomist's young assailant, leaving the axiom of necessary correlation entirely unharmed and exactly as Cuvier had left it.[43] Not knowing his adversary's nascent tenacity, Falconer probably assumed that the matter was now settled. Huxley would have to accept a humiliating peace like that agreed among the victorious European powers in the Treaty of Paris. In July 1856 Falconer even felt able to retire from the fray to the very city where the conflict in the Crimea had been concluded three months earlier. In a letter from Paris "he alluded to the subject as if he had eaten Huxley without salt & left no bones at all," as his old friends Hooker and Darwin noted between themselves with considerably less assurance.[44]

Falconer v. Huxley

Ensconced at Downe, Darwin was "most heartily sorry at the whole dispute," which he worried would "prevent two very good men from being friends." Reluctant to "give up the time to form a very certain judgement to my own satisfaction, in Falconer v. Huxley," Darwin found himself genuinely torn between old and new friends. In a letter to Hooker written over two days in June 1856, he kept veering abruptly—sometimes even midsentence—between their rival claims, stating first that Falconer's "article struck me as very clever," then noting, "I rather lean to the Huxley side" and "I think Huxleys [sic] argument best," before adding, "But to deny all reasoning from adaptation & so called final causes [as Huxley had done], seems to me preposterous," and finally concluding, "I deprecate the contemptuous tone of Huxley." Darwin's own tone of anxious equivocation was at last resolved only in a short postscript, apparently added later, when he told Hooker: "I have just reread your note & it seems to me that there is great justness in your remarks on Huxley & the general question, being discussed as it has been discussed."[45] Although this particular note seems no longer to exist, Hooker's attitude toward the dispute, which apparently decided Darwin's own, can be inferred from another brief missive he sent to Huxley earlier in the same month.

Leafing through "old quarterlies," Hooker had come across a "passage that," as he gleefully informed Huxley, "will amuse you and rile Falconer."[46] The passage appeared in the yellowing pages of a twenty-seven-year-old copy of the *Quarterly Review* and described Cuvier's confidence in his methods as a "delusion."[47] Hooker pointedly underlined this word in his own précis of the passage, emphatically urging Huxley that "it is worth your reading."[48] While Hooker later conceded, "I do not think that Huxley's original lecture [at the Royal Institution] was particularly good at all," he nevertheless came out decisively on his side in the dispute with Falconer, whom, having fallen out with him over his uncouth conduct at the Asiatic Society when the two were in Calcutta in the late 1840s, Hooker was happy to see riled.[49] Hooker's clear stance helped clarify Darwin's own position on the matter, and he soon explained retrospectively, "I thought from the first that he [i.e., Huxley] was right, but was not able to put it clearly to myself."[50] Across the Channel, meanwhile, Falconer was blithely unaware that these two old friends of more than ten years' standing had opted to support, and even furnish with strategically advantageous passages from periodicals, the headstrong young naturalist they had only recently got to know.

Falconer would have been still more surprised because, until now, Darwin had given every impression of being a staunch advocate of Cuvierian methods in paleontology. After all, in his chapter on "Geology" for John Herschel's *Manual of Scientific Enquiry* (1849), Darwin had advised Royal Navy officers that "bones . . . from any formation are sure to be valuable; even a single tooth, in the hands of a Cuvier or Owen, will unfold a whole history. . . . Every fragment should be brought home."[51] The fragmentary bones that Darwin himself brought back from South America had been passed on to Owen, whose "Fossil Mammalia" volume of the *Zoology of the Voyage of H.M.S.*

Beagle (1839) was a veritable textbook of orthodox functional correlation, replete with allusions to Cuvier's "beautiful and justly celebrated reasoning" on a single phalanx bone.[52] Professing his "entire ignorance of comparative Anatomy," Darwin did not challenge the Cuvierian basis of Owen's reconstructions, even while his own, more idiosyncratic approach to reading fossils emphasized environmental rather than functional considerations.[53] In his meticulous research on living species of barnacles, moreover, Darwin actively endorsed Cuvier's distinctive understanding of animal structure, explaining of the *Proteolepas bivincta* that "in accordance with the general law of the correlation of parts, it may be inferred . . . that this abnormal creature was developed within a pupa of the same general structure, and of about the size, as the pupæ whence . . . many other cirripedes are developed."[54] Notably, this inference was published, in the second volume of *A Monograph on the Subclass Cirripedia* (1854), in the very same year that Huxley began his onslaught against the putative law of correlation at the Society of Art's Educational Exhibition. Falconer, in May 1856, had presumably confided in Darwin about his indignation at the manner in which Huxley's subsequent Royal Institution lecture treated Cuvier in the firm belief that his old friend would feel the same way.

Although Huxley represented physiological correlation as ineluctably a quasi-theological doctrine, Darwin remained interested in the nonadaptational correlative changes in an organism's organization that might be effected by natural selection modifying any one part (contra Cuvier's insistence that such complex and holistically coordinated alterations were impossible).[55] In the late 1830s he had jotted in his secret notebook on transmutation, "Thinking of effects of my theory, laws probably will be discovered. of co relation of parts, from the laws of variation of one part affecting another," a principle that, twenty years later, he designated "correlation of growth" in the *Origin of Species*.[56] Darwin's closest confidant in his clandestine evolutionary speculations, Charles Lyell, also discerned a potential support for his friend's provocative theory in Cuvierian correlation, noting in his own private journal on the species question: "Cuvier's observation that each organ of an individual bears a relation to the whole. This supports view of indefinite variation as a tree-climbing creature will acquire all the necessary changes."[57] It is important to recognize that in the mid-1850s, when even Lyell was beginning to entertain the so-called development hypothesis, Huxley still held to a stridently nonprogressive line on species. The apparent impossibility of transitions between species was in fact the one area on which he was in full agreement with Cuvier, with Huxley observing in 1853 that "as CUVIER long ago remarked of the Cephalopoda and Fishes, so we may say of the Cephalous Mollusca in general and other types:—'. . . Nature here leaves a manifest hiatus among her productions.'"[58] There was, at this stage, simply no connection whatsoever between Huxley's fierce antagonism toward Cuvier's paleontology and Darwin's as yet undisclosed, and in any case Cuvierian-inflected, evolutionism. Indeed, the recent contention of both Claudine Cohen and Martin Rudwick that, in the words of the latter, "Huxley criticized Cuvier's claim [regarding correlation] on the grounds that it had been based on a pre-evolutionary conception of each species as an unchanging well-integrated 'animal machine'" is manifestly anachronistic because it does not recognize that Huxley's criticisms of Cuvier predated his conversion to Darwinian evolution.[59] There remained, moreover, important residual differences between Darwin and Huxley over correlation that, as later chapters will show, continued to resurface in subsequent disputes.

The "Law of Necessary Correlation" Is—Nowheres

In the absence of any clear scientific impetus, part of the reason for Darwin's sudden apostasy from Falconer and Cuvierian correlation in June 1856 likely went back to the weekend party in April that Falconer had declined to attend. Darwin had used the occasion to tentatively reveal aspects of his thinking on species mutation to a carefully chosen group of naturalists who could assist him with pertinent questions and objections, and whom he suspected might be susceptible to his circumspect promptings. Had he made the railway journey to rural Kent, Falconer could have provided some much-needed "rudimentary knowledge" of the "Pigeons-skeletons" that littered Darwin's home at this time, and, although "included . . . in the category of those who have vehemently maintained the persistence of specific characters," Darwin was "fully convinced" that, with a little logical persuasion, he would become, as he told him, "less fixed in your belief in the immutability of species."[60] While Falconer later avowed that he had "long enjoyed the privilege of intimate intercourse" with Darwin, and had "been for many years familiar with the gradual development of his views on the Origin of Species," it was instead Huxley who was taken into Darwin's confidence that spring at Downe.[61]

Initially no less opposed to transmutation than Falconer, Huxley was increasingly swayed by the evidence and arguments presented by Darwin, who rejoiced at the "change in . . . Huxley's opinions on species" that was tangible over the course of the weekend.[62] Darwin had been gently insisting on the possibility of organic forms *"generally undergoing* further development" in letters to Huxley for the previous three years, and now, only a month after his first stay at Downe, the erstwhile scourge of those who were "unorthodox about species" forecast that "the 'Theory of Progressive Development' will present by far the most satisfactory solution" to the thorny question of the emergence of new species.[63] Thus, by the early summer of 1856, it was evident that Huxley, with his youthful vigor and rhetorical self-assurance, represented a more effectual ally for Darwin, and his heterodox evolutionism, than the frail old friend who was still socially and intellectually isolated after his long sojourn in India.

Falconer in fact became the source of "splendid joke[s]" among Darwin's inner circle, and his rancorous dispute with Huxley, which coincided with a crucial moment in Darwin's strategic disclosure of his long-harbored speculations, helped forge a sense of solidarity between those who, three years later, would emerge as the principal supporters of the controversial theories presented in the *Origin*. Recent accounts of both Huxley and Hooker, for instance, have emphasized their considerable differences on issues such as gentlemanly conduct, religious authority, and professionalization that were often overlooked by earlier historians who conflated them into a homogenous young guard of self-conscious anticlerical professionalizers.[64] The Cuvierian contretemps, however, established a strong point of connection between the two men that, as has been seen, had them conspiratorially swapping suggestions for riling the unfortunate Falconer.

Although Hooker confessed that he had "only half understood the question before" having read the various contributions to the dispute, he subsequently maintained a virulent and long-standing antagonism to the very Cuvierian axioms that Huxley had alerted him to. Writing to Darwin in 1869, he exulted at his "tremendous upset to Owen's doctrines," exclaiming: "The 'law of necessary correlation' is—nowheres."[65] Even as late as 1895, Hooker maintained that he had "only one criticism" of Walter Lawry Buller's *Illustrations of Darwinism*, "which is as to the necessary correlation" its author invoked to explain the association of the structure of the "huge forms of Palæozoic life with a tropical vegetation."[66] The acrimonious dispute over paleontological method in the opening months of 1856 helped clarify the strategic alliances that would be vital to the defense of Darwinism in the still more bitter controversies to come in the following years and decades. In fact, even though the main participants in what Darwin squeamishly designated "Falconer v. Huxley" were, at the time, equally opposed to species transmutation, and Darwin's own understanding of growth and variation could, in any case, be readily accommodated with aspects of Cuvierian method, an opposition to necessary correlation became enshrined as a key component of the broader agenda of what was increasingly known as scientific naturalism.[67]

This was made particularly evident when, in October 1857, the most prominent public advocate of the development hypothesis, Herbert Spencer, weighed in to the "controversy now going on among zoologists . . . respecting the alleged *necessary correlation* subsisting among the several parts of any organism."[68] With even less experience of actual paleontology than Huxley, Spencer loftily identified the "flaw[s] in Cuvier's principle" by "recourse to a mechanical analogy" more suited to his own theoretical methods, before concluding: "We agree with Professor Huxley. . . . Palæontology must depend upon the empirical method. Necessary correlation cannot be substantiated."[69] Huxley told Spencer that these statements in the anonymous *National Review* had "been ascribed to Huxley himself; 'and that by no less a person than by Dr. Hooker,'" prompting Spencer to boast to his father, "I have heard Huxley say that there are but four philosophical naturalists in England—Darwin, [George] Busk, Hooker, and himself. Thus the article has been ascribed *by* one of the four *to* one of the four."[70] Two years ahead of the *Origin's* publication, opposition to Cuvierian correlation afforded a vital rallying call that brought together the different constituencies of the scientific young guard then emerging in London.

These tactical alliances were only hardened by Huxley's prompt rejoinder to Falconer in the very next number, for July 1856, of the *Annals and Magazine of Natural History*. Far from backing down, Huxley simply reiterated his contention that there was "nothing more erroneous than the popular notion . . . that [Cuvier's] method essentially consisted in reasoning from supposed physiological necessities."[71] Cuvier and his heirs had actually fallen into a logical trap in which "when the result to which a combination tends is obvious, we commonly imagine we can see the reason for that combination." All suggestions that a "correlation is in any case to be called *necessary*" were thus mistaken flights of fancy, given free rein by an unfortunate overemphasis on function rather than structure. At times Huxley did appear to adopt a more conciliatory tone, acknowledging that the "brusque attack from Dr. Falconer" in the previous month's number "caused me at first, I must confess, no slight alarm . . . coming as it did from the pen of a palæontologist of high repute." This was only a tactical retreat, though, and he immediately resumed his assault, insisting that the

> perusal of Dr. Falconer's essay . . . soon relieved me from any real source of uneasiness, by demonstrating very clearly that Dr. Falconer had been far too much in a hurry either to master the real question in dispute, to read what I had written with attention, or to quote me with common accuracy and fairness.

Haste and prejudice had resulted in an "entire misconception of the point at issue," which meant that Huxley "left untouched many points in Dr. Falconer's essay, not because they cannot be answered, but because I conceive they will answer themselves."[72] It was not just Cuvier who was on the receiving end of Huxley's impudence in 1856. Falconer, the discoverer of the world's richest deposits of Tertiary mammals in the Siwalik Hills and a "palæontologist of high repute," was treated with the same laconic disdain by the young tyro still without any paleontological publications of his own.

The aspects of Falconer's article that Huxley did deign to comment on often related more to appropriate conduct than to paleontological method, and Huxley exploited his adversary's long isolation, both on account of his lengthy sojourn in India and because of his frail health and prickly temperament, from the wider scientific community. As Claudine Cohen has observed, Falconer was "a somewhat marginal figure in relation to the scientific establishment of his day," a point that Huxley emphasized with brutal relish.[73] The "singular bad taste" of many passages would, Huxley counseled, "cause Dr. Falconer, in his cooler moments, far more annoyance than they have occasioned to anyone else, except his friends."[74] Hooker, one of those alleged friends, may already have tipped Huxley off about "how ill Falconer behaved" during scientific disputes back in India, where he regularly "gave great offence at the Asiatic Society," for Huxley now insisted, rather disingenuously given Darwin and Hooker's travails over his own election to the Athenaeum, that Falconer had not behaved in the manner expected of participants in metropolitan scientific society.[75]

Oddly, Lyell, who, like Darwin and Hooker, was friends with both Falconer and Huxley, congratulated the latter for "rendering subordinate the personal & controversial part" of their dispute. In any case, Lyell, who when he had initially "read Falconer . . . thought he had the best of the argument," now considered that the "tables are turned." He nevertheless signed off, "I shall ask Falconer what he can say to it," suggesting that he might still be won back to the Cuvierian cause.[76] Others were less tentative in their conversion, with Hooker exclaiming: "I have read Huxley's response to Falconer with eminent gusto—how admirably neatly and clearly he puts the whole question. . . . He has put forth his strength here & will I think startle old Falconer." Darwin concurred, remarking: "I am delighted at what you say about Huxley's answer & I agree most entirely; it is excellent & most clear."[77] Notwithstanding Huxley's staunch and sometimes scornful tone, there was now not even a hint of equivocation in Darwin's approbation for his new friend, whose adversarial manner was beginning to seem potentially very useful.

Not everyone agreed, of course, and James McCosh and George Dickie, a philosopher and a botanist who worked together at Queen's College in Belfast, afford an instructive case in point. In April 1856 they had sent Huxley a copy of their natural theological treatise *Typical Forms and Special Ends in Creation* with an accompanying letter expressing the hope that "it may be a means of bringing into more general notice certain important and admirable views researched by yourself."[78] Despite their enthusiasm for Huxley's early studies of cell structure and marine invertebrates, which, for them, revealed the underlying plan of the divine creation, in the book's second edition a year later, McCosh and Dickie added a lengthy footnote observing that the famous "controversy" between Cuvier and Étienne Geoffroy Saint-Hilaire "has not yet died out. Since the first edition of this work was published, we have this very discussion (in a somewhat confused form) in Huxley's Lecture at [the] Royal Institution . . . Falconer's Examination of this lecture . . . and Huxley's reply." In neither of his contributions, they insisted, had Huxley "succeeded in proving that palæontology, in restoring extinct forms, would be valid, even 'if we knew nothing of final causes or adaptations to purpose.'"[79] Darwin had initially had similar reservations, remarking, as seen earlier, that "to deny all reasoning from adaptation & so called final causes, seems to me preposterous," though these had apparently been resolved by Huxley's reiteration of his argument in the *Annals and Magazine of Natural History*. That McCosh and Dickie were not similarly won over shows how the dispute over paleontological method in 1856 helped instantiate a clear breach between the nascent scientific naturalists and those, such as the authors of *Typical Forms and Special Ends in Creation*, who still endeavored to reconcile the latest findings of modern science, including Huxley's own researches, with natural theology.

More Cuvierian Than Cuvier

Huxley had baited Falconer with the suggestion that "if the master's words be studied carefully, it will be discovered that his followers are more Cuvierian than Cuvier," but in reality the principal defender of necessary correlation in the summer of 1856 was himself not a particularly steadfast advocate of the French naturalist's putative methods.[80] In 1844 Falconer claimed that his identification of an unknown giant tortoise from the Siwalik Hills, which he named *Colossochelys atlas*, had been based on "what is called a Cuvierian restoration founded on . . . the laws of reciprocal connection between special and general structure, which involves the necessary condition that every individual part has a definite relation to the aggregate form."[81] However, his approach to paleontological reconstruction, honed in isolation in the foothills of the Himalayas, was as much self-taught as derived from the pages of Cuvier's writings. As Lyell had explained to the Geological Society when news of the Siwalik fossils first reached London in the late 1830s:

> When Captain [Proby] Cautley and Dr. Falconer first discovered these re-markable remains their curiosity was awakened . . . but they were not versed in fossil osteology, and being stationed on the remote confines of our In-dian possessions, they were far distant from any living authorities or books on comparative anatomy to which they could refer. . . . From time to time they earnestly requested that Cuvier's works on osteology might be sent out to them, and expressed their disappointment when, from various acci-dents, these volumes failed to arrive. The delay perhaps was fortunate, for be-ing thrown entirely upon their own resources, they soon found a museum of comparative anatomy in the surrounding plains, hills, and jungles, where they slew the wild tigers, buffalos, antelopes, and other Indian quadrupeds. . . . They were compelled to see and think for themselves while comparing and discriminating the different recent and fossil bones, and reasoning on the laws of comparative osteology, till at length they were fully prepared to appreciate the lessons which they were taught by the works of Cuvier.[82]

When this autodidactic Cuvierianism improvised in the field was finally compared with Cuvier's long-awaited volumes, it actually showed that the conclusions regarding fossil elephants that the metropolitan-based savant had developed in his Parisian museum were frequently erroneous.

As Falconer and Cautley wrote in their *Fauna Antiqua Sivalensis* (1846), "Notwithstanding his array of authority, we cannot help thinking that Cuvier was premature in his conclusion [that all fossil elephants comprised only one species], and that the identity of forms has rather been assumed against the evidence, than proved by it." There were "not sufficient materials," Falconer and Cautley protested, "for making such a comparison when Cuvier wrote," with "his acquaintance with the Italian fossil Elephant . . . limited . . . to a sin-gle mutilated fragment." It was merely the "great weight of Cuvier's authority [that] has given an undue influence to his statement on this point, which has biased the observations of some later writers directed to the subject."[83] This was strikingly close to Huxley's own complaint, a decade later in the *Annals and Magazine of Natural History*, that the supporters of "Cuvier's law of cor-relation" were "guided more by authority than by right reason."[84] Huxley presumably included Falconer among these slavish adherents to authority, although the latter's more maverick approach to paleontological practice in-stead suggested that there were actually many similarities in their opinions.

Even in his fierce rebuke to Huxley, Falconer acknowledged that the "principle of correlation . . . must not, in practice, be pushed too far in palæontology," conceding that there were "numerous instances on record, in which, in attempting to determine extinct forms from a single bone or tooth . . . very erroneous conclusions have been arrived at; among others, even by Cuvier himself."[85] Nor was Falconer's defense of Cuvier at all inflected with any natural theological agenda; the law of correlation was necessary only in the sense that all other alternatives were inconceivable. The casual invocation of the "Almighty" in the letter to his niece that was quoted earlier was likely only meant to suit the feelings of a pious young woman, and, in any case, elsewhere in the same letter he maintained that most modern scientific developments had initially been "denounced as a heresy opposed to the Bible" and thus "when . . . in a good cause, the imputation of *infidelity* is raised, one need not be ashamed of it."[86] In fact, while Huxley only identified with the Muslim "Prophet-Warrior" Schamyl in print, Falconer had adopted Islamic dress during a period in Afghanistan and "was taken for a pilgrim going to Mecca to worship at the shrine of Mahomet."[87] The truth was that Falconer and Huxley had much more in common than either, in the heat of battle, had been willing to recognize.

Falconer soon came to regret the whole war of words, telling his niece: "I must keep my temper. . . . What a grumpy old uncle you have got. Like an infuriated Toro . . . having an encounter with Huxley." More significant, the self-appointed defender of Cuvierian correlation was himself, in a remarkable volte-face, increasingly converted to his adversary's position. By the early 1860s the "infuriated Toro" had moved on from Huxley and was now "goring Owen, and his jackall [*sic*]."[88] This was a rueful reference to the disputed identification of a mammalian jawbone from the Purbeck beds in southern England, which Falconer assigned to an order of herbivorous rodent marsupialia and Owen to a predatory, carnivorous marsupial similar to the ferocious *Thylacoleo* of Australia. Owen's ostensibly authoritative classification of the so-called *Plagiaulax*, Falconer contended, was merely "an opinion, professing to be founded on the high ground of a connected series of physiological correlations." Both men acknowledged that "comparative anatomy supplies for our guidance fundamental principles," but there was a clear discrepancy in how they applied them. This "conflict of opinion," as Falconer observed, arose "from different methods having been followed by the observers in dealing with the evidence." Unlike Owen's stringently Cuvierian approach, Falconer examined the creature's curious dentition thus:

> Why there should be this plurality of incisors above, and only two invariably occupying the same position below, is wholly unknown to us; but the constancy of the structure makes it certain that there must be a sufficient cause

for it in nature; and we employ the generalization, empirically arrived at, with as much confidence as we do the law of necessary correlation. In many critical cases, where the evidence is limited or defective, the empirical is even a safer guide than the rational law, since it is freer from the risk of errors of interpretation.[89]

This was, of course, a complete reversal of what he had argued in the *Annals and Magazine of Natural History* in 1856, where it had been claimed that it was the "constant effort of every philosophical mind . . . to extinguish the empirical character of the phænomena, and bring them within the range of a rational explanation." Falconer had even mocked Huxley's argument that empirical procedures were more reliable, exclaiming that it was "a rare spectacle to see empiricism chosen by preference" to the rational axioms adumbrated by Cuvier.[90] Now in 1862, it would seem, Falconer had come round to his erstwhile opponent's opinion that simple empirical observations that certain structural phenomena invariably occur together were just as trustworthy as the doctrinaire law of necessary correlation, and, in certain difficult cases, were perhaps even more dependable.

Owen himself certainly considered that this "polemical paper of Dr. Falconer" marked a departure from the "principle which Cuvier laid down as our guide in such dark routes in Palæontology," insisting that "loyalty to our common science compels me to say that he fell into his mistake as to *Plagiaulax* by neglecting its fundamental principle . . . Cuvier's . . . rule . . . of correlation."[91] Falconer was now regarded, by friends and foes alike, as categorically on the side of the anti-Cuvierian cause, with Darwin telling him: "I have not been for a long time more interested with a paper than with yours. It gives me a demoniacal chuckle to think of Owen's pleasant countenance when he reads it."[92] When, two years afterward, Huxley sent Falconer a copy of his *Lectures on the Elements of Comparative Anatomy* (1864), he only had time to "cast a hurried glance over the Classification chapter." He was nonetheless effusive in his praise of the characteristic rhetorical force exhibited in this opening section, telling Huxley: "You write with a sledge-hammer and tick the points with a mace."[93] That Huxley had, in the chapter "On the Classification of Animals," opined that deductions based on the "law of correlation" were "well calculated to impress the vulgar imagination" and then derided "Cuvier, the more servile of whose imitators are fond of citing his mistaken doctrines as to the methods of palæontology against the conclusions of logic and common sense," did not appear to trouble Falconer.[94] Eight years after their acrimonious dispute in the summer of 1856, Huxley's continuing onslaught against Cuvierian correlation now elicited Falconer's admiration rather than his indignation.

Falconer's dramatic conversion to Huxley's empiricism in the early 1860s occurred alongside his increasing acceptance of Darwin's evolutionary theories. As Hooker joked to Darwin, "He seems to me to have just awakened to the fact that there is something in you." The article in which Falconer finally endorsed Darwinian evolution, albeit with some reservations over the adequacy of natural selection, appeared in the January 1863 number of the *Natural History Review*, whose editor in chief was his former antagonist Huxley. That Huxley advised Darwin to look in the "last N.H. Review" to see "a grand paper by . . . good old Falconer" suggests just how far the latter had become accepted as part of the circle of scientific naturalists. This could only happen once Falconer was simultaneously denuded of his doubts over the mutability of species and, no less important, his lingering allegiances to Cuvierian correlation. The closer "good old Falconer" became to Huxley, Darwin, and their inner circle—with even Hooker growing to once more respect him "not only personally, but as a scientific man of unflinching & uncompromising integrity"—the more he renounced his erstwhile Cuvierianism.[95] Once again, and even long after the rancor of the initial skirmishes, the controversy over paleontological method was a crucial barometer of support for a range of broader issues centered around Darwin's evolutionism. While the extent of his apostasy was particularly notable, Falconer was by no means the only naturalist to forsake an earlier adherence to Cuvier's correlative procedures in the wake of Huxley's attack. Before the end of 1856, however, Huxley's criticisms of Cuvier had attracted the attention of a far more formidable adversary.

The Great O. versus the Jermyn St. Pet

Although he was well aware of Owen's fierce loyalty to the Cuvierian law of correlation, Huxley's original Royal Institution lecture had invoked "Professor Owen's determination of the nature of the famous Stonesfield mammal" as a "striking illustration" of how the "whole process of palæontological restoration depends" on nothing more than a knowledge of the "invariable coincidence of certain organic peculiarities."[96] The celebrated mammal Huxley had in mind was the *Didelphis*, which, back in 1838, Owen had confirmed as both mammalian and a marsupial, thereby rebutting Blainville's incessant criticisms of Cuvier's methods. In June 1856 Falconer saw it as another egregious error that "Mr. Huxley holds him [i.e., Owen] up . . . as furnishing a bright example . . . of empirical deduction," adding a slyly flattering footnote observing: "Mr. Owen flies his hawk at a more ambitious quarry in original research; but it is not too much to expect that he may on some occasion record his protest against Mammalian Palæontology being asserted to rest merely on empirical correlation, in a pithy foot-note."[97] A curt aside, however, was not the form in which Owen chose to make his inevitable response to Huxley's provocations, and instead he bided his time through the long summer recess.

The London scientific "season" ran from November to June, and it was only once the metropolitan scientific societies began to reconvene in the autumn that Owen prepared for his long-awaited intervention in the controversy, with the news that he would soon enter the fray setting the scientific community abuzz with gossip. At the beginning of November, Huxley excitedly informed Frederic Dyster, a Welsh doctor with whom he had spent the summer, "There is going to be a set-to at the Geological [Society] on Wednesday—the great O. versus the Jermyn St. Pet on the Methods of Palaeontology," and he promised him: "You shall know the results."[98] But Dyster received no more of his friend's customary mordant gossip on the matter, and instead it was others who swapped hearsay about Huxley's own derisory performance in the much-anticipated "set-to." The news had certainly reached Hooker, who told Darwin: "Owen I hear committed a cutting telling & flaying alive assault on Huxleys [sic] adaptation views at the Geolog. Soc. & read it with the cool deliberation & emphasis & pointed tone & look of an implacable foe.—H. I fear did not defend himself well (though with temper)." While the earlier contretemps with Falconer had come to be regarded as a success for their headstrong young friend, Hooker noted that these continuous "embroglios are very bad indeed & must insensibly have a bad effect upon Huxley—the best natures insensibly deteriorate under such trials."[99] His pugnacious bravado notwithstanding, Huxley had in fact been unwell for much of 1856, and only two days before the Geological Society's Ordinary Meeting, he reported to Dyster that he had "contrived to sit in a draught or something or other so that I have an attack of rheumatism at this moment."[100] Huxley's indisposition allowed Owen's Cuvierian rear guard to snatch back the initiative.

On 5 November Owen brandished before the Geological Society's fellows a small mammalian jawbone recently discovered in the oolitic slate of Stonesfield in Oxfordshire. As central London engaged in the traditionally exuberant and incendiary celebrations of Guy Fawkes night, Owen ignited his own paleontological pyrotechnics. Defying the convention that "papers read to the Ordinary Meetings of the Society were . . . primarily descriptive and fact-based, and it was the discussion that followed that set them within whatever theoretical debate or dispute was current at the time," he made recent methodological controversy the very focus of his talk.[101] The fragmentary fossil,

Owen observed, "excites . . . interest" because it both offered an important "test . . . of the actual value of a single tooth in the determination of the rest of the organization of an animal" and afforded an opportunity, particularly needful "in the present state of Palæontology," to "analyse the mental processes by which one aims at the restoration of an unknown Mammal from a fragment of jaw." Having expertly "conjectured the nearer affinities of the *Stereognathus*," Owen then asked his audience:

> Can this example . . . be justly cited as showing that there is no physiological, comprehensible, or rational law, as a guide in the determination of fossil remains: but that all such determinations rest upon the application of observed coincidences of structure, for which coincidences no reason can be rendered?

This, as Owen well knew, was precisely how Huxley had interpreted the same example at the Royal Institution nine months earlier, and, with the exhausted tyro now sitting across from him on the Geological Society's parliamentary-style benches, he finally responded with emphatic plainness: "I do not believe this to be the case." If the method of determining the structure and habits of the *Stereognathus* had simply been an "empirical one," as Huxley proposed, then the "narrowness of its support from observation" could never "leave the mind free from a sense of the possibility of its being liable to be proved an erroneous conclusion." The "higher law" of necessary correlation, on the other hand, afforded an assurance and certainty that Huxley's indefinite empiricism could never aspire to.[102] Just as his authoritative examination of the *Didelphis* at the Geological Society two decades before had repudiated Blainville's censure of Cuvier, Owen's analysis of another diminutive fossil mammal, in the very same location, now directly challenged, as he told the Dutch naturalist Jan van der Hoeven, the "advocacy of . . . De Blainville by some of our young teachers, in letters in our 'Annals of Natural History'" (although Huxley had not actually invoked Cuvier's embittered successor).[103]

The "small and unfruitful minority" of naturalists who persisted in maintaining the "inapplicability of the law of correlation" were, Owen claimed, closet adherents of the earthly, materialist "tenets of the Democratic and Lucretian schools" formulated in ancient Greece and Rome. Their "insinuation and masked advocacy of the doctrine subversive of a recognition of the Higher Mind" was the result of "some, perhaps congenital, defect of mind, allied or analogous to 'colour-blindness'"—and akin even to compulsive onanism, as Owen's furtive insistence on the absence of "healthy," "normal," or "fruitful" sensibilities implied—which rendered them devoid of humanity's usual "instinctive, irresistible impression of a design or purpose" in nature.[104] Huxley was not named directly in this hyperbolic diatribe, or indeed in the paper as a whole. Owen's auditors, though, would have been fully aware that it was his criticisms of Cuvier that were depicted as not only erroneous, but actually immoral and unhealthy (an accusation likely made more vexing by Huxley's current poor state of health). Spencer defended his friend in the *National Review*, remarking on the "questionable propriety" of the way in which "Professor Owen avails himself of the *odium theologicum* . . . in his defence of the Cuvierian doctrine."[105] What would have been still more questionable for Spencer and his fellow scientific naturalists was that, in direct contrast to Falconer four months earlier, Owen specifically endorsed, in a zealously homiletic idiom, the natural theological implications of necessary correlation. But, for Owen, the religious significance of the Cuvierian method was not restricted to the support it had traditionally lent to a rationalist Paleyite argument concerning the evidence of design in the natural world.

He also invested it with a distinctly fideistic emphasis far removed from the probabilistic approach of conventional natural theology, professing his intuitive "faith in the soundness of the conclusions deduced from the application of such rational law of correlations." This faith rested on an instinctive sentiment, which Owen expressed in unashamedly theistic language: "I feel in the workings of my own mind what I believe to have operated in other minds, an irresistible tendency to penetrate to the sufficient cause of such coincidences—'to know the law within the law.'"[106] This final expression was a quotation from Alfred Tennyson's "The Two Voices" (1842), a poetic dialogue in which the speaker maintains his trust in the human capacity for absolute knowledge despite the suicidal promptings of a skeptical inner voice. In the passage Owen quoted from, the speaker determines

> To search thro' all I felt or saw,
> The springs of life, the depths of awe,
> And reach the law within the law.[107]

Like *In Memoriam* (1850), "The Two Voices" was written in the wake of the crisis in Tennyson's religious beliefs provoked by the death of Arthur Hallam, and it foreshadows the famously fideistic affirmation of faith in that later poem. The literary embellishment of even a brief, and unattributed, allusion to the revered poet laureate helped authorize Owen's own insistence that the "heat of inward evidence," as Tennyson put it elsewhere in "The Two Voices," was sufficient to confirm the truth of the law of correlation.[108]

Having been cast in the same position as the "dull, one-sided voice" that, in "The Two Voices," "make[s] everything a lie," Huxley appeared discomfited when Owen finally sat down, and was unable to muster his usual acerbic wit in the ensuing discussion of the paper.[109] Soon after, he even expressed an uncharacteristic desire to foster a "nobler tone to science" and "set an example of abstinence from petty personal controversies."[110] The campaign against Cuvierian correlation appeared to have run into the doldrums by the end of 1856.

Our British Cuvier in His *True Place*

Owen pressed home his advantage at the beginning of the following year, agreeing to give a course of guest lectures on extinct mammals at the Government School of Mines. This was where Huxley had been employed for the last two years, and, as he observed to Dyster, the "arrangement had been made with my consent" with "only . . . one stipulation which I thought due to myself—viz. that it should be clearly understood that he [i.e., Owen] held no appointment in the School of Mines."[111] Owen had recently stood down as Hunterian Professor at the Royal College of Surgeons, and the lecture series was part of his duties as the new superintendent of the British Museum's natural history collections. In the latest edition of John Churchill's annual *Medical Directory*, however, Owen announced his position as "Professor of Comparative Anatomy and Palaeontology, Government School of Mines, Jermyn-St." This was precisely the title that Huxley, the self-proclaimed "Jermyn St. Pet," considered his own, and he complained to the *Directory*'s publisher that his rival's spurious claim was "calculated to do me injury."[112] Convinced that Owen's "intention of course was to insult me as publicly . . . as possible & to give rise to the idea in the public mind that he had to a certain extent superseded me," Huxley saw no alternative, as he told Dyster, but to once more launch into a "fresh row with the ex–Hunterian Professor."[113]

This infamous spat between Owen and Huxley, during which the latter declared, "I have now done with him, I would as soon acknowledge a man who had attempted to obtain my money on false pretences," has been frequently examined by historians, who have generally considered it to mark the formal termination of any lingering vestige of polite relations between the two men.[114] What has never previously been recognized, though, is that right from the beginning of his contentious course of guest lectures at the Government School of Mines, Owen also continued his defense of Cuvier's axiom of correlation, and this time in Huxley's own institutional stronghold. And just as with their anticlimactic "set-to" at the Geological Society, Huxley again failed to live up to his bellicose imprecations to Dyster that Owen had "reckoned without his host—and will have to eat a large leek," a threat that drew self-consciously on the martial rhetoric of Shakespeare's *Henry V*.[115]

Beginning in late February 1857, four months after his paper to the small group of specialists who attended the Geological Society's fortnightly Ordinary Meetings, Owen addressed a much larger audience containing "many of the most distinguished men in London," who, with the lectures commencing at 2:00 p.m., had "set aside their work at the busiest time of the day in order to be present there." While the early afternoon schedule was "inconvenient" for such "busy men," an engraving of one of the lectures in the *Illustrated Times* suggests that it may have been more expedient for the fashionably dressed women who also flocked to the lecture theater of the Museum of Practical Geology (fig. 6.2). It was here, rather than at the Geological Society, that Owen could defend Cuvierian correlation before the same kind of mixed audience that Huxley had originally condemned it to at the Royal Institution almost exactly a year earlier. Soon after the completion of the lectures, Owen was "arranging with John Murray to publish them," although, despite their manifest success as pedagogic spectacles, no such volume appeared.[116] What the distinguished audience at those crowded weekday afternoons in February, March, and April 1857 heard can nevertheless be partially reconstructed from the printed synopsis, newspaper reports, and private letters.

In the initial lecture Owen evidently spent some of the time considering the recent criticisms of correlation. On the following day, 27 February, Roderick Murchison told a correspondent:

FIGURE 6.2. *Professor Owen Lecturing at the Museum of Practical Geology, Illustrated Times* 4 (1857): 252. This wood engraving accompanying a newspaper report of Owen's lectures on fossil mammals in the spring of 1857 shows the large number of fashionably dressed middle-class women in his audience. Author's collection.

I never heard so thoroughly eloquent a lecture as that of yesterday. . . . It is the
first time I have had the pleasure of seeing our British Cuvier in his *true place*,
and not the less delighted to listen to his fervid & convincing defence of the
principle laid down by his great precursor. Every one was charmed, and he
will have done more (as I felt convinced) to render our institution favourably
known than by any other possible event.[117]

This introduction to the wide-ranging, twelve-lecture course on "Osteology
and Palæontology, or the Frame-work and Fossils, of the Class Mammalia,"
must have contained a great deal besides, but for Murchison it was Owen's
fervid defense of his esteemed precursor's principle that was its conspicuous
highlight.[118] There would, of course, have been at least one notable exception
to Murchison's blithe assumption that everyone was "charmed" by Owen's
"defence." And Huxley would have been still less happy had he realized that
the director of his own institution considered it his archantagonist's "*true
place*," and was directly linking the growth of the Government School of
Mines' public renown to Owen's ardent vindication of Cuvierian correlation.

It is not recorded whether Huxley was actually in the audience for Owen's
lecture, although, even if he opted to stay away, his own words were still heard
in the lecture theater of the Museum of Practical Geology. The section on
mammalia in Owen's entry on "Palæontology" for the *Encyclopædia Britannica* (1858) was based on his lecture series, and in it he observed of the late
attacks on the Cuvierian method:

> It has not only been asserted that the results of such determination are un-
> sound, but that the philosopher who believed himself guided by such law de-
> ceived himself and misconceived his own mental processes! But the true state
> of the case is, that the non-applicability of Cuvier's law in certain cases is not
> due to its non-existence, but to the limited extent to which it is understood.[119]

In the *Encyclopædia Britannica* a footnote informed readers that the precise
source of this fallacious criticism was "Prof. Huxley, 'Lecture on Natural History', &c., Royal Institution of Great Britain, Feb. 15, 1856."[120] Even without
such a printed elucidation, many of Owen's auditors would have been well
aware that he was repeating, and impugning, the very words that the School
of Mines' own professor had acknowledged might seem presumptuous exactly a year earlier.

Owen's second lecture, on 28 February, was on the "Earliest Known Forms
of the Class Mammalia," and, as the *Saturday Review* reported, he declared
that "the first man of science who grappled successfully with the great difficulties which this subject presented was Cuvier, who pointed out the intimate

connexion and mutual interdependence of the different parts of animals." To clarify this point,

> Professor Owen illustrated the relations which exist between the different parts of animals by a diagram which displayed the skull of a lion—the type of the Carnivores—and also his leg. He showed how the lion's teeth are formed—some for killing his prey, and some for cutting flesh—how his leg is armed with formidable claws. . . . So he proceeded through the whole of the limb, pointing out its adaptation to the structure of the head.[121]

Owen's choice of this leonine example was not only on account of its being the "type of the Carnivores." In the previous July's *Annals and Magazine of Natural History*, Huxley had used precisely the same instance when arguing that "the teeth of a lion and the stomach of the animal" did not constitute a "necessary physiological correlation, in the sense that no other could equally fit its possessor for living on recent flesh." The "number and form of the teeth might have been quite different . . . and the construction of the stomach might have been greatly altered, and yet," Huxley insisted, the "function of these organs might have been equally well performed."[122] As Owen's diagram (fig. 6.3) made clear to his audience, "even when the change" in another part of the creature's structure is "comparatively slight . . . we yet find a modification in the teeth," showing that, whatever Huxley's broad-brush approximations of the relationships in an animal's structure, the subtle correlations in the lion's finely balanced configuration were indeed necessary.[123] Owen was dismantling Huxley's arguments against Cuvierian correlation in both his Royal Institution lecture and his article for the *Annals and Magazine of Natural History* point by point.

Conclusion

The course of twelve lectures concluded at the beginning of April 1857 with a "grand party" organized by Murchison and attended by many of the older generation of gentlemen of science such as the astronomer Edward Sabine, as well as the grandees of metropolitan high society, including the recently returned African explorer David Livingstone and the widowed Lady Franklin. Owen reflected contentedly that the "success" of the lectures "exceeded my utmost expectations."[124] A few days after the party, the *Illustrated Times* summarized the main lessons imparted in the now completed course of "lucid lectures," observing that "we saw how . . . an intimate knowledge of anatomy could, from the data of the presence of a muscle, the form of a tooth,

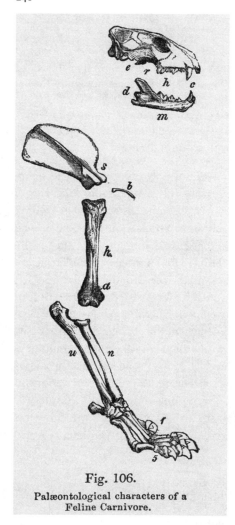

Fig. 106.

Palæontological characters of a
Feline Carnivore.

FIGURE 6.3. *Palæontological Characters of a Feline Carnivore.* Richard Owen, "Palæontology," *Encyclopædia Britannica*, 8th ed., 21 vols. (Edinburgh: Adam and Charles Black, 1853–60), 17:171. Owen used this illustration in his lectures at the Museum of Practical Geology to show the necessary correlation between teeth and feet in the lion's finely balanced configuration. © The Trustees of the Natural History Museum, London.

determine the habits, form, and locality of a creature of which little else remained."[125] Owen must have considered that he had decisively repulsed the previous year's attacks on the central axiom of Cuvierian paleontology, even taking the fight into the very lecture theater and museum that the brash antagonist of correlation considered his own territory. The self-styled "Jermyn St. Pet" had been neutered in his own backyard. Or so it must have seemed to Owen as he accepted the congratulations of the illustrious guests at Murchison's glittering end-of-term soiree.

But if Owen won this particular battle, he had decisively lost the war.

The very arguments in favor of empirical deduction rather than necessary correlation that he presumed had now been expunged had already gained the support, albeit expressed only in private letters at this stage, of Hooker, Darwin, and perhaps Lyell, and, as has been seen, would soon convince even Falconer. After 1856 Falconer ceased his erstwhile flattery of Owen, instead dubbing him "Dirty Dick" and expressing opinions of such vehemence that Huxley concluded only he could "hate him . . . more than good old Falconer does."[126] Spencer too would tie his colors to the mast of empiricism on "this vexed question in physiology" in an article for the *National Review* in October 1857.[127] The Cuvierian controversy initiated by Huxley at the beginning of the previous year had brought together many of the leading lights of what, by 1858, he was proprietorially calling "my scientific young England," providing an empirical and secular approach to organic structure to which they could consent alongside the equally naturalistic mode of species transmutation that Darwin began privately revealing to them at precisely this same moment.[128] In the principal upholder of necessary correlation, moreover, they found an opponent who, at least in Huxley's carefully constructed version of this "servile follower," seemed to embody the corrupt old world of scientific patronage and gentlemanly deference, and whom they would go on to challenge on several other fronts, from simian brain anatomy to museum politics.[129]

Within a few years almost the entire scientific community, with the exception of Owen and a small number of his acolytes, had shifted to the anti-Cuvierian position first outlined by Huxley. This rejection of necessary correlation extended, as will be seen in chapter 9, even as far as the nascent community of émigré paleontologists in Australia. When the disputes of 1856 and 1857 were looked back on at the end of the nineteenth century, it seemed evident that they were a resounding triumph for the young tyro who had taken up paleontology only after assuming responsibility for the Government School of Mines' fossil collection. As the American paleontologist Henry Fairfield Osborn observed in 1897: "Huxley unwillingly entered the field [of vertebrate paleontology], but soon found an opportunity of overthrowing Cuvier's Law of Correlation."[130] This deposition of the great French savant—a veritable paradigm shift in the conduct of paleontology—was both rapid and enduring, as well as geographically far reaching. Or, more precisely, it was among members of the elite scientific community where the tenets of scientific naturalism increasingly held sway. With other groups, however, the story was very different, and this, as the next two chapters will explore, had significant implications for the outcome of Huxley's campaign against correlation.

7

The Problems of Popularization

By the end of February 1857 Richard Owen's guest lectures on extinct mammals at the Government School of Mines were proving so popular that "nominal fees," as Roderick Murchison noted approvingly, had to be "placed on the admission to prevent the mob from entering."[1] Although the mass audience for whom Murchison felt such evident distaste were deliberately priced out, Owen's lunchtime lectures at the Museum of Practical Geology were attended not by specialist savants, for whom the timing was inconvenient, but by stylishly dressed middle-class women and members of the metropolitan glitterati (see fig. 6.2). These fashionable auditors, as was discussed in the last chapter, heard Owen authoritatively repulse Thomas Henry Huxley's recent attacks on Georges Cuvier's law of correlation. In March 1857, exactly halfway through Owen's series of twelve lectures in Jermyn Street, Huxley began his own series of Fullerian Lectures at the nearby Royal Institution on "The Principles of Natural History," which, the printed synopsis promised, would be "*illustrated by familiar examples.*"[2] This emphasis on familiarity had long been a stock feature of much popular science for women and children, which deployed the domestic setting and homely examples of what became known as the "familiar format."[3] While it is unlikely that Huxley's lectures, which were never published, adhered to all aspects of the familiar format, they were advertised as being of particular interest to the "Wives . . . and Sons and Daughters (under the age of Twenty-one) of Members" of the Royal Institution, who could gain entry to all four for a guinea.[4]

The final lecture of the series was on Saturday 4 April and treated "*The Vertebrata*—Fishes—Frogs—Lizards—Birds—Mammals."[5] Significantly, this was just a day after the concluding lecture, on Friday 3 April, of Owen's series at the Museum of Practical Geology in which he had regaled his own popular audience with illustrations of the successful operation of the law of correlation. On the same day as Owen's final lecture, as well as the celebratory soiree organized by Murchison, Huxley told Joseph Dalton Hooker that he would "hold up my head immensely to-morrow when (blessed be the Lord) I give my last Fullerian. Among other things, I am going to take Cuvier's crack case of the 'Possum of Montmartre as an illustration of *my* views."[6] Owen, as seen in the previous chapter, had used his lectures to magisterially dismantle Huxley's arguments against Cuvierian correlation point by point. While he seems to have touched only briefly upon the "opossum" and other "Mammalian life in the overlying beds of the Paris gypsum" in his fourth lecture on 6 March, Owen had long regarded Cuvier's famous discovery of an "Opossum, or true *Didelphys* . . . in the gypsum of Paris" as perhaps the most "striking . . . exemplification of the power of the principle which guided that great anatomist in the interpretation of fossil bones and in the reconstruction of extinct animals."[7] Now, barely twenty-four hours after Owen's triumphant closing lecture, exactly the same instance of correlative reconstruction would be scrutinized by Huxley in order to exemplify what he termed, with strident emphasis, "*my* views."

It is impossible to know exactly what Huxley's views of this particular "crack case" were, although it seems inevitable that he would have disputed Owen's claim that the celebrated prediction of the Parisian marsupial was achieved solely by means of necessary correlation. What is certain, though, is that by the early months of 1857 Owen and Huxley, whose overlapping lectures began, respectively, at 2:00 and 3:00 p.m., when most professional men were otherwise engaged, had opted to continue their bitter technical dispute over the method of paleontology before general—albeit predominantly middle-class and metropolitan—rather than specialist audiences.

Engaging with such nonspecialist auditors was by no means inconsequential to Huxley and other scientific naturalists in their campaign to discredit the famous Cuvierian axiom. While Huxley, as the previous chapter showed, decisively won over his fellow scientific practitioners, he was markedly less successful in persuading the rapidly expanding and increasingly affluent working- and middle-class audiences for science of the latter half of the nineteenth century. They were instead more likely to listen to self-conscious popularizers, whether penny-a-liner journalists, earnest Anglican and Presbyterian clerics, entrepreneurial publishers, or itinerant lecturers, who combined instruction with entertainment to make science an enticing commodity that was eagerly purchased by an unprecedentedly large and diverse set of consumers. The renowned feats of Cuvierian paleontological reconstruction had featured prominently in the initial upsurge of cheap science publishing in the 1830s and 1840s, and, crucially, they continued to be trumpeted in the new forms of popular science—lectures and museum displays, as well as books and journalism—that proliferated in the decades following Huxley's attacks in the mid-1850s.

The popularization of science, of course, ought not to be regarded as a simple top-down process of disseminating authorized enlightenment to a passive and grateful public, and elite practitioners such as Huxley had to be mindful of the success of even the most demotic of popularizers.[8] In fact, the ongoing currency of the law of correlation in popular scientific works posed a direct challenge to the wider cultural authority that Huxley and his colleagues aspired to. As Bernard Lightman has proposed, "Huxley wrote and lectured about science knowing all too well that the explosion of science writing by popularizers who were not practitioners presented some serious problems for realizing the agenda of scientific naturalism."[9] It was necessary to gain the support of the broader public if the scientific naturalists were to garner political and financial backing, as well as influence the agendas of educational institutions and the press. This was the beginning of the period of what Frank Turner has termed "public science," when men of science such as Huxley had to "justify their activities to the political powers and other social institutions upon whose good will, patronage, and cooperation" they depended.[10] Despite having quickly gained the backing of the elite scientific community, Huxley's crusade against Cuvierian correlation needed to be continued amid the numerous competing interests and complex politics of the mid-Victorian cultural marketplace. The triumphant outcome of the disputes of 1856 and 1857 notwithstanding, Huxley had no choice but to take up the cudgels once more.

However, as this chapter will show, after initially failing to win over the middle-class audience of his Fullerian Lectures at the Royal Institution in March and April 1857, Huxley increasingly shied away from repeating the ferocious attacks on Cuvier he had made in the specialist scientific press in his no less acerbic journalism for general periodicals such as the *Westminster Review*. More remarkably, when Huxley did finally broach the issue of correlation before a popular audience, in his lectures to workingmen, he dramatically changed his mind to actually acclaim the certainty of Cuvier's celebrated law. Significantly, the support for correlation that Huxley expressed, seemingly for strategic reasons, in his lectures to plebeian hearers actually came to shape some of his specialist statements on the same subject, and even, potentially, his own paleontological practices.

From the very beginning of the controversy he initiated in 1856, Huxley made the power wielded by inaccurate popularizers a central concern, but his own attempts to offer a more appropriate form of popularization were no less problematic, or inconsistent, than those of the most egregious commercial writers and lecturers. As a consequence, conventional assumptions about the power of the scientific naturalists to simply subjugate older scientific views, and to take the mid-Victorian public with them in their ostensible campaign to reform science and modernize society, become deeply problematic in relation to the law of correlation.[11]

Compilers, and Copyers, and Popularizers

Huxley, at least in the impecunious years following his return from the global voyages of HMS *Rattlesnake* in 1851, was no less subject to the same commercial imperatives as the most pragmatic popularizers. In fact, anxious at being considered a mere journalistic hack while earning his living from the *Westminster Review*, he had disdainfully distinguished himself, in 1854, from those who, on the basis of "mere book knowledge," indulged in "scientific speculations, without the discipline and knowledge which result from being a worker also."[12] "On Natural History, as Knowledge, Discipline, and Power" was the subject of his anti-Cuvierian Friday Evening Discourse at the Royal Institution two years later, and he expressed a similar exasperation at the "compilers, and copyers, and popularizers, and *id genus omne*" who "proceed in the study" of Cuvier only as far as certain passages from the "famous 'Discours sur les Révolutions de la Surface du Globe' . . . and no further," but on whose authority "it is handed down from book to book" that the great savant's reconstructions were accomplished by means of the law of necessary correlation.[13] Although by this time he no longer feared "exile on Grub Street," as James Secord has put it, having secured paying positions at both the Government School of Mines and the Royal Institution itself, Huxley continued to resent the influence of those who had never "performed an experiment or made an observation in any one branch of science," and instead substituted "spurious, glib eloquence" for "original research."[14] The still anonymous author of *Vestiges of the Natural History of Creation* (1844), now in its tenth edition, was the principal peddler of this sham science, though followed closely by the no less fallacious promoters of Cuvier and his purportedly correlative methods.

The same determination to demarcate practitioners from popularizers that characterized Huxley's ferocious reviews for the *Westminster* from 1854 to 1855 was maintained in the following year during the opening salvos of his campaign against Cuvierian paleontology. The situation, however, was now rather different from when he needed only to haughtily catalog the errors and "mere book knowledge" of *Vestiges* or George Henry Lewes's *Comte's*

Philosophy of the Sciences (1853). Huxley, as noted in the previous chapter, actually had little practical experience of working with fossils at this time, and was still yet to make a single original contribution to the study of prehistoric life. In the early and mid-1850s, Huxley "did not care for fossils," as he later acknowledged, and even by the end of the decade his approach to them, as Michael Collie has observed, "relied on others . . . his ideas coming mostly from his reading."[15] Ironically, Huxley's knowledge of paleontology was, at this time, no less dependent on books than that of either the author of *Vestiges* or Lewes.

Hugh Falconer, in his riposte to Huxley's Friday Evening Discourse, muddied the waters still further by conflating the maligned exponents of Cuvierian popularization with the leading experts in the field. In addition to Huxley's disrespectful treatment of Cuvier, Falconer remarked, "later palæontologists are brushed aside with still lighter consideration. They are *les moutons que suivent* 'the compilers, and copiers, and popularizers, and *id genus omne*'. It is some consolation to this *pecus ignobile* to reflect, that Professor Owen has been among their number."[16] Huxley's cavalier ploy of dismissively lumping together all the different categories of writers on Cuvier enabled the deliberately self-deprecating Falconer to include both himself and Owen among the farmyard menagerie of sheepish popularizers. In reality, no ovine characteristics had been ascribed to Owen, and Charles Darwin, as he told Hooker, thought it "a pity that Falconer ludged Owen into Huxley's 'id genus omne' & *not fair*." More significant, Darwin also "deprecate[d] the contemptuous tone of Huxley on compilers &c.," asking Hooker: "What is a lecturer but a compiler?"[17] Even Darwin seemed unsure of the precise difference between popularization and more appropriate scientific activities. The careful distinction between those who produced new knowledge and those who merely disseminated it established in his acerbic contributions to the *Westminster*, which Lightman has designated "Huxley's strategy for preserving his authority as, he hoped, a temporary resident of Grub Street," was beginning to unravel amid the heated disputes over correlation in the summer of 1856.[18]

Even in his supercilious contributions to the *Westminster*, Huxley recognized the necessity of educating the public on scientific matters, and approved of works of popularization such as Charles Kingsley's *Glaucus* (1855), whose "genial pages" were "without the least pretension to scientific lore" and instead accepted, with "conscientious devotion to the truth of Nature," the superior authority of actual practitioners.[19] Nor was Huxley averse to himself endeavoring to shape the public understanding of science, and lecturing to general audiences, especially once Owen's hugely popular lunchtime lectures at the Government School of Mines in the spring of 1857 prompted him to defend his attacks on Cuvierian correlation before the middle-class women and children who were the target audience for his Fullerian Lectures at the Royal Institution. But if Huxley's objective was to direct public opinion away from outdated and quasi-theological axioms upheld only by blind authority, and instead guide it toward the modern secular and naturalistic understanding of the world formulated by practitioners and experts, then his efforts were an outright failure.

Those Industrious "Ants of Science"

The extent of Huxley's inability to direct public opinion is made clear by the attitudes toward the law of correlation expressed in a range of popular writing on science from the two years following his Fullerian Lectures. Although Darwin found it so troubling, Huxley's dismissively catchall category of "compilers, and copyers, and popularizers" encapsulated perfectly the various facets of the career of John Timbs. During the 1830s the then editor of the *Mirror of Literature* pioneered an innovative journalistic practice of reprinting extracted passages from learned transactions and commercial science journals in a separate section entitled "Popular Discoveries in Science." It was this new practice, according to Jonathan Topham, that inaugurated the very concept of *popular science*, a term actually coined by Timbs.[20] Two decades later he was still employing the same technique as "one of those industrious 'ants of science' who garner facts and . . . adapt them for a wider circle of readers than they were originally expected to reach," compiling numerous annual volumes such as the *Year-Book of Science and Art*.[21]

In 1858 Timbs began another venture, *Curiosities of Science*, which "aimed at soundness as well as popularity" and whose "'Curiosities' bear the mintmark of authority." Among the intriguing extracts from authoritative sources compiled by Timbs was an account of Owen's now legendary prediction of the existence of a flightless bird in New Zealand in 1839, to which Timbs appended an explanatory footnote:

> According to the law of correlation, so much insisted on by Cuvier, a superior character implies the existence of its inferiors, and that too in definite proportions and constant connections; so that we need only the assurance of one character, to be able to reconstruct the whole animal. The triumph of this system is seen in the reconstruction of extinct animals, as in the above case of the Dinornis, accomplished by Professor Owen.[22]

Two years after Huxley's vigorous protests against the deadweight of authority in paleontological matters, and only a year following his popular Fullerian

Lectures on the subject, Cuvierian correlation was presented by Timbs in the form of a factual gobbet that readers might have their curiosity stimulated by, but whose veracity they were not encouraged to contest or even consider.[23]

The same year also saw the accuracy of Cuvier's celebrated axiom reaffirmed in other formats of the burgeoning genre of popular science. In January 1858, as Huxley, Hooker, and John Tyndall were considering starting a general *Scientific Review* of their own, Samuel Joseph Mackie began a "popular organ of a Science which has of late years advanced with gigantic strides, and which is daily attracting an increasing share of attention from all classes of society."[24] Mackie was a fellow of the Geological Society and it was his "earnest desire to popularize and to extend the noble science of geology without sacrificing, in any way, its proper dignity."[25] His new journal, the *Geologist*, would, he proposed, be "not only a welcome monthly visitor, but a cherished friend," with its "pages . . . contain[ing] such things as a beginner can comprehend."[26] The opening number caught the attention of Huxley's model popularizer Kingsley, who told Mackie that he had "read [it] with great pleasure and also great hope . . .—and am especially delighted to find that you aim at spreading a popular knowledge of Geology." With so many "artizans desirous of self-instruction, and ladies of rank desirous of instructing their children," the *Geologist*, Kingsley urged, would be "doing a noble work" if "a knot of wise men, and a clergyman as Mr. Brodie among them, would deign to explain to these people their puzzles."[27] Kingsley's proposal for how the *Geologist* might deferentially conduct its work of popularization fitted the mold of Huxley's agenda for maintaining the hierarchical authority of actual practitioners, with one conspicuous exception: the inclusion of a clergyman.

The scientific credentials of Kingsley's choice of an ecclesiastical geologist were indubitable. Peter Bellinger Brodie had been educated at Cambridge under Adam Sedgwick and was elected a fellow of the Geological Society in 1834, before establishing himself as a leading authority on paleoentomology with the publication of his *History of the Fossil Insects of the Secondary Rocks of England* (1845).[28] Brodie, though, was also a pious Warwickshire vicar and thus represented precisely the kind of parson-naturalist whose inclusion among the "wise men" of science Huxley, with his well-known antagonism toward what he called "parsonic influence" on scientific education, could not countenance.[29] To make matters worse, Kingsley's recommendation for the ideal candidate to explain the puzzles of geology to a popular audience was a confirmed Cuvierian.

In only his second contribution to the *Geologist*, Brodie declared that beginning with "disjointed fragments . . . Cuvier and Owen have restored . . . several extinct animals . . . so that magically, as it were, at the bidding of the comparative anatomist, each bone has assumed its right place, and . . . the external form and habits of the creature as living have been fairly demonstrated." This was a paraphrase of the famous passage from the introduction to the third volume of *Recherches sur les ossemens fossiles* (1812) in which, as Brodie would doubtless have recognized, Cuvier invoked the prophetic vision of the valley of dry bones in the book of Ezekiel (37:1–10). In case his autodidactic and juvenile readers were less adept at identifying scriptural allusions (or might possibly have intuited the somewhat irreverent tone, as discussed in chapter 1, of the French original), Brodie added a further theological gloss of his own. The reconstructive capacities of Cuvier and Owen were, he insisted, manifestations of a "wonderful power which God has thus given to man, as the reward of patient study and anxious thought, thus to rebuild the creatures of the past."[30] The ability to utilize the law of necessary correlation was, in complete contradistinction to Huxley's emphasis on quotidian common sense, a divinely bestowed power that enabled only a small elect to discern the underlying design in the natural world. A still more prophetic insight was, shortly afterward, attributed to the same paleontological elect by another parson turned popularizer, with John George Wood proposing in his best-selling *The Illustrated Natural History* (1859) that

> the observer can, in a minute fragment of a bone . . . read the class of animal whose framework it once formed part, as decisively as if the former owner were present to claim his property. . . . Whoever reads these hieroglyphics rightly is truly a poet and a prophet; for to him the "valley of dry bones" becomes a vision of death passed away, and a prevision of resurrection and a life to come.[31]

The law of correlation, in this idiosyncratic Anglican perspective, actually enabled the paleontologist to join the poet and prophet in intuiting the truth of the most fundamental Christian dogmas of resurrection and the heavenly afterlife. Clergymen such as Brodie and Wood were a significant constituency among popularizers of science in mid-Victorian Britain, and their prolific publications, as Lightman has shown, helped maintain a religious framework for scientific discussions that posed a direct challenge to the secularizing agenda of Huxley and his fellow scientific naturalists.[32]

This tension between theological popularizers and expert evolutionists was made especially apparent in the summer of 1859, when Paton James Gloag, a Church of Scotland minister, expounded "in a popular form" the most recent arguments for the harmony of geology and revelation in a concise treatise entitled *The Primeval World*. Shortly before the publication of Darwin's *On the Origin of Species* (1859), Gloag was particularly anxious at the spread of what he called, with evident distaste, the "theory . . . usually known in this country by the name of the *development hypothesis.*" This pernicious conjecture, he insisted, was "based upon mere assumptions and negative statements, and is wholly unsupported by a single scientific fact." The most telling confutation of such transmutationism, Gloag assured his readers, was that it was "in direct opposition to the deductions of comparative anatomy and physiology." After all, he observed:

> It is now an ascertained fact, that all the parts and organs of an animal are so joined together, and so dependent upon each other, that no change can take place on one of them without a corresponding change upon all the rest. . . . Change, for example, the teeth of a tiger into teeth resembling those of an ox, and, in order to [sustain] its existence, the entire form, and organs, and habits, and food of the animal would have to be changed:—a change so great as to amount to a new creation.[33]

This argument, that the holistically coordinated alterations required by the variation of any one part would be impossibly complex, was first articulated by Cuvier in the "Discours préliminaire" to *Recherches,* where it was deployed as an irrefutable bulwark against the transformism propounded by his Parisian contemporaries. With the so-called development hypothesis now finding favor with prominent men of science on the other side of the Channel, Gloag recycled Cuvier's line of reasoning, adding new instances of how such holistic changes could never occur. Significantly, Gloag likely borrowed the example of the tiger from Falconer's riposte to Huxley in the *Annals and Magazine of Natural History,* which had emphasized how a "pure typical digitigrade carnivore like the Tiger . . . rigidly fulfils the terms of the proposition [of the correlation of parts], and every one of the conditions set forth as involved in it."[34] Only months ahead of the *Origin's* publication, the arguments and examples of Cuvier's defenders in the disputes of the mid-1850s were being appropriated by theological popularizers to deny that the mutability of species was actually viable.

Working the Public Up for Science

Although, as seen in the previous chapter, Huxley would soon become extremely particular about the exact title of his first institutional appointment, he had initially given himself the derisive soubriquet of "*non-official* maid-of-all-work in Natural Science to the Government." His salaried post at the Government School of Mines enabled him to turn his back on hack journalism and instead, as his son Leonard later asserted, "devote most of his time to original research." In March 1858 Huxley proudly told his sister Lizzie of his break with the *Westminster*: "I used at one time to write a good deal for that Review. . . . But I never write for the Reviews now, as original work is much more to my taste. . . . Article writing is weary work, and I never do it except for filthy lucre."[35] This state of affairs accorded exactly with his earlier strategy of demarcating expert practitioners such as himself from those who merely disseminated, in popular form, the knowledge they had acquired from books. But the other strand of his strategy, the shaping of public opinion regarding science by such expert practitioners, had proved more refractory, especially when it came to Cuvierian paleontology. With commercial compilers such as Timbs and theological popularizers such as Brodie, Wood, and Gloag all continuing to endorse and celebrate the law of correlation through 1858 and into the following year, Huxley seems to have felt compelled to adjourn his original researches and once more take on the wearying labor of writing for the benighted reviews.

Less than a month after exulting to his sister at having turned his back on the *Westminster*, Huxley accepted the need, as he put it to Hooker, for "one or two of us working the public up for science through the *Saturday Review*," and he agreed to contribute "a scientific article . . . once a fort-night" to the notoriously waspish weekly.[36] Another opportunity for similarly working the public up arose in January 1859, when Whitwell Elwin, editor of the *Quarterly Review*, attempted to commission a contribution from Huxley on a particular topic of the editor's own choosing: "Cuvier & his services to natural history."[37] Like Brodie, Wood, and Gloag, Elwin was another clergy-man concerned to facilitate what he called, in an earlier *Quarterly* article, the "spread of popular science." He differed from his clerical brethren, however, in insisting that popularization ought to be the province of scientific authori-ties who can "adapt the intricacies of science to general apprehension." In a similar manner to Huxley's vituperative contributions to the *Westminster*, Elwin lambasted the "work of persons who, having themselves learnt Natural Philosophy in six lessons, profess to teach it in half-a-dozen—who fill their small phials from another's bottle and adulterate what they steal."[38]

In his editorship of the *Quarterly*, Elwin again seemed to concur with Huxley's hierarchical attitude to science popularization, telling him, "I prize very much the articles which are written upon great subjects by those who are thorough masters of them. Indeed they are the only essays for which person-ally I care two pence." Still more temptingly, Elwin appeared to give Huxley free rein as to how he might approach the subject of his prospective article:

You would be able to treat this question in a way which nobody else could do. What Cuvier was in himself as a naturalist, & his relation to what preceded & what has come after him,—this is a theme of which the public knows nothing & young enquirers probably not much. The repetition of views that you have broached elsewhere is no disadvantage. The only thing of real consequence is that the sketch should be as true, perspicacious, & complete as it can be made under the circumstances.[39]

With the *Quarterly* offering a generous sixteen pounds per sheet, easily surpassing the "12 guineas per sheet (i.e. 16 pages)" he was initially paid at the *Westminster*, Huxley must have considered Elwin's proposal an irresistible opportunity not only to pocket some filthy lucre but also to avail readers of the notoriously Tory journal, which had shifted only to a more moderate conservatism under Elwin, of his own view of the problematic nature of Cuvier's legacy.[40]

Huxley certainly seems to have quickly begun working on Elwin's commission, jotting in his notebook early in 1859: "Whewell's *History of Scientific Ideas*, as a Peg on which to hang Cuvier article."[41] Contributions to the *Quarterly* had still nominally to be reviews of recent books, and with Elwin regularly advising contributors on "the peg on which to hang an article," his editorial influence seems evident in Huxley's approach to writing about Cuvier.[42] The particular peg on which Huxley would hang his article was an abridged reissue of *The Philosophy of the Inductive Sciences* (1840), which the Cambridge polymath William Whewell had published a year earlier, in 1858. In it he had repeated the original's contention that when the

> general scheme and mode of being of an animal are known, the expert and profound anatomist can reason concerning the proportions and form of its various parts and organs, and prove in some measure what their relations must be. We can assert, with Cuvier, that certain forms of the viscera require certain forms of the teeth, certain forms of the limbs, certain powers of the senses.

Whewell, who had received advice from Owen when writing the original, also maintained that the "vast extinct creation which is recalled to life in Cuvier's great work, the *Ossemens Fossiles*," was the profoundest exemplification of the "doctrine of Final Causes" in scientific reasoning, being based on an assumption that the harmonious relation among the different parts of an organism had been designed with a particular end in view. And Cuvier's "extraordinary discoveries," Whewell blithely asserted, were "universally assented to by naturalists."[43] This was, of course, far from being the case even when it had first been published in *The Philosophy of the Inductive Sciences*, but now, eighteen years later, Whewell's refusal to amend or update any of his earlier judgments must have seemed breathtakingly remiss to Huxley. By repackaging such obsolete opinions as if they were the views of contemporary men of science, Whewell's *History of Scientific Ideas* was little different from the defective and pernicious works that popularizers such as Timbs and Brodie also brought out in 1858. Huxley's pen must have dripped with venom as he set to work on his *Quarterly* article.

Oddly, though, Elwin seems not to have bargained on the kind of copy that Huxley would inevitably deliver on the subject of Cuvier, and apparently was not actually aware of the precise nature of the views he had broached elsewhere, or even his characteristically caustic tone. Less than a month after his invitation to Huxley, Elwin condemned the "'Saturday Review' people" for the "vulgarity & flippancy of a large part of the articles," insisting "there will be a reaction against the Saturday Review before long," as its reliance on the "seasoning of abuse . . . palls and disgusts after a while."[44] Perhaps unsurprisingly, the *Quarterly*'s pages never played host to the proposed "Cuvier article" by the *Saturday Review*'s chief science correspondent, and it remained one of the few significant periodicals of the time to which Huxley, who later quipped that he was as "spoiled" by editors "as a maiden with many wooers" and would "remain as constant as a persistent bigamist," did not contribute.[45] It seems likely that the article, whether it was ever completed or not (and no manuscript appears to exist), suffered a fate akin to another contentious contribution by one of Huxley's closest friends.

Like Huxley, Herbert Spencer had been similarly commissioned by Elwin a year earlier, in March 1858, although, as he later reflected ruefully, the article he had "written for the *Quarterly Review* was not accepted by the editor." Spencer could only assume that "possibly its conceptions . . . did not harmonize . . . with his theological system," even though the article, on "Physical Training," was "not conspicuously evolutionary in its doctrines."[46] Early in the next year Elwin would have been particularly alert to anything even remotely evolutionary, having been shown proofs of Darwin's *Origin* by the *Quarterly*'s publisher John Murray and finding that "all kinds of objections & possibilities rose up in my mind." Elwin raised numerous "objection[s] to the publication of the treatise," most crucially its apparent "absence of . . . proofs."[47] A friend of Elwin's later recalled of his opinions at this time: "Darwin he used to speak of as a man who had worked out no theory, but merely put forward a wild hypothesis." Significantly, the friend then added: "Huxley . . . he cheerfully placed in a very back row for ability."[48] Having responded with such circumspection to the proofs of the *Origin*, Elwin would presumably have had similar reservations about a no less naturalistic onslaught against Cuvierian correlation by an associate of Darwin whose intellectual powers he considered even more rash and deficient.

At the end of 1859 it was again suggested that Huxley might contribute to the *Quarterly*, with Murray recommending him as a potential reviewer of the *Origin*. Elwin, however, quickly vetoed the proposal. The task, he told his publisher, required a "really competent & *impartial* enquirer but I have some reason for thinking that Huxley is not that impartial person." Elwin was reluctant to inform Murray of the particular reason for his misgivings, although he was willing to enlighten another rejected reviewer of the *Origin*, stating: "I will tell Hooker my reason for this suspicion when I write to him."[49] Frustratingly, this particular missive seems never to have been written, as Elwin, soon after, "could not use my pen . . . in consequence of a blow on my right arm from the wing of a swan."[50] It was nevertheless almost certainly Huxley's antagonistic attitude to Cuvier that Elwin considered so offensively partisan, precluding him, like Spencer, from the *Quarterly's* circumspect pages, where the job of reviewing the *Origin* was finally given to the hardly unprejudiced bishop of Oxford, Samuel Wilberforce.

Spencer responded to Elwin's rejection by pragmatically sending his problematic paper to the Nonconformist *British Quarterly Review*, where it appeared a year later, in April 1859. Huxley, on the other hand, seems not to have submitted his Cuvier article to any other periodicals, even to what he still called "my favourite organ, the wicked *Westminster*," despite the weary work he had already invested in it and his readiness to generate filthy lucre.[51] He was perhaps mindful of William Benjamin Carpenter's advice—presumably applicable also to Cuvier—that a "critique of Owen . . . would scarcely suit the pages of the 'Westminster', and would be much more suitable to some journal more specially concerned with the subject."[52] Whatever the reasons, Huxley's conspicuous failure to replicate in the *Westminster* or the *Quarterly* (or even his regular contributions to the *Saturday Review*) the intellectual precision and rhetorical force of the anti-Cuvierian lectures and specialist articles that had largely won over his fellow practitioners in 1856 and 1857 left the field open to theological and commercial popularizers, who, regardless of any caveats raised by experts, continued to extol the law of necessary correlation.

Something That People Can Read

Huxley was far from silent on the subject of correlation, though. Over the next decade he repeated his original charges against the famous Cuvierian axiom on at least three further occasions: in a paper at the Geological Society in 1860 on the extinct South American ungulate macrauchenia; in *Lectures on the Elements of Comparative Anatomy* (1864), where his brusque attack on "Cuvier['s] . . . mistaken doctrines as to the methods of palæontology" was commended, as seen in the last chapter, by Falconer; and in *An Introduction to the Classification of Animals* (1869), which again lambasted "Cuvier['s] . . . servile imitators," who continued to adhere to his "methods of palæontology against the conclusions of logic and of common sense."[53] In the latter two works, both of which were based on a series of lectures he gave at the Royal College of Surgeons in the spring of 1863, Huxley extended his attack on necessary correlation to also encompass Cuvier's celebrated natural system of classification, in which the *embranchements* of vertebrates, articulates, mollusks, and radiates were founded, as was adumbrated in *Le règne animal* (1817), on their constituent creatures' distinct functional adaptations to the conditions necessary for existence. The four *embranchements*, in other words, represented each of the systems of possible correlations available in nature.[54]

For Huxley, "classification . . . acquires its highest importance as a statement of the empirical laws of the correlation of structures; and its value is in proportion to the precision and the comprehensiveness with which those laws . . . are stated." The accuracy and therefore the usefulness of any classificatory system was predicated on how precisely the laws of correlation were comprehended. Above all, as Huxley had been urging since the mid-1850s, it needed to be "carefully borne in mind, that, like all merely empirical laws, which rest upon a comparatively narrow observational basis, the reasoning from them may at any time break down." Without the underpinning of functional correlations that were necessary and a priori, Cuvier's system of classification, even though it had "taken such deep root," could be considered neither unimpeachable nor natural. In *Lectures on the Elements of Comparative Anatomy*, Huxley eagerly pointed out "how far" his own empirical understanding of classification "departs from that embodied in the opening pages of 'Règne Animal,'" avowing that the "Cuvierian *Radiata* is, in my judgement, effectually abolished."[55] Huxley's invalidation of necessary correlation simultaneously undermined the whole edifice of Cuvierian classification.

At the close of the 1860s the *Popular Science Review* reflected that "it is strange in how many lecture-theatres the old Cuvierian classification still flaunts in diagrammatic form on the walls, and how many teachers, in defiance of the intelligence of their pupils, cling to the barbarous *omnium gatherum* it expresses."[56] Huxley's criticisms of this enduring classificatory system were in sore need of reaching a wider audience. The combined attacks on necessary correlation and the natural system of classification that Huxley published in the 1860s, however, remained inaccessible—or simply incomprehensible—to most nonspecialist readers. His *Introduction to the Classification of Animals*, for instance, was intended "to be useful as a text book to lecturers, and students," and cost six shillings, a relatively hefty price for a slim volume of only 130 pages.[57] As for his *Lectures on the Elements of Comparative Anatomy*, one nonspecialist reader, Emma Darwin, declared of it, "I don't call that a Book. . . . I want something that people can read," prompting her husband to suggest that Huxley should instead "write a 'Popular Treatise on Zoology.'" Responding curtly, "Tell Miss Emma that my last *is* a book," Huxley confided to Darwin:

> I wish I could follow out your suggestion about a book on Zoology. . . . But I assure you that writing is a perfect pest to me unless I am interested—and not only a bore but a very slow process—I have some popular lectures . . . which have been half done for more than a twelvemonth & I hate the sight of them.[58]

Such disenchantment with the banalities of writing was a familiar rhetorical posture among mid-Victorian literary professionals, men of science as much as journalists and novelists, and Huxley's self-consciously jaded missive to Darwin, echoing his earlier complaint that "article writing is weary work," does not negate the integral role that writing played in his scientific practice (fig. 7.1).[59] However, while Huxley both continued and extended the anti-Cuvierian campaign he had begun in the mid-1850s into the following decade, and indeed apparently could not desist from again and again broaching the issue of correlation, his continuing reluctance to write popular accounts of his scientific researches, whether in general journals such as the *Westminster* or in the kind of demotic textbooks favored by Emma Darwin, meant that he did so only in a narrow range of specialist forums and publications. This left ill-informed teachers and popularizers, as the *Popular Science Review* implied, to continue peddling an outdated understanding of classification and animal structure.

Hard-Handed Fellows Who Live among Facts

If Huxley considered the prospect of writing popular works, or still worse seeing them through the press, an unendurably tedious and time-consuming "bore," his letter to Darwin suggests that he was more amenable to addressing nonspecialist audiences if it could be done verbally. The Fullerian Lectures in the spring of 1857 were, as seen earlier, "*illustrated by familiar examples*" that explicated what Huxley called "*my* views" on Cuvier's purported use of necessary correlation. The audience for these familiar lectures, though, was largely confined to the guests or relations of the Royal Institution's members, and

FIGURE 7.1. Theodore Blake Wirgman, *Thomas Henry Huxley*. Pencil, 1882. National Portrait Gallery, London. Despite his hyperbolic claims to find it tiresome, writing was an important component of Huxley's scientific practice. © National Portrait Gallery, London.

Huxley, as he told Frederick Dyster, had grown "sick of the dilettante middle class." Instead, he wanted the "working classes to understand that Science and her ways are great facts for them," and determined "to try what I can do with these hard-handed fellows who live among facts."[60] From the mid-1850s Huxley gave regular courses of lectures to London's artisans and autodidacts as part of the program established in 1852 by Henry De La Beche at the Government School of Mines. Crucially, these lectures, part of a general endeavor to ensure the inclusion of the working classes in the improving agenda of science in the wake of the Chartist uprisings of the 1840s, were held after the end of the working day rather than in midafternoon. On weekday evenings in Jermyn Street's Museum of Practical Geology, Huxley spoke in a "theatre [that] holds 600, and is crammed full," and he soon found that the "fellows are as . . . intelligent as the best audience," enabling him to make them "participants in my train of thought—not . . . shove information down their throats as if they were Turkeys to be consumed."[61] Having already expressed his contempt for popular science writing in the *Westminster Review*, Huxley now avowed: "*Popular* Lectures I hold to be an abomination unto the Lord." His own addresses to the masses were instead to be titled "People's Lectures," in which he would eschew the stolid style of mere popularizers, who, constrained by their reliance on books rather than practical experience, could offer only derivative didacticism and force-fed indoctrination.[62]

Yet when he came to tackle the vexed issue of correlation in front of the massed ranks of eager plebeian hearers, Huxley assumed precisely the same celebratory and almost reverential tone as the very popularizers he most derided. The "People's Lectures" that he so vehemently distinguished from vulgar "*Popular* Lectures" in fact represent an extraordinary departure—even a complete volte-face—from his customary statements on Cuvier's correlative methods in specialist journals and lectures to more elevated audiences. In the autumn of 1858 Huxley delivered the last of the School of Mines' annual series of Lectures to Working Men on "Objects of Interest in the Collection of Fossils." The sixpence tickets for the lecture, as the *Builder* reported, had been "disposed of to *bonâ fide* workmen, and it is impossible to speak too highly of the attention and intelligence of the audience." Reaching the conclusion of the lecture, Huxley addressed his attentive auditors directly, telling them that he was "most desirous not to be misunderstood" on a subject on which he had "allowed myself a longer interval for reflection" than usual.[63]

Like other mid-Victorian science lecturers who successfully maintained the interest of working-class audiences, Huxley generally spoke without a prepared script, relying only on brief notes.[64] His practice, as he later recalled, was to "make notes on two or three or four slips of note paper" and "leave the words to come at call while I am speaking."[65] But such was the significance of this particular peroration that he departed from his customary impromptu manner, as "contrary to my wont, I have written down in full, and will read, what I have to say." The carefully elaborated lesson that Huxley read aloud was that the "many justifications" of scientific methods of reasoning "teach us what implicit and absolute faith we may place in the conclusions of the human intellect." Foremost among recent examples of the remarkable powers of man's unfettered intellect was the nascent recognition that "successive races *must* have proceeded from one another in the way of progressive modification." Still more provocatively, Huxley added, "And I *do* include man in the same category as the rest of the animal world," before going on to observe:

> Such doctrines are supposed to be antagonistic to religion, or rather, to be opposed to certain traditions handed down to us with our religious beliefs. . . .
> If it be *really* true that science is opposed to religion, all I can say is, so much the worse for religion. . . . For science, and the methods of science, are the masters of the world.

Huxley was boldly apprising his exclusively plebeian audience of the profound implications of evolution, ideological as much as intellectual, more than a year ahead of the *Origin*'s publication.

Other workingmen beyond those crammed into the lecture hall in Jermyn Street were soon alerted to Huxley's incendiary claims when the weekly artisanal trade journal the *Builder* was "urged by several of his hearers that the publication of these remarks in our columns would be useful."[66] The report, which appeared in January 1859 under the title "Science and Religion," also attracted the attention of more genteel readers unaccustomed to paying the fourpence cover price, with Darwin telling John Lubbock: "I shd. much like to see the Builder, with Huxley, which shall be returned."[67] The interest generated, among readers new and old, by the *Builder*'s account of Huxley's contentious yoking of support for evolution with an attack on the "bigotry" of those who "wear the tonsure of a priest" inevitably meant that the other instances he used of the "sufficiency of the faculties with which man is endowed, to unravel . . . the mysteries by which he is surrounded" went largely unnoticed.[68] But to any of the *Builder*'s readers—or, still less likely, listeners to the original lecture—who were aware of the tenor of Huxley's specialist contributions to the study of fossils over the previous few years, they would nevertheless have been striking.

As key examples of the "winning of every new law by reasoning from ascertained facts" and the "verification by the event, of every scientific prediction," Huxley listed:

> Donati's comet lately blazing in the heavens above us at its appointed time; the first quiver which betrayed to the anxious watcher of the telegraphic needle on the other side of the Atlantic, that an electric current would follow, even under such strange conditions, the laws which man's wit and industry had discovered; the bone which, laid bare by Cuvier's chisel, justified his trust in the law of organic correlation which he had discovered.[69]

Having disdainfully lambasted "those who, guided more by authority than by right reason" mistook the invariability of a particular empirical supposition as a necessary axiom and "denominated it Cuvier's law of correlation" two years earlier in the pages of the *Annals and Magazine of Natural History*, Huxley now assured his artisanal auditors that the very same "law of organic correlation," which Cuvier had apparently discovered and was fully justified in trusting, was actually as unerring as the fundamental principles of physical sciences such as astronomy and physics.[70]

When, three years later, an enterprising reporter taking shorthand notes at a subsequent series of the School of Mines' Lectures to Working Men asked

Huxley "to allow him, on his own account, to print those Notes," he "will-ingly accede[d] to this request," even though he had "no leisure to revise the Lectures, or to make alterations in them."[71] Huxley's readiness to permit the publication of this unofficial transcript of his lectures *On Our Knowl-edge of the Causes of the Phenomena of Organic Nature* (1863), which appeared in weekly parts under the imprint of the popular science publisher Robert Hardwicke, suggests that he similarly had no problems with the *Builder's* own unauthorized extract from an earlier lecture, a cutting of which he kept among his private papers.[72] In fact, Huxley voiced none of the concerns about being misrepresented that, by the 1880s and 1890s, came to blight his relations with certain sectors of the popular press, and his only regrets at this time were financial, with the unexpected success of his serialized lectures making him "lament I did not publish them myself and turn an honest penny by them as I suspect Hardwicke is doing."[73] But if the *Builder's* report is taken as an accu-rate (if similarly unremunerative) record of the original lecture, the question remains why Huxley, who always vaunted his prioritization of intellectual truth over all other considerations, so dramatically altered his attitude toward Cuvierian correlation when addressing a plebeian audience.

That the peroration was read from a carefully prepared script suggests that this seeming endorsement of the law of correlation was not merely an instance of when, as Huxley later reflected, the "strange intoxication which is begotten by the breathless stillness of a host of absorbed listeners weakens the reason and opens the floodgates of feeling," inducing "what bitter reflec-tions" in the lecturer "when the report of his speech stares him in the face next morning."[74] Nor was it a repentant retraction of the fierce views that had prompted the controversy with Falconer and Owen in 1856, for in the year following the *Builder's* account of his apparent approval of "Cuvier's . . . law of organic correlation," Huxley was once more regaling the specialist audi-tors of the Geological Society with a new "refutation of the doctrine that an extinct animal can be safely and certainly restored if we know a single impor-tant bone or tooth."[75] Huxley, it would seem, was treating the topic of cor-relation in different ways according to the educational and social background of his audiences.

Huxley's principal aim in his scripted peroration was to win a fair hear-ing for evolution among his artisanal auditors, and in order to achieve this, he needed to align it with other scientific principles that, at least to an audi-ence more usually addressed by popularizers than practitioners, had been established as entirely indubitable. The comet first observed by Giovanni Battista Donati in the summer of 1858 had quickly become a standard subject of popular science writing. Mary Ward, in her *Telescopic Teachings* (1859), observed that the "beautiful spectacle of Donati's comet has awakened a new interest in these singular bodies," and also proposed: "We may fairly assume that all those who read these pages have seen the great comet of 1858." On the issue of the uncannily accurate predictions of the Italian astronomer, Ward addressed her autodidactic audience directly, asking: "Reader, do we trace a slightly incredulous expression in your countenance, as you read these statistics, so confidently given? We cannot now enter into any demonstration of their truth, but shall merely say that this comet in its subsequent journey, strikingly verified the mathematical theory of its motion."[76] Although Ward, a prolific female popularizer who benefited from her aristocratic family ties to the astronomer William Parsons, declined to discuss the exact proofs of Donati's forecast of the comet's trajectory, her popular readership were assured that they could take its remarkable character on trust.[77] As seen earlier with Timbs's *Curiosities of Science* (whose 1858 edition included an insertion noting that "while this sheet was passing through the press, the attention of astronomers, and of the public generally, was drawn to the fact of the . . . Comet . . . first discovered by Dr. G. B. Donati"), popularizers similarly presented Cuvier's axiom of correlation as a settled fact whose veracity could not be challenged.[78] For Huxley, anxious to win the ear of the working classes on the subject of evolution, the popular renown of these two putatively perfect laws, and the celebrated predictions they had facilitated, enabled him to imply that the hugely contentious theory of "progressive modification" was actually no less certain and incontrovertible, even if this entailed appearing to affirm the very law of correlation that he had so vociferously disputed in more specialist forums.

Calculated to Impress the Vulgar Imagination

Entering the Museum of Practical Geology, a building styled on an Italian palazzo and symbolizing middle-class respectability by closely resembling the gentlemen's clubs in nearby Pall Mall, most of the artisans who paid sixpence to hear Huxley's lecture on "Objects of Interest in the Collection of Fossils" would have headed straight through the resplendent entrance hall to the lecture theater at the back.[79] However, at least some of this exclusively plebeian audience would, on other occasions, have already ascended the grand staircase to enter the actual museum on the first floor (fig. 7.2). Since 1856, the Museum of Practical Geology had designated Monday to Wednesday as "public days" and offered free admission between 10:00 a.m. and 4:00 p.m. This policy had quickly almost doubled attendance, with 13,055 having visited in the whole of 1855, and then 2,200 coming through the doors in each

FIGURE 7.2. *The Museum of Practical Geology—The Great Hall, Illustrated London News* 18 (1851): 446. From 1856 the public was admitted to the museum from Mondays to Wednesdays, and thousands of men and women of all classes, as with the middle-class families depicted here, took the opportunity to enter its grand premises and view the specimens accumulated by the Geological Survey. Reproduced by permission of the University of Leicester.

of the first months of the following year.[80] By the mid-nineteenth century, especially following the Museums Acts of 1845 and 1850, museums were increasingly seen as another important route, alongside popular science writing and workingmen's lectures, for bringing rational recreation to the laboring classes. The Museum of Practical Geology was just one of many scientific institutions, both in London and elsewhere, that made their collections more accessible to an eager public.[81]

The collections housed in the Museum of Practical Geology had been accumulated since the 1830s by the Geological Survey, and in its fashionable Mayfair location the museum had constantly to balance the requirements of specialist researchers studying its various rocks, minerals, and maps with accessibility to casual visitors. Huxley's employer, the Government School of Mines, was also located within the museum's grand palazzo, a circumstance that Sophie Forgan and Graeme Gooday see "shaping the direction of his scientific work" in this period.[82] His specialist publications for the museum, no less than his workingmen's lectures, certainly reflect its uneasy balance between research and recreation, showing that it was not just Huxley's attempts to communicate with new audiences for science that were importantly shaped by the continuing currency of Cuvierian correlation among popularizers.

With personal guided tours such as those given by Owen at the Hunterian Museum in the 1830s and 1840s (see fig. 3.3) increasingly unfeasible after midcentury, all types of museum visitors relied on printed books, whether exhaustive catalogs or inexpensive guidebooks, to help them understand and interpret the collections.[83] The Museum of Practical Geology recognized the need to cater to both ends of the market, with Roderick Murchison pledging in early 1856 that a "catalogue of the fossils in the Museum has been commenced, and is in an advanced state; and a popular catalogue is in due course of preparation."[84] While the avowedly popular catalog, Robert Hunt's *A Descriptive Guide to the Museum of Practical Geology* (1857), duly appeared in the following year, it would be almost another decade before its specialist equivalent, *A Catalogue of the Collection of Fossils in the Museum of Practical Geology* (1865), was finally published. This massive tome, which had been continually delayed by the sheer amount of specimens to be cataloged, was a collaboration between the Geological Survey's paleontologist Robert Etheridge and Huxley, with the latter also contributing an "Explanatory Preface" summarizing the present state of paleontology.

This "Explanatory Preface" had actually been drafted by at least December 1857, when the "proof Sheets" were read by Darwin, who commended it as "simply the very best Resume by far, on the whole science of Natural History, which I have ever seen" (although he was less effusive behind Huxley's back, telling Hooker it was "admirable, but very brief").[85] With the *Catalogue* growing increasingly extensive and detailed as ever more important specimens came into the museum, the eight-year hiatus between the initial composition of the prefatory essay and its eventual publication affords a reason why it often seems out of sync with the specialist *Catalogue* it ostensibly introduces. Although Hunt, at the beginning of his *Descriptive Guide*, warned that the "volume must not be mistaken for a Catalogue of the Museum," there were much better reasons why Huxley's "Explanatory Preface" to the specialist paleontological *Catalogue* might itself be mistaken for a popular introduction to the fossil collections.[86]

Huxley certainly gave meticulous consideration to the particular structure and style necessary for such an introduction, proffering detailed suggestions on the matter to the curator of the Warwickshire Natural History Society's geological collections. Explaining, in October 1859, that "I am sorry to say that I can as yet send you no catalogue of ours. . . . Only the introductory part of my catalogue is written," Huxley then advised:

If I may make a suggestion I should say that a catalogue of your museum for popular use should commence with a sketch of the topography and stratig-

raphy of the county, put into the most intelligible language. . . . After that I
think should come a list of the most remarkable and interesting fossils, with
reference to the cases where they are to be seen . . . and under the head of each
a brief popular account of the kind of animal or plant which the thing was
when alive.[87]

The recipient of Huxley's epistolary recommendations was none other than
Peter Bellinger Brodie, the same clergyman and popularizer who, as seen earlier, had emphatically endorsed the religious credentials of the law of correlation in the *Geologist* only a year before receiving Huxley's letter. Huxley, as
Paul White has shown, was willing to enter into pragmatic coalitions with the
same clerics whose continuing cultural authority he elsewhere lamented, and
his conspicuously friendly advice to Brodie indicates another crucial aspect
of his attempts to render his own "Explanatory Preface" suitable for "popular
use" by what he called "person[s] of intelligence, unversed in science."[88] As
with his 1858 workingmen's lecture on "Objects of Interest in the Collection
of Fossils," Huxley again adopted a very different attitude toward the vexed
issue of correlation than he had in more unambiguously specialist writings
and lectures, once more appearing to change his mind on an issue that, as
seen in the last chapter, was a crucial barometer of support for the broader
agenda of scientific naturalism.

Applauding the "great law of the invariable correlation of organic peculiarities," and insisting that the "magnificent researches of Cuvier first prac-
tically . . . showed that the laws of correlation of parts . . . hold good to a
wonderful extent among the extinct forms," Huxley went as far as to propose:

> If the morphological law which expresses this invariable co-existence, or cor-
> relation, of organic peculiarities has been as regularly verified by our experi-
> ence as the astronomical law [that the earth turns from west to east], we may,
> for all practical purposes, reckon as securely upon the one relation as upon
> that of the other.[89]

Like the unerring astronomical principle that ensured the accuracy of the
calculation of a comet's elliptical orbit made by Donati, the still more fundamental law of the direction of the earth's rotation was posited as potentially
equally reliable as the Cuvierian axiom of correlation, even if the ultimate
arbiter of both tenets was a posteriori experience of their invariable operation. As with the workingmen's lecture in which he invoked Donati's famous
prediction, Huxley once again followed Cuvier and subsequent generations
of his popularizing disciples in aligning the French savant's anatomical doctrine with the fixed mathematical laws of physical sciences such as astronomy
and physics.

With more space than in the succinct peroration to his workingmen's lecture, Huxley's "Explanatory Preface" did caution that "we should be led into most erroneous conclusions by reasoning without hesitation from these data," and urged that "we must . . . be content to regard . . . the great laws of the construction of animals . . . as only approximatively correct." But such mild caveats were easily missed amid the hyperbolic adjectives and general tenor of approbation for correlation—with its "great law," "magnificent researches," and "wonderful extent"—that led Huxley to eulogize Cuvier's "famous gypsum quarries which furnished so many occasions for the display of his genius and knowledge." The discovery in 1804 of an "opossum" in the quarries at Montmartre was presented as "one of the most remarkable examples of . . . successful prediction." In fact, as Huxley observed approvingly, "Cuvier was so confident . . . that he invited some friends to witness the picking away of the stone from the region where he believed the marsupial bones would be found; and the result verified his expectation."[90]

Later, in his specialist *Lectures on the Elements of Comparative Anatomy*, Huxley would tetchily demythologize exactly the same instance of Cuvier's renowned predictive abilities, insisting that

> deductions . . . such as that made by Cuvier in the famous case of the fossil opossum of Montmartre . . . are well calculated to impress the vulgar imagination; so that they have taken rank as the triumphs of the anatomist. But it should be carefully borne in mind, that . . . the reasoning from them may at any time break down.[91]

The jubilant tone in which Huxley recounted the particular anecdote in his "Explanatory Preface" was entirely at odds with this censorious bluntness, and instead more closely resembled that of Owen in his *History of British Fossil Mammals* (1846), where he related how Cuvier "called together a few friends . . . and, predicting the result of his operations, commenced the removal of the Montmartre fossil . . . and soon brought into view the . . . precise form and proportion of those of the Opossums." This celebrated episode, according to Owen, was a "striking . . . exemplification of the power of the principle which guided that great anatomist," an opinion that Huxley's quotation of Cuvier's own view that it was a "very singular monument of the force of zoological laws" gave every impression he concurred with.[92] If the famous case of Cuvier's discovery of the Parisian opossum was, as Huxley acknowledged, "calculated to impress the vulgar imagination," then it was a strategic calculation that, at least in his popularizing "Explanatory Preface," he was prepared to make himself.

Huxley's correspondence with Darwin suggests that he drafted much of the "Explanatory Preface" in the summer and autumn of 1857.[93] Notably, this was only months after the concluding Fullerian Lecture on "The Principles of Natural History," given on 4 April, in which, as seen earlier, Huxley used the "crack case of the 'Possum of Montmartre" to contrast what he called "*my views*" on correlation with those of Owen in his own series of lectures on fossil mammals. Huxley was apparently willing to give, almost simultaneously, entirely contradictory accounts of exactly the same concepts and events, both savaging and then confirming Owen's laudatory account of the role of correlation in Cuvier's discovery of the opossum. In drafting his overtly popularizing "Explanatory Preface," it was acceptable, perhaps even necessary, for Huxley to conceal the profound reservations over Cuvierian correlation he had articulated in other forums.

Exact Scientific Writings

But, crucially, by the time the *Catalogue of the Collection of Fossils in the Museum of Practical Geology* that Huxley's "Explanatory Preface" introduced was eventually published in early 1865, it was much more detailed and considerably longer—at nearly 450 pages, categorizing over thirteen thousand extinct species—than had originally been envisaged. To meet the increased production costs, the *Catalogue* was priced at four shillings. The popular *Descriptive Guide*, by contrast, cost a mere sixpence, and whereas Hunt's pocket-size manual had already sold out two editions of five thousand copies and would soon require a third, Huxley and Etheridge's bulkier tome was never reprinted. Huxley still considered that, as he told Darwin when sending him a copy, it was "not much worth your reading," proposing instead that Darwin's daughter Henrietta might "act as taster" and "read the explanatory notice & give me her ideas thereupon." It is unlikely, though, that a "young lady" without such a wealthy or well-connected family would have had similar access to Huxley and Etheridge's hefty *Catalogue*, which was reviewed exclusively in specialist geological journals rather than in general periodicals.[94]

The *Catalogue* had, as a relieved Murchison acknowledged when it finally came out, "long been a desideratum" for researchers using the Museum of Practical Geology.[95] The president of the Geological Society, William John Hamilton, agreed, urging in his annual address that the volume "must find its way into the hands of every British geologist."[96] Tellingly, when, three decades later, Michael Foster and E. Ray Lankester collected together Huxley's *Scientific Memoirs* (1898–1902) following his death in June 1895, they included the "Explanatory Preface" among these "exact scientific writings," which were considered as clearly distinct from "his more popular writing" that "might safely be entrusted to the usual agencies of publication" offered by commercial publishers.[97] Whatever Huxley might have intended when he began drafting it in the mid-1850s, his "Explanatory Preface" was received as an unequivocally specialist publication by most of its readers.

This, of course, meant that the decidedly more sanguine approach to correlation that, for a variety of reasons, Huxley adopted in his attempts at popularization during 1857 and 1858 came to inform what, by the middle of the following decade, was no longer considered as merely popular exposition. The demarcations between popularization and practice that Huxley had sought to maintain since he began reviewing for the *Westminster Review* in the early 1850s were never quite as clear-cut as he had initially supposed. Both Darwin and Falconer, as seen earlier, had questioned the young tyro's distinction between the production and dissemination of knowledge, and now, by the 1860s, the putative change of heart on correlation that was apparently reserved only for his popular writings could not be prevented from leaching across into publications regarded as part of Huxley's more specialist oeuvre. As he later acknowledged when reflecting on the pitfalls of popularization, "I have not been one of those fortunate persons . . . who keep their fame as scientific hierophants unsullied by attempts . . . to be understood of the people."[98] His apparent endorsement of the law of correlation in his attempts at popularization certainly adulterated the precision of the anti-Cuvierian arguments he had made in the methodological disputes of the mid-1850s, preventing Huxley's technical criticisms from gaining recognition beyond the small cadre of experts and fellow scientific naturalists whose support he had already secured.

What Huxley perceived as the defilement of having to adapt his tone and attitude in order to be "understood of the people" might nonetheless have also had productive consequences for his work as a self-styled scientific hierophant. It was, after all, only during 1858 that Huxley, whose tenacious participation in the disputes on paleontological method belied his very limited practical experience, finally took charge of a consignment of fragmentary fossil bones sent to the Museum of Practical Geology from northern Scotland. The most immediate of the new "practical problems Huxley faced," as Michael Collie has suggested, "was how to study *incomplete* fossil remains." Collie has made a detailed examination of Huxley's paleontological practices when working with these reptilian fossils from the Morayshire coast near Elgin, and he concludes that "what Cuvier recommends [in the "Discours préliminaire" about reconstructing the whole from a part] is what Huxley did. He applied the principles enunciated by Cuvier to the Elgin fossils" as the "best method he knew at the time." Never having visited Elgin and with only limited knowledge of comparable prehistoric reptiles, Huxley, Collie avows, had "to implement the Cuvier method of comparative anatomy . . . in the strictest, most disciplined way imaginable."[99]

It is striking that Huxley's emergence as a strict and disciplined Cuvierian in his actual paleontological practice, at least in Collie's account, only occurred once he had already amended his erstwhile antipathy toward correlation when addressing the nonspecialist implied readers of his draft "Explanatory Preface" and the artisanal hearers of his lecture on "Objects of Interest in the Collection of Fossils." Just as the expectations of such popular audiences induced Huxley to adopt a different, markedly more positive, attitude toward Cuvier's famous axiom, so the demands of actually working with perplexingly incomplete fossil remains compelled him to bring that new approach into his paleontological practice. Huxley's skillful identification of the crocodilian characteristics of the Elgin fossils, during which, as Collie claims, his "complete confidence in the methods he adopted . . . was fully vindicated when he realised that large conclusions could indeed be drawn from fragmentary evidence," earned him the role of secretary of the Geological Society.[100] Intriguingly, the particular approach to the fragmentary remains that occasioned this prestigious appointment seems to have been shaped by the equivocations and compromises necessitated by his attempts to negotiate the requirements of popularization and to be "understanded of the people."

Huxley's work on the Elgin fossils, as well as his travails with the inadvertently specialist "Explanatory Preface," afford an especially conspicuous instance of how, as Roger Cooter and Stephen Pumphrey have contended, "popularization . . . reconfigures the nature of science itself," demonstrating the existence of what Ralph O'Connor has termed "influence in both directions between elite and nonelite practices, including the often underappreciated processes by which popular science shapes the practice of elite science."[101] O'Connor's acknowledgment that such processes are "often underappreciated" suggests that historians have been too ready to follow the self-serving assumptions of Huxley himself that the two activities were entirely distinct. Instead, they need to recognize, in line with Darwin and Falconer as much as recent historians such as James Secord, that such sharp distinctions between practice and popularization, and the "making and communicating of knowledge," are untenable.[102]

Conclusion

It is usually assumed that Huxley was a hugely effective communicator of science, even, as John Carey has proposed, the "greatest Victorian scientific popularizer."[103] At the end of the 1860s, however, Huxley reflected on the potential for "spreading a knowledge of science among the people," and conceded ruefully: "I have long looked upon any spreading of science from above downward as utterly hopeless."[104] His attempts to articulate his intense opposition to the law of correlation in ways that were appropriate for nonspecialist audiences were certainly far from successful. Ironically, Huxley had made fallacious popularization one of the defining issues of his anti-Cuvierian campaign from its very beginning in February 1856. In the following months and years, though, he increasingly found himself unable—and often tetchily unwilling—to provide a more worthy alternative. While a survey of *Journalistic London* (1882) later reflected that Huxley's pellucid prose was such a prized commodity that "editors of periodicals . . . will pay him anything to write for them," his antagonistic opinions of Cuvier were vetoed from the pages of the *Quarterly Review* by Whitwell Elwin, and, seemingly impotent in the face of the editor's embargo, he declined to again broach the subject in general periodicals for several decades to come.[105]

This uncharacteristic reticence in the press coincided with a still more remarkable willingness to tailor his approach to suit the expectations of audiences more usually addressed by popularizers than practitioners, and actually reverse, at least temporarily, his vaunted condemnation of Cuvierian correlation. Notably, Huxley only committed this strategic volte-face when engaging in popularization under the aegis of the Museum of Practical Geology, an environment where, as Forgan and Gooday have suggested, his authority and independence were constrained by larger institutional demands.[106] Whatever the local factors that might have been involved, Huxley's conspicuous failure as a popularizer of his anti-Cuvierian argument in the late 1850s and early 1860s had important consequences for perceptions of the law of correlation that would last for the rest of the nineteenth century.

One of these corollaries was a widespread misunderstanding of Huxley's own distinctive position on the issue. To the astronomer and popularizer Richard Proctor, for instance, Huxley seemed to be no less of an adherent of the celebrated Cuvierian axiom than even his archopponent Owen. In *Other Worlds Than Ours* (1870), Proctor amalgamated their very different stances, asserting that "a great naturalist like Huxley or Owen can tell by examining the tooth of a creature belonging to some long extinct race, not only what the characteristics of the race were, but the general nature of the scenery amongst which such creatures lived." Notwithstanding their ferocious clashes over paleontological method during the preceding decade and a half, Huxley and Owen shared a faith, according to Proctor, in reconstructive procedures that were so infallibly perfect, as well as necessitating such exalted powers of genius, that they were analogous to how "a single grain of sand or drop of water must convey to the Omniscient the history of the whole world of which it forms a part."[107] While reviewers often complained that Proctor was prohibitively prolific and churned out books strewn with errors and misunderstandings, his evident misapprehension of Huxley's and Owen's respective attitudes toward the reliability of Cuvierian correlation was not necessarily the result of mere haste.[108] Such confusion was instead the inevitable consequence of Huxley's own obfuscation of his actual opinions in both overtly popular as well as more specialist publications, and Proctor, as will be seen in later chapters, was far from alone in this misunderstanding.

But, for Huxley, it was not just the large and diverse constituency of popularizers who continued to endorse Cuvier's famous axiom, and sometimes mistakenly portrayed Huxley himself as an enthusiastic advocate of it, that presented a problem. Even those who actually concurred with Huxley's denunciation of necessary correlation could be no less—and perhaps still more—of a hindrance to his attempts to render a secular and naturalistic understanding of the world acceptable both to his fellow men of science as well as to broader public opinion, as the next chapter will show.

Unfortunate Allies

In early 1856 Thomas Henry Huxley was preparing the Friday Evening Discourse "On Natural History, as Knowledge, Discipline, and Power," which he would soon deliver, as seen in chapter 5, before a fashionable audience at the Royal Institution in Mayfair. Across London at exactly the same time, an elderly, one-eyed Scottish anatomist was completing a translation of Henri Milne-Edwards's *Cours élémentaire d'histoire naturelle* (1841–42) in his shabby lodgings in Hackney. Despite their contrasting circumstances, this hack translator, whose meager income came largely from his writing, would have agreed fervently with the scornful words then being penned by the ambitious and well-connected young lecturer. As he recounted six months later:

> I accidently met with a passage or two in the Manual of Zoology, by Milne Edwards, (a work I have recently translated), which showed me that there is a class amongst modern zoologists, rigid followers of M. Cuvier, who believe that the anatomical method invented and applied by him to the classification of the existing fauna, and to the restoration of the extinct, was actually elevated by him, first, to the dignity of an exact science—almost geometrical: and, second, that he, the great anatomist, trusted to it solely or mainly in his celebrated restorations of the extinct species. . . . I was not aware, until I translated the valuable work . . . that the doctrine had been received unreservedly.[1]

This was precisely the same complaint that Huxley made at the Royal Institution against the "compilers, and copyers, and popularizers" whom he also viewed as overly rigid disciples of Cuvierian orthodoxy. Indeed, only a month before the above passage was published in the *Lancet* in August 1856, Huxley had likewise argued that "if the master's words be studied carefully, it will be discovered that his followers are more Cuvierian than Cuvier."[2] Along with their mutual antipathy for the French savant's slavish devotees, Huxley also had a personal connection with the impecunious translator, having been trained at Charing Cross Hospital under Thomas Wharton Jones, who had previously served as the elderly Scot's assistant when he had been Edinburgh's most successful extramural anatomy lecturer three decades before.[3] Huxley's favorite tutor had long since forsaken his erstwhile employer, though, and it would have been unthinkable for Huxley to acknowledge that his opposition to Georges Cuvier's law of correlation was shared by the disgraced and infamous Robert Knox.

Knox's once glittering reputation had never recovered from his involvement in the so-called West Port murders of 1828, when, unwittingly so he claimed, he had purchased cadavers for dissection in his anatomy classes from two Irish immigrants posing as grave robbers, William Burke and William Hare. While body snatching from cemeteries was rife in Edinburgh, Burke and Hare had expedited the process by simply murdering their vulnerable and impoverished victims and then delivering their bodies to Knox's lecture theater. Impressed by the unusual freshness of the corpses, Knox's assistants, including Huxley's tutor Jones, had generously remunerated the two killers, encouraging them to provide more, and eventually purchasing the cadavers of sixteen murder victims.[4]

After Hare turned king's evidence and Burke was hanged in January 1829, Knox was officially exonerated of any deliberate involvement. Suspicions of complicity nonetheless clung to him. The influential "Noctes Ambrosianæ" column in *Blackwood's Edinburgh Magazine* alleged that "Dr Knox stands arraigned at the bar of the public. . . . He stands, at this hour, in the most hideous predicament in which a man can stand—in that of the *suspected* accomplice or encourager of unparalleled murders."[5] He was, the *Medical Times* later reported, "hourly searched for by the rabble . . . and it was only by the utmost precaution that Knox escaped Lynch law at the hands of the populace, or being hanged across a lamp-post." For months the streets of Edinburgh resounded with cries of:

> Hang Knox, Burke, and Hare,
> Burn the College and Surgeons' Square.
> Burke's the murderer, Hare's the thief
> And Knox the butcher who bought all the beef.[6]

Even if he had not connived in the murders, the Burke and Hare scandal tarnished Knox's professional reputation for the rest of his career. The "stigma cast upon Dr. Knox by the West Port atrocities," as his devoted student Henry Lonsdale later acknowledged, "damaged his name and destroyed his prospects."[7] By the end of the 1830s, another—albeit less lurid—scandal had barred Knox from lecturing in his native city, and in 1842 he moved to London.[8] Living in the obscurity of Hackney, Knox was still unable to find employment as an anatomy lecturer, and he relied on translations and journalism to make ends meet.

Even before his ruinous involvement in the Burke and Hare scandal, Knox had already affronted many among the Edinburgh medical establishment with the discourteous belligerence of his lectures, in which he advanced Étienne Geoffroy Saint-Hilaire's transcendental understanding of anatomy and sarcastically impugned all teleological explanations of organic structure. His outspoken lectures, as the *Medical Times and Gazette* reflected, made no attempt to disguise just how much "Dr. Knox loathed teleology, or the utility principle."[9] Taking his lead from Henri de Blainville, with whom he had studied in Paris, Knox particularly derided the paleontological claims, based on teleological final causes, of Cuvier's British followers in his lectures during the mid-1820s. He then continued his campaign into the next decade in the militant medical journal the *Lancet*.

Significantly, the fiery and avowedly atheistic Knox recommenced his prolonged onslaught against the alleged excesses of Cuvierian functionalism, once more in the *Lancet*, in the summer of 1856, just as Huxley was himself making similar criticisms in more respectable publications such as the *Annals and Magazine of Natural History*. With Knox living in poverty and having no institutional affiliations, his writings were, as Evelleen Richards has observed, "overshadowed by his notoriety and questionable 'morality,'" especially once his provocative views on the inevitability of racial conflict were adopted by an aggressive and self-consciously iconoclastic clique within the Ethnological Society who later designated themselves the Cannibal Club.[10] These were precisely the kinds of incendiary associations that the emergent scientific naturalists, endeavoring to purge secularism and evolution of their earlier connotations of scurrilous political radicalism and immorality, needed to avoid.[11]

Already struggling in his attempts to challenge popularizers who still espoused correlation, Huxley had also to urgently distinguish his own particular strain of anti-Cuvierianism from Knox's older, but much less reputable, version of the same argument. Although Knox's attacks concurred almost entirely with Huxley's own denunciation of the principle of necessary correlation, even employing the same examples, line of argument, and rhetorical tone, he presented no less of a problem—and actually perhaps still more of one—than the ranks of popularizers who, as seen in the last chapter, continued to endorse Cuvier's famous axiom. A further complication was that while the Cannibal clique within the Ethnological Society, soon to secede and form the rival Anthropological Society, adhered closely to Knox's views on race, its members were, paradoxically, at odds with him on correlation. Knox's protégé Charles Carter Blake emerged as a particular thorn in Huxley's side with his ardent defense of Richard Owen's paleontological methods. Even as Huxley's opposition to the law of correlation was steadily winning over much of the elite scientific community, the interventions of the one-eyed hack translator tainted by the Burke and Hare scandal and now laboring in a Hackney garret threatened to jeopardize the entire anti-Cuvierian campaign. His paradoxical supporters in the Anthropological Society, meanwhile, constituted a nascent scientific faction who, throughout the 1860s, fiercely resisted both Huxley's authority and the broader agenda of scientific naturalism.

Cuvierian Mania

In his final years in Edinburgh, as attendance at his extramural lectures slumped, Knox supplemented his income by producing annotated translations of Continental texts on anatomy. In 1839 Thomas Wakley commissioned him to translate the initial installment of Blainville's *Ostéographie* (1839–64) for the *Lancet*. In his anatomical lectures Knox was notorious for "always saying the truth without reference to the feelings or prepossessions of his auditory," and for his "sly, satirical wit," and he quickly adapted to the combative and often vituperative tone of Wakley's campaigning weekly.[12] Even in a translation, Knox regularly interposed notes and addendums that conformed with the *Lancet*'s distinctively irascible character as well as enabling him to advance his own heterodox and self-consciously controversial outlook. In fact, the *Lancet* advertised its forthcoming translation of Blainville's new work as effectively a collaboration between "two of the most learned and accomplished anatomists of France and Scotland," with the "English form, edited by Dr. ROBERT KNOX, of Edinburgh . . . very greatly enriched by remarks and annotations from the pen of that celebrated anatomist, the accumulation of twenty years of observation and research."[13]

As Cuvier's successor at the Muséum d'histoire naturelle, and, since their personal estrangement in 1816, his most prominent and persistent critic in France, Blainville had lost little time in denouncing the law of necessary correlation, with its apparent "degré de prévision" from "qu'un seul os," in the opening number of his part-issued *Ostéographie*.[14] In Knox's translation this appeared as:

> To believe that the science of osteography is sufficiently advanced, or can ever be so, to enable us ever to reach such a degree of prevision — . . . that a single

bone, or single facette of a bone, being examined, it would be possible for any person to reconstruct or pourtray the whole skeleton . . . is to encourage a pretension which will appear the more exaggerated and the more extraordinary in proportion as any one may himself have studied the science, as well *à priori* as *à posteriori*. In my opinion, no one has ever made good this pretension.[15]

To this, Knox appended a fierce note adding his own opinion. He commended the "common sense and truth again vindicated by the powerful pen of M. De Blainville" in his attack on the "Cuvierian mania." Significantly, Knox also observed that it was "in Britain, where almost universally Cuvier's writings have been substituted for the book of Nature, that we find the best specimens of the manifold absurdities which are so happily criticised by M. De Blainville."[16] Knox's translation and its accompanying note were published in the *Lancet* on 9 November 1839. The distinctively British "Cuvierian mania" he diagnosed must have seemed verified when, just three days later, Richard Owen announced to the Zoological Society, as discussed in chapter 3, that "so far as a judgement can be formed of a single fragment," he was "willing to risk" his entire "reputation" on the prediction of the past existence of an unknown struthious bird in New Zealand.[17]

Having previously castigated Knox as the notorious "receiver of the bodies of the sixteen unfortunate creatures butchered at Edinburgh," the *Lancet* now strategically avoided any allusion to its translator's awkward entanglement in the scandalous events in West Port a decade earlier.[18] Instead, it warmly recommended Knox to readers on the grounds that "in 1821–2, he studied in Paris, in communication with DE BLAINVILLE, CUVIER, and the elder GEOFFROY (ST. HILAIRE), on terms of daily intercourse, such as is rarely permitted in that country to foreigners." Even now, the *Lancet* pointed out, he "continues an intimate acquaintance" and is the "constant correspondent of M. DE BLAINVILLE."[19] Knox dated the emergence of his skepticism regarding correlation to precisely the time he spent in Paris and his growing friendship with the *Ostéographie*'s author (which had quickly superseded his initial reverence for Cuvier), and he related how he had soon brought such anti-Cuvierian sentiments back with him to Scotland. "As early as the year 1821," he avowed,

> I denounced the pretensions . . . of the Cuvierian school and its followers, as being without any foundation in truth; being, in fact, absolutely absurd. Subsequently,—in a series of lectures on comparative anatomy, delivered here [i.e., in Edinburgh] to large and distinguished classes from the year 1825 to 1828 . . .—I endeavoured, to the utmost of my power, to stem the torrent of assertion, amounting to absolute nonsense, which had set in, in this country, on the subject.[20]

Knox's lectures propagating the opposition to Cuvier's functionalist account of anatomical structure he had imbibed in Paris—which seemingly concluded in the very same year that his nefarious transactions with Burke and Hare came to light—were largely extemporized, and, with the fees paid by his audience of up to five hundred students affording a lucrative income, he had no need to publish them (fig. 8.1). The only permanent trace of Knox's reasoning in the course of lectures he gave each summer between 1825 and 1828 appeared in the *Transactions of the Royal Society of Edinburgh* exactly a year after Burke's public execution.

In January 1830, in a paper on the idiosyncratic stomach of the Peruvian llama, Knox proclaimed that the "high authority which would persuade us, that from a small portion of bone we may determine the form, the anatomy, the natural history, the antiquity of an unknown animal, I altogether disregard." This "seeming neglect of such well-earned reputation" was predicated,

FIGURE 8.1. *Robert Knox.* Line engraving, circa 1830. Knox, here in the act of lecturing on the structure of the hand but positioned so as to make his missing left eye less apparent, expounded his vehement opposition to the law of correlation in his lectures on comparative anatomy in Edinburgh. Wellcome Trust, London.

he claimed, on the "strong conviction," acquired during his own "anatomical inquiry into the structure of almost every kind of animal," that all the "possible combinations of form have not been fully determined." As such, to "declare *à priori*" that any particular correlation of structure was so certain, or even necessary, "as not to admit of refutation, or at least of doubt" was "not to use a harsher style of criticism, eminently imaginative and fantastic." Alongside the Peruvian llama, Knox gave a further instance of the inability of fossil evidence to afford absolute knowledge of a creature's anatomical structure, observing that

> the molar teeth of bears are not carnivorous molar teeth; and it is by the observation of the living species only that we have become aware of the frugivorous habits of some, and of the strictly carnivorous habits of the polar species. To speculate from such facts as these as to the anatomy and natural history of the extinct Ursus spelæus, must, to every reflecting mind, appear exceedingly ridiculous.

It was just such caveats, Knox acknowledged, that he had "been in the habit of alluding to . . . in my summer course of lectures on comparative anatomy."[21] In the audience for at least some of these lectures was Hugh Falconer, who had come to Edinburgh in 1826 to study medicine after graduating from the University of Aberdeen, and whom Lonsdale later included among "those who occupied Knox's benches" (Falconer gained his MD in 1829 so likely dissected bodies supplied by Burke and Hare in Knox's anatomy classes).[22] Ironically, three decades later, and long after Knox had become persona non grata even to his former students, Falconer was compelled to argue that precisely the same ursine example did not refute the Cuvierian axiom of necessary correlation after all.

Falconer might well have recalled Knox's crowded summer lectures of the mid-1820s when, in the spring of 1856, he read the printed version of Huxley's Friday Evening Discourse at the Royal Institution. In it, Huxley asked:

> What difference exists in structure of tooth . . . between the herbivorous and carnivorous bears? If bears were only known to exist in the fossil state, would any anatomist venture to conclude from the skull and teeth alone, that the white bear is naturally carnivorous, while the brown bear is naturally frugivorous? Assuredly not.[23]

Huxley implied that this particular example of brown and white bears was a case that Cuvier himself had chosen to exemplify the precision of correlative methods, but Falconer knew better, insisting that the "case in question is *not* that of Cuvier's selection," and exasperatedly maintaining "How then is Mr. Huxley warranted in asserting, that the Bears were '*the case of Cuvier's own selection*'?"[24] Huxley's apparent sleight of hand, as Falconer may have recognized, strategically distanced the example from what seems to have been its actual source, Knox's 1830 paper in the Royal Society of Edinburgh's *Transactions*, although Falconer, even in his ill-tempered response to Huxley, likewise avoided any mention of his disreputable former tutor. Philip Rehbock has observed of Knox that "because of the stigma attached to his name . . . scientists may have been reluctant to cite him as an authority," and this certainly seems to have been the case in 1856 with both Huxley and Falconer.[25] While steering clear of any direct reference to Knox, the contretemps with Huxley in 1856 gave Falconer the opportunity to, finally, offer a riposte to his vehemently anti-Cuvierian lecturer of thirty years earlier, and, contra both Knox and Huxley, he contended that it was possible "in the cases of certain extinct fossil bears, to form conclusions as to their habits of diet upon the osteological evidence."[26]

Although it could not be acknowledged, the parallels between Knox and Huxley's opposition to correlation went much further than their mutual insistence that the differences between brown and white bears were not discernible from fossils. Knox, as Evelleen Richards has proposed, "was clearly influenced by secular naturalism. . . . He was concerned with subjecting the whole of nature and society to the sway of natural law and opposing such naturalistic or 'scientific' explanations to traditional theological modes of explanation." Richards suggests Knox be included among the exponents of "scientific naturalism," the very label that Huxley came to use to describe his own position.[27] Neither Knox nor Huxley had any faith in final causes or divine design, and both considered empirical observation the sole means of attaining reliable knowledge of organic structures. They recognized that, with regard to Cuvier, this position might suggest a "seeming neglect of such well-earned reputation" (Knox) or "appear to be not a little presumptuous" (Huxley), but nonetheless they contended that his "master-mind misconceived its own processes" (Huxley), allowing the "anatomist [to] deceive himself" (Knox).[28]

Despite his own reliance on hack journalism, Knox also shared Huxley's contemptuous distaste for the "compilers, and copyers, and popularizers" who traded merely on their "spurious, glib eloquence" and further exacerbated Cuvier's own misconceptions.[29] Knox lambasted such "literary pirates," alleging that

> carefully excluding from their compilations all elevated and correct views of science, they have, by their anecdotic and quasi-popular style, contributed to debase the works of the most eminent zoologists to such an extent, that the . . . researches of Cuvier . . . can scarcely be recognised. Their views are anti-scientific and anti-educational.[30]

Even the belligerent manner in which they presented these opinions had conspicuous parallels, with a critical notice of Knox in the *Athenæum* noting the "Doctor's sledge-hammer style," and Falconer, in a private letter, similarly telling Huxley, "You write with a sledge-hammer."[31] Huxley's examples, line of argument, and rhetorical tone in the disputes over necessary correlation in 1856 all had distinct echoes of Knox's invectives against Cuvier from the 1820s and 1830s.

Viciously Aiming Arrows of Withering Scorn

In neither his Friday Evening Discourse to the Royal Institution nor his rejoinder to Falconer in the *Annals and Magazine of Natural History* did Huxley acknowledge any antecedents for his objections to Cuvier's law of correlation. Not even Blainville warranted a single mention. Regardless of this, Owen had privately alluded to the "advocacy of . . . De Blainville by some of our young teachers, in letters in our 'Annals of Natural History'" during his imperious response to Huxley's provocations.[32] He subsequently shifted his position in a confidential letter to the Council of the Royal Society, expressing his incredulity that Huxley seemed not to recognize that the arguments on which his purported "refutation" of Cuvier's method was based "have been propounded before . . . shortly after the decease of the great Reconstructor of extinct forms." This had been by "Prof. de Blainville," and in revealing the "fallacy" of Huxley's attacks, Owen wearily insisted, "one repeats the refutation to which the original assailant of Cuvier had exposed himself."[33]

Huxley instead presented his outspoken views as emerging, seemingly sui generis, from a rigorous and disinterested scrutiny of the facts of natural history. He had "examine[d] this principle [of necessary correlation]" by, first, "taking . . . one of Cuvier's own arguments and analyzing it" and then "bringing other considerations to bear." This scrupulously impartial approach, even if it risked contradicting the time-honored views of established authorities, afforded a "gymnastic for the intellectual" as well as the "moral . . . faculty," showing that the "moral faculties of courage, patience, and self-denial, are of as much value in science as in life."[34] Such assertions regarding paleontological method were an early version of the doctrine, articulated more explicitly in Huxley's 1866 lecture "On the Advisableness of Improving Natural Knowledge," that "scepticism is the highest of duties; blind faith the one unpardonable sin," and that "natural knowledge, in desiring to ascertain the laws of comfort, has been driven to discover those of conduct, and to lay the foundations of a new morality." It was precisely this new moral code based on scientific skepticism, supplanting Reformation dogma by insisting on "justification, not by faith, but by verification," which underpinned the claims of Huxley and other scientific naturalists to exercise a wider cultural authority in Victorian Britain.[35]

The only problem was that this avowedly moralistic scientific agenda, at least in its initial anti-Cuvierian incarnation, had unmistakable correspondences with the equally naturalistic views of Knox, whose entanglement in an infamous homicidal scandal had long cast considerable doubt on the ethical claims of scientific practitioners. *Blackwood's* had labeled Burke and Hare's crimes "experimental murders" on account of Knox's apparent complicity, while another contemporary account of the scandal, *The Murderers of the Close* (1829), observed that "'science' had, somehow or other, mixed itself with the most dreadful criminality that ever disgraced the name of man."[36] Nor was Knox's own strain of scientific skepticism much less problematic, with the *Medical Times* observing in 1844 that "Knox may have been too free in expressing his contempt for . . . professors of Christianity," and as a consequence "some have not hesitated on the score of his private conversation, to denounce Knox as an atheist, Deist, infidel, &c."[37] By the 1850s Knox's pungent secularism was also accompanied by an increasingly outspoken radicalism, which was imbued with the disenchanted cynicism he had felt since his fall from grace in Edinburgh. He vehemently denounced British colonial policy, all forms of philanthropy and reformist legislation, especially on sanitation, and even attempts to prohibit prostitution, on the grounds that it restricted the right of women to use their bodies as they wanted.[38] Knox advocated a "savage radicalism," as an obituary in the *Medical Times and Gazette* later put it.[39]

The pressing need to earn money from his writing often compelled Knox to rush such impetuous opinions into print, to the dismay of many reviewers. The *Athenæum* complained in 1852 of "our meeting Dr. Knox so often now in literature, whom we used to hear of only as a lecturer," remarking that "whoever reads a few pages . . . will feel that Dr. Knox is wrong-headed . . . pouring the vials of his wrath over all that authority has established."[40] Knox occasionally included Cuvier among the authorities he denounced in his increasingly scattershot invectives. In *Great Artists and Great Anatomists* (1852), he determined to "assail the reputation of Cuvier," lambasting the "law of the co-relations of structure" and insisting that the "facts of anatomy do not offer many fitted to form the basis of *a priori* reasoning."[41] The *Literary Gazette* noted the book's "abuse of everybody and everything jumbled together," and accused Knox of "viciously aiming arrows of withering scorn or savage ridicule at men of renown." It observed in particular that it was Knox's "intention to disparage Cuvier" and rebuked his "unreasonable sneers," suggesting that the "Doctor is too hardened a scribe . . . to sit down seriously and earnestly to the working out" of the revered savant's actual words and arguments.[42] The implication was clear: the disgraced penny-a-liner's haughty attacks on Cuvier were both laughably inappropriate and lacking in any scientific rigor. Worst of all, they were tastelessly couched in the sensational language of hack journalism.

Huxley must have been aware of the risks of being similarly perceived by the press as indulging in mere scorn and derision rather than proper science, especially as the *Literary Gazette*'s anonymous notice was penned by his close friend Edward Forbes.[43] It was, after all, Forbes's advice and canny institutional maneuvering, according to Adrian Desmond, "which had made the public Huxley" in the early 1850s.[44] Forbes, who, like Falconer, was a former student of Knox's anatomy school, had even introduced Huxley to paleontology by having him appointed his replacement at the Government School of Mines after he moved back to Edinburgh in May 1854. His untimely death six months later was a grievous loss to Huxley. Forbes, as the *Literary Gazette*'s editor reflected, "was a bold thinker, but he used a guarded pen; and a cautious style became natural to him," although the same style had not yet rubbed off on his less diplomatic protégé.[45] Less than two years after Forbes's death, it was Huxley's own "attack upon Cuvier," in 1856, that, like Knox's, was being derided in the press for its unscientific "special pleading . . . beside the real scope of the argument," and desire "to produce effect instead of merely giving instruction" by willfully taking on "established reputations" in words of "strong seasoning." Only "rarely in the history of science," as Falconer remarked in the same tone of withering incredulity Forbes had used in the *Literary Gazette*, had "confident assertion been put forward, in so grave a case, upon a more erroneous and unsubstantial foundation."[46] Just as Cuvier's ebullient rhetorical style in the "Discours préliminaire" had, as seen in chapter 1, a decisive impact on the triumph of the law of correlation half a century before, the belligerent "sledge-hammer style" that Huxley shared with Knox might now derail his attempts to halt the same law's continued success. Even without Forbes's circumspect counsel, the dangers of his argument being regarded as merely another version of Knox's fanatical and unwarranted antipathy toward Cuvier must have been worryingly apparent to Huxley.

No Faith in "the Method"

To make matters worse, after remaining virtually silent on the subject for more than fifteen years, with the brief allusion in *Great Artists and Great*

Anatomists the only exception, Knox chose to recommence his prolonged and bitter onslaught against Cuvierian correlation in August 1856, only weeks after Huxley's unrepentant rejoinder to Falconer's exasperated incredulity and at the very height of the dispute that had been ignited in February. In that same month Knox had completed his translation of Milne-Edwards's *Cours élémentaire d'histoire naturelle*, which, like his earlier rendition of Blainville's *Ostéographie*, was "undertaken with the Author's full approbation," and in which, as Knox avowed, "I have scrupulously avoided omitting any fact or idea or opinion of the author."[47] Whereas, in 1839, Knox's translation of Blainville had enthusiastically approved the original's demolition of the law of correlation, his vaunted meticulousness now obliged him to render into English Milne-Edwards's contrary insistence that there were "évidente les indices d'un principe de coordination."[48] Knox's translation of Milne-Edwards's textbook asserted that

> between every part there reigns the strictest mutual dependence. . . . Some of these harmonies are so obvious and striking that the naturalist may, from the observation of a single organ—a tooth for example, deduce nearly the whole natural history of the animal. . . . In fact, from this single organ may be deduced nearly the whole structure . . . *à priori*, or without having ever seen it.

This effusive affirmation of the Cuvierian true faith was unaccompanied by any comment from Knox. Even in his own introduction and annotations, he held back from directly criticizing correlation, although he could not resist a wryly condescending observation that it was only in "England, where Cuvier and his supposed views had become fashionable."[49]

The completed translation, published as *A Manual of Zoology* in the spring of 1856, was envisaged by Knox as a "safe text-book in all schools and colleges," and he was no doubt well aware that the widespread adoption of the volume in such educational establishments would afford a considerable supplement to his meager earnings from journalism. In return for such tacit respectability and the welcome remuneration that went with it, Knox had to suppress, or at least disguise, his own fierce hostility toward Cuvierian correlation. This was confirmed by a review in the *Leader* that advised that, even in its relatively muted form, the "reader will do well to pass over unread all that Dr. Knox furnishes . . . and devote himself to the *Manual* Dr. Knox has translated."[50] Knox's own opinions were evidently neither reputable nor lucrative, but he could not contain them for long.

Having kept silent in his translation of Milne-Edwards's *Manual*, Knox's antipathy for Cuvierian correlation was given free rein less than five months later in the accommodating pages of the *Lancet*. His series of articles "On

Organic Harmonies," published in four weekly installments between August and September 1856, had apparently been prompted, as was noted earlier, by his experiences translating Milne-Edwards's paean to the "principe de co-ordination" in the *Cours élémentaire*. But that work had been in print continually since the early 1840s, and with Knox boasting of his long-standing connection with "my distinguished friend, M. Edwards," it seems unlikely that he had not read it until beginning his translation in the mid-1850s.[51] Instead, it is more plausible that as a freelance journalist, as Richards has suggested, Knox's "writings were perforce commodities . . . and their production was tailored to his pressing financial needs. . . . They ranged over topics that a fickle public might find of interest."[52] It was certainly the case that in the months since the publication of the calculatedly uncontentious *Manual*, Huxley, followed by Knox's former student Falconer, had brought Cuvier's paleontological method firmly back into the scientific spotlight, and made it a subject that would likely be of interest to the *Lancet*'s readers.

Knox had long considered himself the initial and still the most significant British adversary of correlation, and now eager to "prove . . . incontestably that" as far back as 1830 "and indeed long before, (as early as 1820 — that is, ten years prior), I had no faith in 'the method,'" he entered the fray.[53] He did not allude to either Huxley or Falconer, who had themselves failed to acknowledge his own contributions to the subject, and instead invoked French authorities with whom, notwithstanding his scandalous past, he remained on equal terms. It would nevertheless have been apparent to readers of the *Lancet* who were also aware of the contents of the previous month's number of the *Annals and Magazine of Natural History* (which regularly addressed medical readers by carrying notices of books that it could "recommend both to the medical student and amateur botanist") that the objections that Knox had long held against what he pointedly called "the method" were strikingly similar to those that Huxley had recently presented as a new, and potentially morally improving, direction in science.[54]

In the same acerbic and highly personal style he had honed in his many previous contributions to the *Lancet*, Knox regaled its readers with details of how, while having the "good fortune to have made the personal acquaintance of Cuvier" in Paris, he had been unable to comprehend "how so cautious an observer . . . could ever have penned certain passages in the first volumes of his works, claiming for his *method of restoration* the merits of a fixed science." The explanation, according to the ever-cynical Knox, was nakedly pragmatic and political. Cuvier's marginal "position in the Academy in France" had compelled him to willfully brush aside any complications and instead "raise the anatomical method to the dignity of a fixed science," placing the anatomist on the same "footing with . . . the illustrious mathematicians" who dominated the Académie des sciences. Of course, as Cuvier must have recognized, "living matter . . . is wholly different" from numerical equations, and "there are no laws by which any structure can be foretold *à priori*; they cannot be reduced to a scientific formula" in the same way as mathematical phenomena.

With his customary brusqueness, Knox then interrogated the most celebrated—and mythologized—statement of the so-called method, asserting:

> "Give me the teeth", said Cuvier, "and I will describe the stomach, the feet, the skeleton". Now, given the teeth of the dolphin, porpoise, seal, bear, pig, how stand the co-relative structures?—what has become of them? Given the paddle-shaped limbs of the fossil and extinct saurian *alone*, no anatomist that ever lived could possibly have conjectured the truth of a fragment of the rest.

There were, Knox acknowledged, some "obvious co-relations of forms in certain well-marked characteristic animals," and in cases where there exists "a *constant coincidence* . . . we give the name of an anatomical co-relation." But these had not "been discovered by genius," and it was instead by mere "common sense" that they were "rendered obvious," even to the "untutored observer." Such commonplace deductions were simply the "result of direct comparison with other similar objects—in fact, of immediate observation. They are strictly empirical observations, and not scientific laws."[55] This was precisely the same argument that Huxley had reiterated only weeks before in the *Annals and Magazine of Natural History*, with Knox even placing the same emphasis as Huxley on the entirely quotidian and commonplace nature of the empirical method.

There were other areas, though, in which Knox's argument and tone went well beyond even Huxley's self-conscious impudence. The most deleterious consequence of the misplaced confidence in Cuvier's specious method, he alleged, had been the "*restoration mania* in the hands of some of the theologico-geologists of England," who had merely "guessed at the fossils" and indulged their "highly fanciful imaginations." Nor was Knox afraid to specify his targets, asserting that "in Britain . . . the *method* was hotly taken up by amateurs, of whom the Reverend Dr. Buckland was the type."[56] While Huxley, at the Royal Institution, had been careful to restrict himself only to generalized insinuations against the long-dead "[William] Paley and the natural theologians," the dean of Westminster had succumbed to brain disease barely a month before Knox publicly derided him—in the past tense, notably—in the *Lancet*.[57]

Already, within two weeks of William Buckland's death on 14 August 1856, Knox, although without directly naming him at this stage, had cruelly ridiculed the ersatz restoration of "troops of hyenas (which never live in troops), that scoured the fields of ancient Britain, dragging to their caverns everything they could collect, from the mouse to the rhinoceros!"[58] The particular target of Knox's scorn was Buckland's famous investigations in the 1820s of the miscellaneous fossil bones at Kirkdale Cave in Yorkshire, which were just then being celebrated once again in respectful obituaries, although he had long been equally dismissive of what he derisively called the "Bilgewater Treatises."[59] In July's *Annals and Magazine of Natural History*, Huxley had emphasized the importance of gentlemanly conduct in questioning the "singular bad taste" of Falconer's indignant defense of correlation, but by August, in the pages of the infamously irascible *Lancet*, his own side of the argument had become a means to traduce a highly esteemed member of the scientific establishment whose body was barely cold in the grave.[60]

Knox's deliberately disrespectful style of sensational journalism, along with his increasingly savage radicalism and the baggage of the Burke and Hare scandal he brought with him from Edinburgh, were now firmly, and very publicly, tied to the same anti-Cuvierian argument that Huxley had presented as his own before the grandees of the Royal Institution only seven months earlier. In fact, Knox's articles in the *Lancet* during August and September, whose full title was "On Organic Harmonies: Anatomical Co-relations, and Methods of Zoology and Paleontology," would have had around sixteen times as many readers as Huxley's own contribution on "The Method of Palæontology" in the July number of the *Annals and Magazine of Natural History*, with their respective circulations approximately eight thousand and five hundred.[61] Knox's resumption of his decadeslong antagonism toward the Cuvierian method at precisely this moment must have presented a particularly awkward problem to Huxley, eager to avoid any scandalous or ideological taint that might threaten his scientific credibility and nascent cultural authority.

It may explain his otherwise inexplicable reluctance, following Whitwell Elwin's apparent rejection of his contribution to the *Quarterly Review* discussed in the previous chapter, to take his own crusade against correlation into the general press in the late 1850s, where it would have been still more likely to become entangled with Knox's disreputable version of the same argument. With other respectable men of science such as Falconer also shunning Knox, Huxley, by confining himself to specialist periodicals, could render it less likely that he and his argument would be perceived as in any way

Knoxite. In such circumstances, leaving the field to commercial and theological popularizers might have seemed an acceptable price to pay.

For Knox, on the other hand, Huxley's instigation of a high-profile dispute over necessary correlation both helped him generate some additional income, with the *Lancet* willing to pay for four successive installments on the suddenly topical issue, and even gave his long-standing anti-Cuvierian views a newfound modicum of respectability. When a second edition of his translation of Milne-Edwards's *Manual of Zoology* was commissioned in the early 1860s, Knox felt able to render it "*au courant* with the present state of zoological knowledge" in ways that he had not risked in the initial edition. The translated passage in which Milne-Edwards elaborated the implications for paleontology of the "mutual dependence" among parts that had been left without comment in the first edition was now accompanied by a peremptory parenthetical warning: "[The law of organic harmonies must be applied with the greatest caution. Cuvier seldom trusted to it; it cannot be made with safety the basis of *à priori* reasoning]."[62] The sentiments that Knox had evidently not felt able to express in a textbook at the beginning of 1856 were now, six years later and with Huxley, Falconer, and other prominent scientific authorities having taken up the same issue in the intervening period, much more permissible. Having begun raising objections to what he called Cuvier's "method" more than forty years before, in the early 1820s, Knox now came to the end of his life making the very same complaints, with him dying, at the end of 1862, while still in the midst of correcting proofs of the *Manual*'s new, modified edition.

Cuvier, the Saxon

The proofs of Knox's new *Manual* were completed by Charles Carter Blake, who, motivated by "respect to the memory of the late Dr. Knox," made "as few alterations as possible" and "left the manuscript untouched." While Blake concurred with many of Knox's most contentious opinions and had "enjoyed the privilege of the personal acquaintance of the venerable anatomist until the period of his decease," he must have read the proof pages containing the new notes that Knox added to Milne-Edwards's textbook with gritted teeth. Blake was among the staunchest advocates of the provocative understanding of race and racial conflict that Knox had begun to promulgate in the 1840s, acclaiming him as "the great master" on the subject, but he combined this with a no less steadfast support for the Cuvierian law of necessary correlation, frequently requesting "Professor Owen to give his *imprimatur*" to his own work on comparative anatomy, and regarding him as no less a "great master of the science" than Knox was on questions of race.[63] Nor was Blake alone in mixing a specifically Knoxite approach to race with an acceptance of the infallibility of correlation that would have repulsed the recently deceased anti-Cuvierian. In fact, this paradoxical amalgam became a distinctive feature of the new Anthropological Society, which was founded after its leaders, including Blake, seceded from the Ethnological Society only weeks after Knox's death.[64]

Knox contended that racial traits were immutable because biologically determined in the embryo, rendering the different races of man as distinct as separate species and therefore structurally unable to merge through miscegenation. The society and culture of each race were shaped distinctively by their own innate characteristics, leading Knox to proclaim: "Race is everything: literature, science, art—in a word, civilization, depends on it." It was inevitable that these discrete civilizations would struggle against one another, and Knox identified almost all wars, from ancient conquests to the recent tensions that would soon lead to conflict in the Crimea, as direct results of an unavoidable racial antagonism. That "race in human affairs is everything," he avowed, "is simply a fact, the most remarkable, the most comprehensive, which philosophy has ever announced," although he also acknowledged that this purported truth was "wholly at variance with long-received doctrines, stereotyped prejudices, [and] national delusions."[65] British imperialism, for Knox, was inherently flawed, and white settlers would soon be usurped by fierce indigenous races better suited to their native climate and conditions.

When Knox published these challenging views in *The Races of Men* (1850), he felt the time had also come for "declaring the æra of Cuvier at an end." Knox's determinist approach to race invalidated Cuvier's understanding of organic structure in two ways. First, the "permanent varieties of men . . . originate . . . by the laws of unity of the organization, in a word, by the great laws of transcendental anatomy," with the emergence of immutable racial traits occurring in the embryo as it developed through all the possible stages of life, in the mode identified by Geoffroy Saint-Hilaire, rather than as the adaptations to function that Cuvier proposed.[66] Second, even Cuvier's own inability to appreciate the transcendental anatomy of Geoffroy in the famous dispute that Knox had witnessed while in Paris during the early 1830s could itself be attributed to racial characteristics. In *Great Artists and Great Anatomists*, Knox alluded to Cuvier's German descent by calling him "Cuvier, the Saxon, and therefore the Protestant," and noting that "in personal appearance he much resembled a Dane, or North German, to which race he really belonged."[67] Along with the apparently inevitable Protestantism, this racial pedigree also entailed mental habits of practicality and common sense. As

Knox observed in *The Races of Men*, "In a vertebra the matter-of-fact Saxon mind sees merely a vertebra; beyond this it seldom proceeds," thereby overlooking the vertebra's status as the ideal type of all organic structures. With the "transcendental doctrine . . . a theory originating unquestionably with the mixed Slavonian and German race, inhabiting Southern Germany," it was inevitable that it would be "resisted to the last by Cuvier," whose "true Saxon" brain was predetermined not to apprehend its "imaginative, romantic" subtleties, and instead to cleave to the "uninventive, unimaginative" mechanical contrivances of functionalism. Knox also noted that "nor is the Celtic mind very gifted in this respect." It was therefore a direct result of the two nations' particular racial makeup that Cuvier's "narrow . . . view of the philosophy of animal beings" had for so long been "embraced [by] spiritual France and imitative England."[68] The acceptance or otherwise of Cuvierian correlation was, according to Knox, determined by race.

A reprinted edition of *The Races of Men* was soon published in America, and Knox's laws of ineluctable racial segregation and antagonism were widely adopted by proponents of the Southern states' "peculiar institution" of slavery, especially as the debate over black emancipation became more heated in the years leading up to the Civil War. With Knox still largely shunned at home, it was in "influential American proslavery texts," as Richards has pointed out, that his views on race "were reimported into England."[69] Significantly, however, these repackaged versions of Knox's views on race, which took advantage of the absence of transatlantic copyright laws, often either disregarded or actually reversed his vaunted opposition to Cuvier.

In *Negroes and Negro "Slavery"* (1861), John H. Van Evrie, a New York physician and publisher who supported the Confederacy, followed Knox in asserting the "fixed and indestructible *facts*" that "whites, Indians, and negroes" were "different creatures, different *species* of men, with different bodies, and different minds." But in order to demonstrate this Knoxite polygenism, he deployed an extended analogy that could never have appeared in *The Races of Men*. Van Evrie proposed:

> Cuvier, the great French zoologist, it is said might pick up a bone of any kind, however minute . . . and from this alone determine the species, genera, and class to which it belonged. This at first seems almost incredible, but a moment's reflection shows not only its practicability, but the ease and certainty with which it may be accomplished. . . . That great fundamental and eternal law of harmony or adaptation which God has stamped on the organic and material universe permits of no incongruities or contradictions to mar its beauty or deface its grandeur. Thus an anatomist, who had given a certain amount of attention to the subject, might select the smallest bone . . . and determine

from among millions of similar ones, whether it was that of a white man or of
a negro with perfect certainty and the greatest ease. . . . The analysis of a single
bone or of a single feature of the negro being is thus sufficient to demonstrate
the specific character or to show the diversity of race.

The anatomical distinctions among races, with, for instance, the "negro . . .
incapable of an erect or direct perpendicular posture" and "the whole structure . . . thus adapted to a slightly stooping posture," were sufficiently clear to
be inferred with absolute certainty using Cuvier's now legendary law of correlation.[70] The structural traits that could be identified from merely a single
bone made it clear that slavery, far from being an iniquitous imposition, was
actually the natural condition of the biologically inferior Negro.

Another recommendation for the Cuvierian axiom was that it had also
been used to similarly spectacular effect by one of Van Evrie's fellow Northern
sympathizers with the South. He remarked:

We have recently witnessed a still more remarkable instance of this tracing the
life and defining the relations of organized beings from a minute and remote
point. Agassiz has been able, from a single scale of a fish, to determine the
specific character of fishes, and those, too, which he had never before seen!

The Harvard ichthyologist Louis Agassiz, whom Van Evrie acclaimed as "unquestionably the greatest of American naturalists," had lectured in the Southern slave port of Charleston during the 1850s on the distinctness of the human
races, and was nauseated by his encounters with black slaves in the neighboring plantations.[71] He lent his considerable imprimatur to the overtly racist
science that endorsed the continuation of Southern slavery.[72] Agassiz's earlier
accomplishments in identifying extinct fish from fragmentary fossil remains,
largely conducted in Switzerland and France before he came to America in
the late 1840s, continued to augment his scientific celebrity and were now a
further reason why correlation was attractive to American polygenists who
were otherwise strongly influenced by Knox. The segregationist laws adumbrated in *The Races of Men* were readily repackaged as concomitant with the
correlative methods of Cuvier and Agassiz in the incendiary circumstances of
America in the early 1860s.

Van Evrie's juxtaposition of Knoxite racialism and Cuvierian correlation
came to Britain in James Hunt's *On the Negro's Place in Nature*, a paper read
before the Anthropological Society in November 1863 and subsequently published as a pamphlet. Hunt applauded "Dr. Van Evrie, of New York, who has
paid considerable attention to the character of the Negro, and had ample
opportunities for observation," before quoting the section of his inflamma-
tory treatise beginning "The analysis of a single bone or of a single feature

of the negro being is thus sufficient to demonstrate the specific character."[73] Van Evrie returned the favor, reprinting Hunt's pamphlet in New York amid the "slaughter and destruction around us," with the recommendation that it "collected all the reliable modern authorities, and demonstrates what every unperverted American *knows*—that the Negro is a different and subordinate species or race," a "fundamental *fact*" that when "clearly apprehended and accepted" would provide a "starting point for . . . the restoration of peace, Union and harmony."[74] The version of Van Evrie's argument that was reimported back across the Atlantic in the penultimate year of the Civil War was nonetheless subtly attuned to Hunt's own particular circumstances in Britain.

Hunt's strident polygenism, as he acknowledged in 1866, had been "imbibed from the late Dr. Knox," whom he had first got to know in the mid-1850s, when Knox was reviving his long campaign against Cuvier and the law of correlation.[75] By 1860 Hunt, as secretary, had ensured Knox's election to the Ethnological Society, where he had previously been blackballed by the Quaker monogenists who dominated the society. Knox was also the intellectual mentor of the breakaway group, led by Hunt, who, bridling at the religious and philanthropic agenda of the Ethnological Society, formed the avowedly polygenist, and often savagely racist, Anthropological Society in January 1863.[76] In *On the Negro's Place in Nature* later in the same year, Hunt quoted extensively from Van Evrie's account of how Negro traits could indeed be inferred from just "the analysis of a single bone," but he carefully elided any of the parts of the quotation that alluded to Cuvier's, or even Agassiz's, "incredible" accomplishments. Van Evrie's repackaging of Knox's abrasive racialism was itself repackaged to render it less discordant with Knox's no less strident antagonism toward Cuvierian correlation.

A Silly Empty-Headed Young Man

When Hunt presented some of the material contained in *On the Negro's Place in Nature* at the annual meeting of the British Association for the Advancement of Science earlier in 1863, his "statement of the simple facts was received with . . . loud hisses." Amid the melee created by "those 'outer barbarians' whose habit it is to sneer at any views opposed to their own," Hunt noted, "my friend Mr. C. Carter Blake ably supported me, but the audience also favoured him with strong marks of disapprobation."[77] The title of Hunt's pamphlet was an unsubtle parody of Huxley's *Evidence as to Man's Place in Nature* (1863), a work condemned by Blake in the *Edinburgh Review* for its evolutionary vindication of the "unity of the human race." Blake scorned the "so-called 'Theory of Development,'" agreeing with Knox that racial dimorphism occurred only in the embryo, after which races were stable and immutable and could not be altered by transmutation.[78] Huxley, who resigned his honorary fellowship of the Anthropological Society and instead joined the council of the rival Ethnological Society, responded in a series of lectures at the Royal College of Surgeons, in which, according to the *Reader*, he "pronounce[d] the system of slavery to be root and branch an abomination—thus making his physiological definition of the Negro's place among men equivalent to an earnest plea for Negro emancipation."[79] Huxley actually had little sympathy for the black slaves and instead his principal concern with slavery was its moral and economic consequences for Southern whites, with whom he had close family ties.[80] The violent antipathy between him and Blake, however, extended beyond their views on race and the conflict in America, and also embraced correlative methods of paleontology, once more complicating the relation between Knox's resolute racialism and his uncompromisingly anti-Cuvierian stance.

Blake liked to present himself as if he were another of the "great English minds that owe their first scientific *impetus* to Knox's influence and personal teaching," although, having been born in 1840, he would have been only seven when Knox's license to teach anatomy was withdrawn by the Royal College of Surgeons, and barely thirteen when Knox's final attempt to give unlicensed lectures at the Royal Free School of Medicine was ignominiously thwarted in 1853.[81] Whatever training Blake might have received from Knox, it must have been wholly informal. His official instruction in anatomy was instead undertaken with Owen, to whom Blake dedicated one of his publications "as a testimony of the respect and friendship of his pupil."[82] At the age of sixteen, Blake had attended Owen's course of lectures at the Government School of Mines in the spring of 1857, where, as seen in chapter 6, he had vigorously defended Cuvier against Huxley's attacks. Later that year, the precocious Blake wrote to Owen requesting a "ticket of admission . . . for the Osteological Department of the British Museum" so that he could "carry . . . out the directions given by you to zoological students in your lectures" regarding the "examination of . . . actual specimens." When, at the end of 1860, Huxley again began disputing the reliability of the law of necessary correlation, four years after both he and Knox had separately traduced the famous principle, Blake unsurprisingly took the side of the tutor to whom he felt "under a great obligation" rather than that of his unofficial mentor.[83]

Back in November 1856, Owen had unleashed what Joseph Dalton Hooker called a "cutting telling & flaying alive assault on Huxleys [*sic*] adaptation views at the Geolog. Soc."[84] Exactly four years later, Huxley returned to the very same location to offer the response he had so conspicuously failed to

make at the time. Where Owen had once suggested that opposition to correlation was a "congenital, defect of mind" indicative of an absence of "healthy
and normal" sensibilities, Huxley now pointed to an "important palæontological moral" of his own.[85] At the end of a paper on a new species of an
extinct South American ungulate mammal that had been variously likened to
camels and llamas, he concluded:

> *Macrauchenia*, alone, affords a sufficient refutation of the doctrine that an
> extinct animal can be safely and certainly restored if we know a single impor-
> tant bone or tooth. If, up to this time, the cervical vertebra of *Macrauchenia*
> only had been known, palaeontologists would have been justified by all the
> canons of comparative anatomy in concluding that the rest of its organization
> was Camelidan. . . . Had, therefore, a block containing an entire skeleton of
> *Macrauchenia*, but showing only these portions of one of the cervical ver-
> tebræ, been placed before an anatomist, he would have been as fully justi-
> fied in predicting cannon-bones, bi-trochanterian femora, and astragali with
> two, subequal scapho-cuboidal facets, as Cuvier was in reasoning from the
> inflected angle of the jaw to the marsupial bones of his famous Opossum. But,
> for all that, our hypothetical anatomist would have been wrong; and, instead
> of finding what he sought, he would have learned a lesson of caution, of great
> service to his future progress.[86]

Owen, in his portion of the *Zoology of the Voyage of HMS* Beagle (1839), had
classified the macrauchenia remains Charles Darwin brought back from
Patagonia using the accustomed Cuvierian method. He identified the idiosyncratic creature as "referrible [*sic*] to the Order Pachydermata" with certain close "affinities to the Ruminantia, and especially to the Camelidæ,"
although with other features that "deviate remarkably from those of the *Camelidæ*."[87] Owen later came to recognize that those deviations meant that the
macrauchenia was not a ruminant after all, and he revised his classification
accordingly.[88] Huxley, presented with new bones from Bolivia that showed
still more clearly the macrauchenia's divergence from the camel, strategically
yoked Owen's awkward change of heart to "Cuvier's crack case of the 'Possum,'" which, as seen in the previous chapter, he had already taken on in his
Fullerian Lecture on the vertebrata of April 1857, to teach a mortifying "lesson in caution" over the same parliamentary-style benches where Owen had
humiliated him in front of the Geological Society's fellows four years before.[89]

By the early 1860s, as was noted at the end of chapter 6, Huxley's opposition to necessary correlation had begun to win over the scientific community, with even Falconer retracting his previously steadfast support in
1862. Quickly losing the backing of his peers, Owen, as Falconer observed a
year later, had "now got hold of a silly empty-headed young man—Carter

Blake—to work upon—for his dirty work."[90] Whether vacuous or not, Blake was certainly doing an underhand favor to Owen when he wrote to the editors of the *Annals and Magazine of Natural History* in May 1861, noting that "Prof. Huxley, impugning the philosophical laws of 'correlation of structure' as defined by Cuvier and Owen, suggests that, upon the Cuvierian method of induction, a palaeontologist, reasoning alone from the cervical vertebræ of *Macrauchenia*, would have confidently predicted its Cameloid affinities." Having given Owen equal billing with Cuvier as the definer of the law of correlation, Blake explained why the case of the macrauchenia had no bearing whatsoever on its efficacy or reliability. He wrote:

> But when Prof. Huxley founds an argument, put hypothetically into the mouth of an ideal adversary, upon a structure so liable to variation as the perforation by a blood-vessel of a cervical vertebra, it can hardly be accepted as a correct exemplification of the principle which Cuvier has so successfully applied. The non-perforation of a cervical vertebra by an artery is certainly not such a character, subserving an important purpose, and denoting ordinal distinction, as the presence of a marsupial bone in an Opossum, with which Prof. Huxley compares it. The analogy which it is attempted to deduce, as adverse to the principles of correlation, therefore totally fails, whilst this high law of comparative anatomy, "*aussi certaine qu'aucune autre en physique ou en morale*", remains unimpaired by the re-discovery of Macrauchenian remains in the Andes.[91]

Huxley's hypothetical scenario was invalid because it assumed the use of inconsequential bones and variable characters that the properly Cuvierian paleontologist would never rely upon. The impudent young tyro of 1856 was now himself being upbraided by an assured and well-informed assailant who was barely in his twenties.

Blake's poise and incisive understanding of anatomical technicalities doubtless arose from having been carefully primed by his erstwhile tutor. Owen, as he told the Council of the Royal Society, was well aware of what "Prof. Huxley would have the readers of his Paper on *Macrauchenia* . . . believe to be Cuvier's 'Doctrine.'" Like Blake, he insisted that this "attack seemingly on the principle on which the founder of Palaeontology based the science" was, in reality, a "misstatement of Cuvier's Law" and a "travesty of that principle." The "fallacy" Huxley's paper perpetuated, Owen explained, "lies in propounding any or every bone or tooth for those truly light giving & guiding parts which Cuvier was careful to define." Frustrated at Huxley's persistent deprecation of "Cuvier's 'Law of Correlation'" regardless of his own limited practical experience of working with fossils, Owen noted that "those who are loudest in decrying it are also those to whom Palaeontology owes least for the reconstruction of an extinct animal from a single bone or tooth." Attributing Huxley's attack to "either ignorance or bad faith," Owen acknowledged that with a "Paper in which the author indulges in polemical topics . . . if one leaves them to their fate, silence may be misconstrued as assent."[92] Despite this risk, Owen preferred to confide his indignation only within a private missive. It was left to Blake to make exactly the same refutation of Huxley's arguments in public.

Notwithstanding the assured and knowledgeable tone honed by Owen's briefings, Huxley's friends remained brusquely dismissive of Blake's scientific abilities. Alongside Falconer's withering contempt for the "empty-headed young man," Hooker told Darwin: "I wish Huxley would not go out of his way to pick quarrels with such cattle as Carter Blake . . .—who he thus magnifies greatly."[93] Hooker's correspondent, though, seems to have considered Huxley's adversary far from bovine. In January 1862 Blake told Darwin that he was "very much gratified by the reception of your note of the 1st June last, in relation to my article on Macrauchenia in Bolivia."[94] Darwin's seemingly approving note to Blake has not been found, but there is good reason why, despite the animadversions of Falconer and Hooker, he might have appreciated the younger man's letter to the *Annals and Magazine of Natural History*. The macrauchenia's putative affinities with the camel had afforded Darwin, when jotting down theoretical speculations in his Red Notebook during the late 1830s, a valuable instance of an intermediate form that suggested the ostensibly separate mammalian orders of ruminants and pachyderms might actually be related. When, in subsequent years, the creature's classification was altered, Darwin, as Stan Rachootin has commented, "had sufficient motivation to keep seeing a camel when none remained."[95] Blake's incisive missive rebutting Huxley's demolition of the macrauchenia's cameloid affinities might therefore have piqued Darwin's interest, even if it also risked reviving the residual differences with Huxley over correlation that had been concealed, although not necessarily resolved, by the strategic alliances forged during the heated disputes over paleontological method in 1856.

Scientific Recognition of the Confederate States

No less perturbingly, Blake's defense of correlation had reached a much broader audience than just Darwin, as he had cannily submitted the very same letter that attracted the evolutionist's admiration when it appeared in the *Annals and Magazine of Natural History* to the *Geologist*.[96] Samuel Joseph Mackie's monthly magazine, as seen in the last chapter, endeavored to "popularize and to extend the noble science of geology," and had already played host to confirmed Cuvierians such as Peter Bellinger Brodie.[97] In 1858 Mackie and his journal had also been involved in founding the Geologists' Association, which, with the "'Geological Society' . . . too far advanced in the strict course of scientific method," aimed to provide a "common means of communication among those who, while not devoting their lives to the pursuit, yet take an active interest in the facts and teachings of Geology."[98] Blake was again quick to utilize the Geologists' Association to uphold the value of correlative methods before a much larger and more diverse audience than Huxley could address in specialist forums such as the Geological Society.

In a lecture "On the Distribution of the Fossils of South America," delivered to the Geologists' Association in January 1863, Blake only briefly alluded to the contested macrauchenia, which, he conceded, was "allied to the tapirs, but with a stiff neck like the llamas." Instead, he gave an "illustration . . . of the mode employed by palæontologists in reasoning by the aid of the 'correlation law', as applied to the reconstruction of the skeleton of the Megalonyx from the solitary unequal phalanx originally discovered by the President of the United States, Jefferson."[99] This was the celebrated instance, back in 1797, when the polymathic American politician Thomas Jefferson inferred the existence of a large leonine creature from just an outsize claw and a few other smaller bones, although without any reference to Cuvier's nascent law of correlation (which at that stage was entirely unknown in America).[100] Eschewing the continuing controversy over the use of correlation in classifying the macrauchenia, Blake shrewdly deployed a classic—albeit distorted and much mythologized—instance of the success of the famous inductive method, which would leave his popular audience in no doubt as to its enduring effectiveness whatever the specialist naysayers might claim.

As well as Jefferson's status as an icon of the correlative method, Blake, whom the *Evening Star* labeled a "Confederate physiologist," had other reasons for dwelling on America at this time.[101] Introducing "American palæontology" to the "minds of the British public" in the *Geologist* in September 1862, he noted ruefully that the "mighty hemisphere" was "now the seat of political convulsions." In the same month that more than three thousand men died at the ferocious Battle of Antietam, Blake suggested that impartial attention to paleontology and other forms of natural history might actually afford a means of resolving those same political tensions without the need for further bloodshed. He proposed:

> An argument for the scientific recognition of the Confederate States might be founded upon the fact that their flora and fauna differ essentially from those of their more northern antagonists in the less fertile country north of

the Ohio. The term "fauna of the United States" conveys no meaning to the scientific mind.[102]

The Mason-Dixon Line, according to Blake, was a natural border rather than an arbitrary political demarcation, giving an apparently objective and scientific rationale to the South's demands to secede from the Union.

Blake saw science supporting the Confederacy in other ways too, especially Knox's polygenist account of race. For Knox, new species, including the various human races, were created by embryonic change rather than transmutation in the adult animal. This view contradicted the arguments of evolutionary monogenists such as Huxley that the different races shared a common descent from the same primitive ancestors, and thus slavery could only be regarded as an abomination that contravened the essential unity of man. As Blake noted in his review of Huxley's *Man's Place in Nature*, the "opinion of one of England's best anatomists, the venerable and recently departed Robert Knox, was vehemently opposed to that of the modern Darwinite transmutation school." In particular, Knox, as Blake recalled, "took every opportunity of denouncing the hippocampus minor controversy as a 'silly dispute,'" considering Huxley's criticisms of Owen over the structure of apes' brains as missing the point that common descent of the human races from simian forebears was so unfeasible that tiny anatomical parallels were meaningless. Blake's anonymous notice of Huxley's book for the *Edinburgh Review* unequivocally tied its colors to the Knoxite mast, thundering: "We entirely agree with Dr. Knox in his opinion of the extreme absurdity of this quarrel."[103] It was once more in the popularizing *Geologist*, however, that this vehemently Knoxite racialism became entwined with Blake's endorsement of necessary correlation.

In a signed article on "Fossil Monkeys" for the *Geologist* in May 1862, Blake explained the paleontological evidence that opposed the "alleged origin of the human race from a transmuted gorilla." He quoted from a paper in the *Proceedings of the Zoological Society*, "On the Gorilla," from 1859, in which Owen adumbrated the clear differences between simian and human anatomy. As Blake approvingly observed: "Professor Owen sums up by stating,— 'There is no law of correlation, by which, from the portion of jaw with the teeth of the *Dryopithecus*, can be deduced the shape of the nasal bones and orbits . . . or any other cranial characters determinative of affinity to Man.'" The infallibility of the law of correlation, even if on this occasion yielding negative evidence, decisively proved that "we have thus amongst the fossil species of *Simidæ* no form sufficiently allied to Man to have served as his ancestor."[104]

Correlation, the so-called method that Knox had so zealously disputed and derided since the 1820s, was now utilized by his self-proclaimed protégé to bolster his own rejection of the Darwinian doctrine of the common descent of the human races from a shared ancestor.

In Blake's paradoxical amalgamation, Cuvier's famous law underwrote a determinist approach to race that, as was just then being reiterated in the second edition of *The Races of Men*, published in 1862, considered Cuvier himself too limited by his matter-of-fact Saxon brain to advance beyond drearily mechanical functionalism and properly grasp the transcendental unities of anatomy. At the same time, with Blake using the quotation from the Zoological Society's *Proceedings* to disabuse the "disciple[s] of unity of descent" who proclaimed the equality of blacks and whites and supported the abolitionist North in the American Civil War, Cuvierian correlation was invoked on behalf of the Confederacy and the cause of slavery.[105]

A Jackal of Owen's

Clearly perturbed by Blake's relentless attacks in both the scientific and the general press, Huxley portrayed his predacious adversary as "a jackal of Owen's."[106] To the bohemian wits of *Punch* he appeared even more rapacious, and the comic weekly reported in 1865 that at the "last meeting of the Anthropological Society there was a delightful discussion of Cannibalism. MR. CARTER BLAKE fired up over this subject to such an extent, that several members, noticing the presence of the 'devouring element' in his speech, felt slightly uncomfortable."[107] Reveling in such provocative and distasteful discussions, the inner coterie of the Anthropological Society, which included the outspoken explorer Richard Burton and the sadomasochistic poet Algernon Charles Swinburne, dubbed themselves the Cannibal Club, as a counterpart and rival to the more respectable X Club established by Huxley and his fellow scientific naturalists.[108] With Blake and Hunt the leaders of both this Cannibal clique and the larger society itself, as the founding secretary and president respectively, it was inevitable that their mutual, and equally paradoxical, juxtaposition of Knox's laws of racial segregation with Cuvier's axiom of correlation would be reflected in the wider activities of the Anthropological Society.

At times, the Cuvierian law appeared to be more highly regarded among the society's members than even the Knoxite laws of race. Hunt oversaw and personally financed an extensive program of publications. In the *Anthropological Review*, one of two periodicals that he edited, it was proposed, by Luke Owen Pike in 1868, that when the bone of any known animal is discovered, the anatomist . . . can tell us to what kind of animal the bone belonged. If there really are well-marked classes of language, philology ought to be able to do as much for us with a fragment of any class, as anatomy can do with a single bone.

Pike, a leading barrister and fellow of the Anthropological Society, was contesting the assumption that "race and language must be co-extensive," and the inability of the correlative method to identify distinct linguistic identities among the so-called Teutonic races showed only the indeterminacy of these purportedly pure races rather than any flaws in the method itself (Knox himself actually agreed that race could not be inferred from a single bone, but inevitably blamed the failure on the flawed Cuvierian "method" rather than the indistinctness of racial differences).[109] In fact, Pike insisted, contra Van Evrie, that in determining race, even "our best anthropologists . . . would not feel any confidence in giving an opinion on a single skull."[110] Two years earlier, in his *The English and Their Origin* (1866), Pike had observed, "Dr. Knox assures us that . . . races . . . are unchangeable," but he insisted that the "evidence adduced in support of this great law is of the weakest possible character," invalidated by the "monstrous contradiction" of the "inhabitants of America, who are of . . . mixed European descent."[111] Even with Knox, as Richards has contended, "indubitably the Anthropological Society's intellectual mentor," his polygenist principles were seemingly less sacrosanct and incontrovertible than the anatomical techniques of his narrow-minded Saxon nemesis.[112]

At least part of the reason why, notwithstanding Knox's enduring antipathy, the much-contested principle of correlation appealed to the members of the Anthropological Society, and especially to its inner Cannibal coterie, was simply that it had been so persistently opposed, for more than a decade, by Huxley. In 1868, the same year as Pike's rejection of Knoxite racial laws and endorsement of the law of correlation, Huxley was elected president of the Ethnological Society, augmenting his status as the principal adversary of Hunt and Blake's rival organization. Although Huxley, as Efram Sera-Shriar has recently argued, actually agreed with many of the Anthropological Society's reforms to the methodology of ethnological research, and his dispute with Hunt and Blake was largely a professional competition for control of the new science, their animosity toward each other was nonetheless vehement.[113] The predatory Cannibals of the Anthropological Society had sought to undermine Huxley's scientific standing in whatever ways they could since the early 1860s, when Blake had traduced *Man's Place in Nature* in the *Edinburgh* and Hunt had tastelessly parodied the same book's title in his controversial pamphlet *The Negro's Place in Nature*. With Huxley's installation as the Ethnological Society's president at the end of the decade, they now had even more reason to find common cause with his enemies. Indeed, when Huxley brusquely rebuffed the Anthropological Society's representatives at the British Association meeting in 1870, they responded by giving Owen, Huxley's long-standing opponent on correlation and several other matters, an especially "rousing welcome." The bemused Owen, as one participant later recalled, "never knew to what he owed the extraordinary warmth of his reception."[114] In the bitter and increasingly petty institutional conflict between the Anthropological and Ethnological Societies, support for Cuvierian correlation was sometimes merely a convenient means to irritate Huxley rather than a matter of sincere intellectual allegiance.

In fact, by February 1871, less than eight years after aligning himself with his two mentors, Knox and Owen, as "vehemently opposed to . . . the modern Darwinite transmutation school," Blake was assuring Darwin that in a forthcoming review of *The Descent of Man*, "I shall have no other object than fiat justitia ruat cœlum [let justice be done, though the heavens fall], & if I have to modify some of my previously published opinions, the shame will merely attach to myself."[115] In his anonymous notice for the *British and Foreign Medico-Chirurgical Review*, Blake not only avowed that the "evidence in favour of the working of Darwinic principles . . . is already strong enough to prove that Mr. Darwin's are 'veræ causa,'" but contrasted the Darwinian mode of evolution with the "derivation hypothesis of Professor Owen" (discussed in the next chapter), which, he insisted, was "without a *vera causa* to rest upon."[116] The ferocious "jackal of Owen's" was now disloyally sinking his teeth into his own erstwhile master, and, as with the similar apostasy of Falconer discussed in chapter 6, such an acceptance of Darwinism generally entailed the repudiation of the axiom of necessary correlation.

Conclusion

Intriguingly, Blake's dramatic change of heart seems to have been related to his own parlous financial situation. He wrote again to Darwin in 1879, telling him that he had fallen into a "great state of pecuniary distress & privation" and asking for a loan of "five pounds till some brighter days come for me." Blake then pragmatically invoked his review of the *Descent*, reminding Darwin that "though at one time I may have been opposed to the theory of Natural Selection, I have long since . . . acknowledged the theory," before asserting bluntly (and far from convincingly): "I have given you my support, &

do not ask you to buy it."[117] Darwin was sufficiently persuaded by Blake's protestations, and, according to his account books, provided the five-pound loan, which, tellingly, appears never to have been repaid.[118]

No matter how cynical it might have been, Blake's expedient conversion to Darwinism, like the earlier disdain for his intellectual capacities expressed by Falconer and Hooker, does not diminish the very real challenge that he, along with Hunt and their fellow Cannibals, posed to Huxley and his scientific naturalist agenda during the 1860s. In the competitive metropolitan scientific scene of that decade, as Adrian Desmond has proposed, they represented a "rival faction which threatened the Darwinian hegemony . . . leaving Huxley's clique feeling more besieged than ever."[119] The Anthropological Society's membership and resources far outstripped those available to the Ethnological Society, even if its leaders deviously counted nonpaying members in the figures and, during the American Civil War, drew upon a secret slush fund set up by the Confederate government.[120] This potent rival group was no less naturalistic than Huxley and his allies, and its members were even less accommodating of religious views of race, but they vigorously repudiated Darwinian evolution while also embracing Cuvierian correlation, precisely the two issues, as seen in chapter 6, that had first brought together the emergent scientific naturalists in the mid-1850s. With Blake unstintingly defending Owen throughout the following decade, necessary correlation received the approval of a still more youthful scientific young guard.

Crucially, the Cannibal clique, and Blake in particular, was also far more adept at utilizing new methods of popularization to uphold the efficacy of correlation than Huxley, as seen in the last chapter, had been in advancing his anti-Cuvierian argument. Knox's long adherence to the same argument may, of course, have been an important factor in Huxley's reluctance to criticize correlation in such popular forums, and the disgraced anatomist's paradoxical supporters in the Anthropological Society, having already contested Huxley's claims in the specialist scientific press, took full advantage of his continuing failure to defend his position before the much larger audiences for popular lectures and journals.

Huxley's apparent triumph in the methodological disputes discussed in chapter 6 has, in the light of the last two chapters, come to seem increasingly hollow, at best merely a pyrrhic victory. The chapters in the following part will examine the longer-term consequences of this disparity between the elite scientific community's relinquishment, or at least fundamental modification, of the law of correlation, and its continued currency among several other groups during the final decades of the nineteenth century.

PART IV

Afterlife, 1862–1917:
Missing Links and Hidden Clues

Evolutionary Modifications

On 17 July 1862 George Robert Waterhouse returned to London after a short trip to the Bavarian town of Pappenheim. Waterhouse was the keeper of the British Museum's geological department, and he went immediately to the home of the superintendent of the museum's natural history division. It was at Richard Owen's behest that Waterhouse had traveled to Germany, and he now disturbed his superior's domestic tranquility to deliver bad news. Waterhouse, as Owen's wife, Caroline, recorded, had "been in treaty for the collection of fossils, in which is the curious fossil with the alleged feathered vertebrate tail. The old German doctor is obstinate about the price, and Mr. W. has come away empty-handed. We ought not to lose the fossil."[1] Owen, sharing his wife's sentiments, authorized Waterhouse to match the valuation of the stubborn physician who owned the strange feathered fossil, and, as the *Times* later reported, "It was made a *sine quâ non* to purchase the collection of which this was a part, and the sum paid for the whole amounted to not less than 750*l*."[2] This was soon corrected in the gossip column of the *Athenæum*, which had been "informed on authority" that the price was "not more than 400*l*.," although even this was regarded by many as an extravagant misuse of public finances, especially as the British Museum's board of trustees had not given its approval.[3] Defending the controversial purchase, the art critic John Ruskin, a close personal friend of Owen's, revealed that, as well as paying "four hundred pounds at once," Owen had "himself become answerable for the other three!" and Ruskin complained that the "public will doubtless pay him eventually, but sulkily . . . only always ready to cackle if any credit comes of it." There is no other evidence to corroborate Ruskin's claim (he prefaced it by observing, "I state this fact without Professor Owen's permission: which of course he could not with propriety have granted"), but whether or not

Owen did actually underwrite the acquisition with his own money, his determination to bring the enigmatic Bavarian fossil to Britain seemed to have no limits.[4]

The incomplete skeleton that Owen coveted with such intensity, which had been excavated from the fossiliferous limestone quarries of Solnhofen, combined feathers and other ornithic characters with a tail that was distinctly reptilian. In 1861 the German naturalist Andreas Wagner, albeit without having actually seen the fossil, recognized that "nothing more surprising and odd could be imagined" than a "skeleton with such a combination of characters" and proposed that it was a "singularly constructed Saurian," an anomalously feathered reptile that he named *Griphosaurus problematicus*. Following the publication of Charles Darwin's *On the Origin of Species* (1859) two years earlier, such intermediate forms, so-called missing links that might provide tangible evidence of evolution in action, had assumed a new pertinence. Wagner, though, was careful to "ward off Darwinian misinterpretations of our new Saurian," insisting that such "strange views upon the transformation of animals" were "fantastic dreams, with which the exact investigation of nature has nothing to do."[5]

Even if the peculiar creature's curious admixture of taxonomic characters was not attributed to evolution, it still had important, and potentially fatal, consequences for an earlier approach to animal structure. As the *Popular Science Review* reported of the enigmatic fossil in January 1862:

> Dr. Wagner . . . thinking it a reptile, gave it the name of Griphosaurus. Should the latter supposition prove correct, it will furnish one more illustration of the fallacy of Cuvier's doctrine that an unknown creature can be restored from a single bone; for, had the tarsus only been found, it would . . . surely have been referred to a bird.[6]

If there was no clear taxonomic connection between particular component parts and the overall structure from which they derived, then the Cuvierian law of correlation, and the celebrated paleontological inferences that it had facilitated over the last six decades, could no longer be considered reliable. Atypical and unstable intermediate forms such as Wagner's feathered reptile, as the *Popular Science Review* quickly realized, would be the final nail in the coffin of Georges Cuvier's vaunted capacity to reconstruct from merely a single bone.

Wagner's interpretation of the fossil was not the only one from a German naturalist. Hermann von Meyer, who had previously identified an ornithic species he named *Archaeopteryx lithographia* on the basis of a single fossil feather from the Solnhofen quarries, proposed that the partial skeleton was likely to be another specimen of the same prehistoric bird (although, like Wagner, without having seen the fossil). He nevertheless acknowledged that "in one and the same creature, the most different types may occur together," conceding that his single "fossil feather . . . even if agreeing perfectly with those of our birds, need not necessarily be derived from a bird." Meyer had in fact been a critic of the law of correlation for almost as long as either Cuvier's estranged successor Henri de Blainville or the Scottish scourge of functionalism Robert Knox, and, in light of the incongruous feathered fossil, he insisted: "As early as the year 1834 I indicated the danger to which we expose ourselves in palæontology by drawing logical conclusions in accordance with Cuvier's theory, from the similarity of particular parts as to the similarity of other parts or of the whole."[7] It was in his *Palaeologica* (1832) that Meyer had initially contended that "conclusions drawn from one part of the skeleton to the structure of the entire animal have turned out to be erroneous," and Thomas Henry Huxley gleefully noted that "in his later works this eminent palæontologist constantly repeats this much-needed warning against the current exaggerations of Cuvier's unguarded phrases" (Huxley evidently had in mind Cuvier's hyperbolic rhetoric in the "Discours préliminaire" from *Recherches sur les ossemens fossiles* [1812], discussed in chapter 1).[8] Both of the rival German interpretations of the feathered fossil from Solnhofen, translations of which were quickly published in the *Annals and Magazine of Natural History*, had equally problematic implications for those who still fostered such alleged exaggerations.

By dispatching Waterhouse to secure the contentious fossil regardless of the price demanded by its uncompromising Bavarian owner, Owen was not merely acquiring a "museum icon" that would add luster to the British Museum's natural history collections, as Nicolaas Rupke has suggested.[9] He was also taking personal control of the particular specimen that, while it remained in Germany, most challenged his own long-standing advocacy of Cuvierian paleontological methods. "British gold," as the *Times* remarked, had "deprived" the "German philosophers . . . of the means of completing the investigation which they had so successfully commenced."[10] Within a month of the fossil's eventual arrival in London in October 1862, Owen took advantage of his exclusive proprietorial access to a fossil that neither Wagner nor Meyer had actually seen to make his own authoritative interpretation of it. At a crowded meeting of the Royal Society, Owen insisted that the feathered creature was indubitably a bird, a classification that enabled him to powerfully reaffirm the efficacy of the law of correlation.

Both Wagner and Meyer had been equally insistent that their rival conclusions regarding the Bavarian feathered fossil also refuted the "supposed
birds' footprints" found in the Triassic sandstone of the Connecticut River
valley, which, as Wagner proposed, should now be considered "reptile-
tracks."[11] The New England geologist Edward Hitchcock's long-standing
interpretation of the footprints as the only surviving remnants of unprecedentedly ancient struthians relied on the same assumptions that enabled
Cuvier to infer ruminant feeding habits from the imprint of a cloven hoof,
and had been dramatically corroborated in the early 1840s by Owen's iconic
exemplar of the Cuvierian method, the discovery of the dinornis. Owen's capacity to disavow Wagner's reptilian *Griphosaurus* and appropriate Meyer's
archaeopteryx as unequivocally ornithic was now vital to uphold Hitchcock's much-disputed explanation of the tracks, and with it the wider validity of the law of correlation. This was especially needful as Huxley, whose
initial antagonism toward Cuvier in the mid-1850s had been unrelated to
Darwin's views on species mutation, deployed the evolutionary affinities between birds and reptiles as a potent new weapon in his ongoing campaign
against correlation.

In the wake of Darwin's *Origin of Species* and the fossil remains of transitional forms that soon after came to light, Cuvierian correlation faced a profound crisis. Rather than simply being eclipsed by evolution, however, it was,
as this chapter will show, often modified and adapted to the new scientific
priorities of the post-*Origin* era. Darwin himself actually appropriated key
elements of Cuvier's law in his own law of correlated variability, to which he
devoted an entire chapter of *The Variation of Animals and Plants under Domestication* (1868). Crucially, Darwin's detailed elaboration of his own version
of correlation denuded the original of its erstwhile antievolutionary potential. Owen, at last breaking cover on the incipient evolutionism he had cultivated since the 1840s, publicly disputed Cuvier's views on the fixity of species,
explaining the complex functional adaptations of the Madagascan aye-aye as
the result of a preordained law of evolution. At other times, though, Owen
still adhered to an unreconstructed Cuvierian orthodoxy, endeavoring to
impose the French savant's paleontological methods on obdurate colonial
naturalists in Australia who complained that the law of correlation could not
account for the peculiar forms of the indigenous fauna and was therefore not
universally applicable. Amid the transformative scientific developments of
the 1860s and 1870s, the famous principle Cuvier had first formulated more
than half a century before was both modified and contested, but it was far
from forgotten.

Preening the Plumage of *Archeopteryx*

In January 1863 the *Geologist* excitedly reported that the "singular fossil" Waterhouse had purchased from the stubborn Bavarian physician was now "placed in the Gallery of the British Museum, where geologists who feel an interest in this remarkable discovery—and many unscientific persons, too, attracted to it by the notoriety it has attained—have flocked to inspect the blocks of lithographic limestone" (fig. 9.1).[12] While the *Geologist*, as discussed in chapter 7, had been an innovative commercial organ of popular science when it was founded in 1858, by the early 1860s the market for specialist periodicals that made science accessible to middlebrow reading audiences had become increasingly crowded.[13] The sensational "fossil Bird with the long tail" that Darwin considered "by far the greatest prodigy of recent times" was an opportune source of exciting journalistic copy that might entice readers to these new popular science periodicals, with the *Geologist's* editor already aware that "accounts of this extraordinary fossil have appeared in almost every daily and weekly journal."[14] At the end of 1862, the *Intellectual Observer*, which competed for readers with both the *Geologist* and the *Popular Science Review*, responded to the prospect of the "unique fossil" being "open to the observation of all the world" with an insider's account of the specimen penned by an ambitious assistant in the British Museum's geological department.[15] Henry Woodward would later reflect that Owen was "eager to conceal his treasure from the curious and inquiring eyes of youthful aspirants," and he clearly had much less access to the coveted fossil than his possessive superior.[16] Despite these constraints, Woodward's popularizing article cast doubt on whether the "feathered enigma" could be considered a "bird at all," and, like Wagner, he concluded that a "flying reptile *might have been endowed with feathers.*"[17]

Significantly, Woodward's contribution to the *Intellectual Observer* began by observing:

> It has always been held that the form and proportions of any symmetrical object may be inferred from the inspection of a part or fragment of the whole. The disciples and admirers of Cuvier have often asserted that it was possible for those who (like the great anatomist) possessed the requisite knowledge, to reconstruct the entire frame of any extinct animal from a single bone, or tooth, or claw. These notions went, doubtless, far beyond the pretensions of the illustrious founder of the science of comparative osteology, and they have led to mistakes and disappointment; but they serve to show the strength of the conviction which long ago found expression in the proverb, "*ex pede Herculem*".[18]

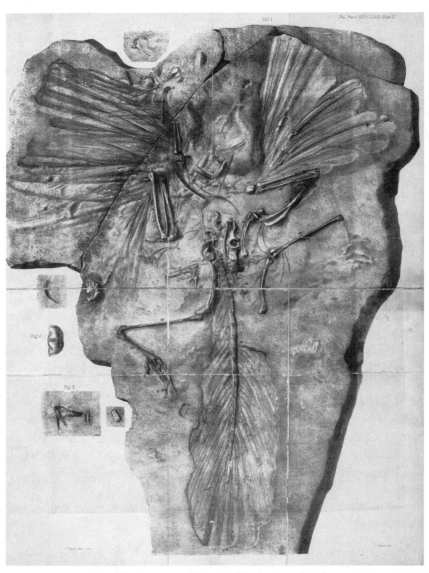

FIGURE 9.1. Joseph Dinkel, *The Moiety of the Split Slab of Lithographic Slate, Containing, with the Impressions of the Feathers, the Major Part of the Fossilized Skeleton of* Archeopteryx. Richard Owen, "On the *Archeopteryx* of Von Meyer," *Philosophical Transactions of the Royal Society* 153 (1863): plate 1. This lithograph—ironically of the very type of stone from which lithographs were printed—shows the slab of matrix containing the feathered skeleton that Owen obtained, at great expense, from Germany and then employed to authoritatively classify the archaeopteryx as avian at the Royal Society. It attracted thousands of eager visitors when it was subsequently put on public display at the British Museum. © The Trustees of the Natural History Museum, London.

Although Woodward was less scornful than the *Popular Science Review* had been in January, he too implied that the notorious fossil soon to be put on public display would reconfirm the mistakes and disappointment made inevitable by Cuvier's vaunted reconstructive abilities, no matter how strongly his disciples and admirers maintained their flawed convictions. Having paid such a lavish fee for the feathered fossil (some of which, potentially, even came from his own pocket), Owen now saw it used by a junior member of the British Museum's staff to impugn the law of correlation. Even worse, the perfidious criticism appeared in the kind of popular periodical where the Cuvierian law, as seen in chapter 7, had hitherto remained largely impervious to the ferocious criticisms unleashed by Huxley in the previous decade.

Woodward completed his callow copy for the *Intellectual Observer* on 10 November 1862, acknowledging that, with the journal not being published until the traditional "magazine day" at the end of the month, even "before the issue of this present number" the feathered fossil would already have been "described by Professor Owen before the Royal Society."[19] Only ten days after Woodward set down his pen, Owen rose to address what one attendee called "one of the largest meetings of the Royal Society we remember to have seen."[20] The massed ranks of the Royal Society's eminent fellows were apprised of the "true class-affinity of the *Archeopteryx*," with Owen demonstrating the "positive proof of the ornithic proportions" of its skeleton, which, he insisted, afforded "no evidence of a reptilian structure" whatsoever. Although the head of the specimen appeared to be missing, Owen concluded that

> the best-determinable parts of its preserved structure declare it unequivocally to be a Bird, with rare peculiarities indicative of a distinct order in that class. By the law of correlation we infer that the mouth was devoid of lips, and was a beak-like instrument fitted for preening the plumage of *Archeopteryx*.[21]

Not only was the archaeopteryx not an incongruous jumble of taxonomic characters that would inevitably stymie Cuvier's long-standing method of reconstruction, but the law of correlation could even be deployed to infer the distinctively ornithic structure of its absent head. With Wagner now dead— apparently "killed" by his "spite at Darwinianism"—and Meyer having less opportunity to view the fossil than even the lowliest plebeian visitors to the British Museum, Owen's interpretation of the archaeopteryx authoritatively rebuffed the emerging consensus that such apparently intermediate forms invalidated the fundamental premises of Cuvierian correlation.[22]

Accounts of Owen's particular interpretation of the archaeopteryx soon also appeared in the new commercial popular science journals that had previously implied that the feathered fossil would vitiate the law of correlation.

Barely a month after the Royal Society's thronged meeting and before the publication of Owen's own paper in the *Philosophical Transactions*, his ornithic classification of the archaeopteryx was reiterated in an anonymous article for the *Geologist* written by its editor Samuel Joseph Mackie. In particular, Mackie noted that

> an article in our contemporary "The Intellectual Observer", written with much care and a complete acquaintance with the bibliography of the subject, by Mr. Henry Woodward, of the British Museum, would seem to leave an open doubt that the Archæopteryx might have reptilian affinities.

Slyly insinuating that Woodward's principal acquaintance was with the "bibliography of the subject" rather than the actual fossil specimen, Mackie thundered: "Neither of these surmises are tenable."[23] Woodward evidently took the hint and soon fell into line, reflecting in a further popularizing account of the archaeopteryx, this time for the *Popular Science Review*, on how the "assistance of Professor Owen, disposed of the difficulty arising from the law of correlation, which requires that a beak and feathers should be associated together."[24] The *Geologist*, as seen in the previous chapter, had already carried several articles defending Cuvier's paleontological methods against Huxley's trenchant criticisms, and its editor, who elsewhere admitted that he personally felt it "imperative to show distinctly that it [i.e., the archaeopteryx] does not possess reptilian characters," now helped Owen to contain the concerns over the law of correlation raised by Woodward in a rival popular journal.[25] More established foes of the same Cuvierian law could not be put off so easily, though.

An Advocate Urging a Plea for the Birds

Since the early 1830s, the Royal Society's *Philosophical Transactions* had employed an innovative process of peer review. Refereeing Owen's paper on the archaeopteryx, Hugh Falconer acknowledged that it was "on a subject of high interest and importance in Palæontology—giving a description of one of the most remarkable and least expected forms that has ever turned up in the fossil state." The author, Falconer observed, had "seized upon the indisputably strong evidence of the Bird" and "argues the case ably from the bird-aspect throughout." Despite the submission's manifest strengths, though, there was, Falconer complained, a "strong leaning upon one side so that the paper has assumed somewhat the character of an advocate urging a plea for the birds rather than that of a well-balanced exposition of the case."[26] Falconer's confidential referees' report was dated 20 December 1862, only six months after he had finally recanted his erstwhile endorsement of the law of correlation and, as seen in chapter 6, joined forces with Huxley and Darwin in their vitriolic antagonism toward Owen. Falconer had recently disputed Owen's pompous profession that his classification of a carnivorous marsupial was "founded on the high ground of a connected series of physiological correlations," and his accusation that Owen's partisan interpretation of the archaeopteryx was "more specious than convincing" now once more challenged his adversary's invocation of the same elevated terrain.[27]

At the end of his censorious referees' report, Falconer proffered a casual afterthought, "I observe in some of the weekly and monthly periodicals that Mr. John Evans is reported to have detected the cast of the cavity of the skull &c.—previously unobserved—on the slab of matrix containing the fossil," before nonchalantly adding: "The statements have, doubtless, attracted the attention of the author."[28] But the apparent insouciance of this addendum was no less calculating than anything in Owen's allegedly legalistic paper, for Evans, a geologist and antiquary, was a close friend of the peer reviewer.[29] In a private letter, Falconer had responded to his friend's discovery of the missing skull—comprising a jaw encased in the main slab of the matrix and a negative impression of the brain cavity on the counterpart slab—with undisguised glee:

> Hail, Prince of Audacious Palæontologists! Tell me all about it. I hear that you have to-day discovered the *teeth* and jaws of the *Archæopteryx*. To-morrow I expect to hear of your having found the liver and lights! And who knows, but that in the long run, you may get hold of the *fossil song* of the same creature, impressed by harmonic variation on the matrix.[30]

Evans's location of these parts only weeks after the original paper at the Royal Society was not only a chastening correction of an embarrassing interpretative oversight but also had the potential to immediately negate the ornithic version of the mouth that Owen had inferred by means of the law of correlation.

After all, the jaw Evans found in the matrix containing the remainder of the archaeopteryx's skeleton, although distinctly birdlike in most respects, showed the presence of teeth, which, as the exultant italicization in Falconer's missive suggests, could not be accounted for on Owen's exclusively ornithic classification. There were instead, as Evans later reflected, "many analogies with the jaws of some species of fish," leading him to conclude that, while it was "of course contrary to all our existing notions to suppose that a jaw, such as this, armed with teeth, could belong to a creature so truly bird-like in most respects as the Archæopteryx," the "presence of feathers does not of necessity imply that the beak with which to preen them should be edentulous."[31] This was precisely what Owen had inferred of the archaeopteryx's beak at the end of his Royal Society paper, with Evans even echoing the very terms of his correlative supposition that the toothless beak was "fitted for preening" feathers. Once again, it was contended that the prevailing assumptions underpinning Cuvierian correlation were rendered invalid by the unpredictability of such intermediate forms.

Although Evans had immediately alerted Owen to the neglected skull in the matrix on display at the British Museum and seems to have deposited a "valuable . . . unpublished document" with Falconer, he refrained from publishing his own account of the discovery. It was nevertheless not long before news of the archaeopteryx's misplaced head appeared in the press, as Falconer had noted in his referees' report. These journalistic reports, however, did not have Evans's approval, nor, crucially, did they concur with his own particular interpretation of the skull. As Evans's wife, Frances, recorded in her diary in February 1863: "His brain was mentioned in an article in the *Geologist* of 1st Jan.—without his permission—his jaw is spoken of in the 1st Feb. *Geologist* without any mention of his name at all, which is almost worse."[32] When the first of these illicit allusions was published, Falconer told Darwin:

> A very ludicrous event has turned up. John Evans appears to have hit upon the very obvious cast of the interior of the skull—undetected by the describer, and before Owen's paper is out, we have Mr. Mackie describing the hemispheres—and optic lobes of Archæopteryx! Look at the Geologist.[33]

If he followed Falconer's excited exhortation, Darwin would soon have noticed that the *Geologist*'s editor had not worked alone in describing the contents of the newly discovered cranium.

In addition to slyly denigrating Woodward and dutifully reiterating the arguments of Owen's Royal Society paper, Mackie's anonymous article also proclaimed that "with the assistance of Mr. Carter Blake, we believe we have decisively made out the actual parts of the brain . . . for the apt means of the determination of which too much praise cannot be given to Mr. Evans."[34] Charles Carter Blake, as seen in the previous chapter, was a tenacious—and often predacious—protégé of Owen, whom Falconer derided as a "silly empty-headed young man" that Owen had "got hold of . . . for his dirty work."[35] Unsurprisingly, the analysis of the archaeopteryx's brain in which Blake assisted upheld a conclusion entirely different from that of Evans, who had hesitated over reaching a classificatory conclusion from the slight traces of the brain and spurned any "hypothesis built on such slender foundations."[36] In Blake and Mackie's view, on the other hand, the "presence or known existence of the head . . . prevented any reptilian mystery," affording clear "evidence" that "goes far to support the admirable inferences of Professor Owen, as the fossil brain presents true bird's characters, and can thus be perfectly distinguished from the very peculiar form of brain in reptiles."[37] Blake, a founding member of the Anthropological Society's Cannibal Club whom Huxley dubbed "a jackal of Owen's," was once more sinking his teeth into anyone whose opinion differed from that of his mentor.[38]

When, in the following month's *Geologist*, the vexed issue of the archaeopteryx's jaw was broached, Evans was not only not named, but his opinion of the teeth was silently expunged. Instead, it was proposed that the jaw had nothing to do with the archaeopteryx and was actually the "upper part of the head of a lepidoid fish" that had become accidentally mixed up in the matrix.[39] This convenient conclusion seems to have originated in unofficial briefings from within Owen's own institutional stronghold, where, as Falconer informed Darwin, "Waterhouse I am told pronounces it to be a *fish's* jaw."[40] Mackie too was told the same thing by "Mr. [William] Davies, of the Palæontological Department of the British Museum," and, although he later confessed that this "'fish-head' theory did not . . . rest easily on my mind," it was Mackie's own contributions to the *Geologist* that ensured that the oral speculations of the British Museum's staff reached a wider audience.[41] Even in the mid-1860s, Owen could evidently still rely on the same informal contacts with editors that he had utilized so adeptly in earlier decades. Indeed, Falconer quickly recognized how Evans's discovery was being distorted and deliberately denuded of its potential to harm Owen's correlative inferences regarding the archaeopteryx's edentulous beak, and he advised his friend: "Keep your own counsel, and do not let yourself be blarneyed out of it. Have nothing to say to the *Geologist*."[42]

The Grand *Darwinian* Case of the *Archæopteryx*

When Evans did finally publish his own account of the archaeopteryx's toothed jaw, he was adamant that its "extreme importance as bearing upon the great question of the Origin of Species must be evident to all."[43] Falconer agreed with his friend on the "grand *Darwinian* case of the *Archæopteryx*," telling Darwin that "had the Solenhofen quarries been commissioned—by august command—to turn out a strange being à la Darwin—it could not have executed the behest more handsomely." Darwin himself "particularly wish[ed] to hear about the wondrous Bird," assuring Falconer that the "case has delighted me."[44] Huxley, by contrast, was noticeably less excited than his friends and allies. In fact, having recently argued, as seen in the last chapter, that the South American pachyderm the macrauchenia did not link together ruminants and pachyderms while nonetheless stymieing Cuvierian correlation, he now similarly combined a categorical refutation of Owen's correlative interpretation of the archaeopteryx's beak with a more ambivalent stance on the peculiar creature's status as an intermediate form.

Back in January 1863, the *Geologist* had concluded its account of Owen's paper by complaining that the "palæontologists who were silently present at the Royal Society's meeting, or who were 'conspicuous by their absence', whose opinions we should be glad to know, have maintained a significant silence."[45] This, according to Adrian Desmond, was a surreptitious reference to Huxley, who either did not attend or sat through Owen's detailed description of the archaeopteryx and the subsequent discussion of his conclusions without offering any comments.[46] Huxley's reticence on the subject would persist for another five years, lasting until January 1868, when he too addressed the fellows of the Royal Society on the "unique specimen . . . which . . . is undoubtedly one of the most interesting relics of the extinct fauna of long-past ages."[47]

Huxley's intention, he declared, was simply to "rectify certain errors which appear to me to be contained in the description of the fossil in the Philosophical Transactions for 1863." The most egregious fault in Owen's paper had come in the conclusion, and Huxley responded to it with characteristic asperity:

> I may remark that I am unaware of the existence of any "law of correlation" which will enable us to infer that the mouth of this animal was devoid of lips, and was a toothless beak. The soft tortoises (*Trionyx*) have fleshy lips as well as horny beaks; the *Chelonia* in general have horny beaks, though they possess no feathers to preen; and *Rhamphorhynchus* combined both beak and teeth, though it was equally devoid of feathers. If, when the head of *Archæopteryx* is discovered, its jaw contains teeth, it will not the more, to my mind, cease to be a bird, than turtles cease to be reptiles because they have beaks.[48]

Notably, Huxley did not dissent from Owen's ornithic classification of the archaeopteryx, which fitted with his own understanding of the long persistence of birds as a distinct taxonomic class.[49] It was instead Owen's imperious assumption that, notwithstanding the manifest evidence to the contrary afforded by both living and extinct fauna, there was a necessary and invariable correlation between feathers and edentulous beaks that provoked Huxley's ire. What the archaeopteryx showed most clearly, as Huxley later reflected, was that "animal organization is more flexible than our knowledge of recent forms might have led us to believe; and that many structural permutations and combinations, of which the present world gives us no indication, may nevertheless have existed."[50] It was in fact the very absence of such rigid correlations that meant that the archaeopteryx could combine teeth and feathers without ineluctably representing a "missing link" between separate orders. Unlike Evans, Falconer, or even Darwin himself, Huxley reflected guardedly: "So far as the specimen enables me to judge, I am disposed to think that, in many respects, *Archæopteryx* is more remote from the boundary-line between birds and reptiles than some living *Ratitæ* are."[51]

Only a month later, Huxley elaborated on this singular view in a Friday Evening Discourse at the Royal Institution. He averred that the "precious skeleton of the *Archæopteryx*" was merely a "very early bird," and contended that "those who hold the doctrine of Evolution (and I am one of them)" needed to look elsewhere for an "approximation to the 'missing link' between reptiles and birds." While Huxley conceded that "all existing Birds differ thus definitely from existing Reptiles," he maintained that "one comparatively small section comes nearer Reptiles than the others. These are the *Ratitæ*, or struthious birds." Huxley then advised his audience that the "extinct Reptiles which approached these flightless birds" would only be found "if we cast our eyes in what at first sight seems to be a most unlikely direction." Indeed, it was the "*Dinosauria*, a group of extinct reptiles . . . for the most part, of gigantic size" that, Huxley avowed, "appear to me to furnish the required conditions."[52] Rather than the celebrated archaeopteryx, it was a dinosaurian specimen from the very same limestone quarries in Solnhofen, the diminutive *Compsognathus*, that, for Huxley, constituted a more plausible "missing link" (and he later also proposed the *Hesperornis*, an aquatic struthian with teeth discovered in Kansas in the early 1870s, as the closest intermediate link from the bird side).[53]

What made this surprising conclusion viable was that "from the great difference in size between the fore and hind limbs, Mantell, and more recently Leidy, have concluded that the *Dinosauria* (at least *Iguanodon* and *Hadrosaurus*) may have supported themselves, for a longer or shorter period, upon their hind legs."[54] In complete contradistinction to the bulky quadrupedal dinosaurian models constructed for the Sydenham Crystal Palace in 1854, the American paleontologist Joseph Leidy, only four years afterward, proposed that the hadrosaurus, a fossil skeleton of which had recently been excavated in New Jersey, was bipedal, something that Gideon Mantell had intimated might be the case with the iguanodon as far back as the early 1840s. Ironically, it was Benjamin Waterhouse Hawkins, the artist who had worked with Owen on the stout-limbed brick-and-mortar models at the Crystal Palace, who, in November 1868, first articulated the skeleton of the hadrosaurus in Leidy's pioneering bipedal posture.[55] Huxley, nine months earlier, had utilized the conclusions of Mantell and Leidy to claim that the "hind quarters of the *Dinosauria* wonderfully approached those of birds in their general structure."[56]

This new conception of the former existence of "reptiles more like birds" also entailed a revised understanding of the renowned "sandstones of Connecticut," which, as Huxley reflected, "have yielded neither feathers nor bones" but contain "innumerable tracks which are full of instructive suggestion." Disregarding the prevailing interpretation of these three-toed footprints as exclusively ornithic, Huxley contended that the "important truth which these tracks reveal is, that . . . bipedal animals existed which had the feet of birds, and walked in the same erect or semi-erect fashion. These bipeds were either birds or reptiles, or more probably both." The intermediate creatures that had left such hybrid footprints would, in all likelihood, "obliterate completely the gap . . . between reptiles and birds."[57] While Huxley's skepticism that the archaeopteryx represented the "dawn of an oncoming conception à la Darwin" meant that he remained largely indifferent to the incongruous creature that several commentators, both in Germany and in the new middlebrow popular science journals, implied would decisively overthrow the law of correlation, his own alternative interpretation of the transitional bipedal dinosaurs that left birdlike tracks in Connecticut had no less problematic implications for Cuvier's increasingly beleaguered axiom.[58]

The Efficacy of a Foot-Print

To understand these awkward implications, it is necessary to return briefly to the period when the fortunes of the Cuvierian law of correlation were at their apogee. The fossil footprints in the Triassic New Red Sandstone of the Connecticut valley that Huxley now considered so suggestive were first noticed by laborers at the beginning of the nineteenth century, although it was a further three decades before their significance became apparent to local naturalists (fig. 9.2).[59] In 1836 the *American Journal of Science* published the initial conclusions of Edward Hitchcock, a Congregationalist pastor who had left the ministry to become professor of natural history at Amherst College. He confidently announced that the two-footed tracks "could not have been made by any other known biped, except birds," and consequently named them "*Ornithichnites.*" By examining the "character of the foot, and the length of the step," Hitchcock identified several new ornithic species, which, as well as being considerably older than any other birds then known, must also have been much larger than their modern counterparts. This capacity to "determine the character of an animal from its footmark" was the principal facet of the "new branch of paleontology" specializing in fossil footprints that

FIGURE 9.2. Francis Arthur Lydston, *The Moody Foot Mark Quarry, South Hadley*. Edward Hitchcock, *Ichnology of New England* (Boston: William White, 1858), plate 1. This small quarry by the side of the main highway in the Massachusetts town of South Hadley contains the tracks of a bipedal animal Hitchcock identified as a bird and named *Otozoum moodii*. © The British Library Board, 7203.e.31.

Hitchcock inelegantly named "Ichnolithology," prompting William Buckland, across the Atlantic, to coin the more concise alternative *Ichnology*.[60] Having read the American reprint of the translation of Cuvier's "Discours préliminaire" that Samuel Latham Mitchill had edited in 1818, Hitchcock recognized that the "sagacious mind of Cuvier has anticipated this principle" in

the passages where he boldly asserted that he could surmise the entire structure of a ruminant from merely a single footprint with more certainty than
Voltaire's Zadig.[61] Hitchcock's insistence that it "required only to extend this
principle to other tribes of animals to constitute ichnolithology in its present
state" explicitly aligned his own examination of the New England footprints
with some of the most notoriously hyperbolic passages in Cuvier's oeuvre.[62]
Hitchcock, as Claudine Cohen has recently proposed, was self-consciously
working "in the wake of the great Cuvierian tradition."[63]

The "adoption in America of the theoretical framework established by
Georges Cuvier," as Patsy Gerstner has proposed, occurred during the mid-
1830s, but notwithstanding his conspicuous invocation of the French savant's
putatively infallible methods, both the unprecedented antiquity and the
enormous size of Hitchcock's sandstone birds ensured that naturalists, on
both sides of the Atlantic, were wary of endorsing his ornithic interpretation
of the footprints.[64] These apparent "difficulties" were soon "unexpectedly
removed," as Hitchcock exultantly observed, by Owen's "discovery of the
Dinornis" in early 1843, whose gigantic proportions appeared to spectacu-
larly affirm his own inferences. Although Owen had himself actually doubted
the ornithic character of the Connecticut tracks, Hitchcock effusively lauded
him as the "man on whom so deservedly the mantle of Cuvier rests," and
who, having been "able to construct the Dinornis from a single fragment of
the shaft of a bone," would now similarly "place before us some Dinornis of
sandstone days."[65] Conveniently disregarding his own erstwhile skepticism,
Owen even implied that Hitchcock's bold surmises regarding the giant birds
of the Connecticut valley had assisted him in first reconstructing the dinornis, telling him: "Your beautiful discovery of the Ornithichnites has always
been in my thoughts while working out the New-Zealand bones."[66] If, as
Owen brooded, "a single bone has been deemed insufficient to give the entire
animal, with more reason may we doubt the efficacy of a foot-print," but
equally the manifest success of either technique helped confirm the other.[67]
As Owen appreciated, the parallels between the colossal birds of New England
and New Zealand mutually reinforced the same Cuvierian method that both
he and Hitchcock had employed.

By the late 1850s, however, Hitchcock had encountered new footprints
that, because of the accompanying marks of anomalous tails and heels, he
struggled to accommodate with his previous exclusively ornithic classifications. This, as he recognized in 1863 despite the distractions of the "disastrous
war" that had engulfed America, would be no less disastrous for the validity of
the Cuvierian method on which his original interpretations had been based.
Explaining that the "number of phalanges, and their order in the toes of liv-

ing birds, enables the anatomist to distinguish them from other animals," Hitchcock ruefully acknowledged that if "in the fossil foot-marks, birds cannot be distinguished from quadrupeds by the number of phalanges," then consequently the "law of correlation among living animals would seem not to have been true with the fossil."[68] In particular, it was the amorphous genus he named the *Anomoepus*, which combined the forefoot of a bipedal bird with the hindfoot of a lizard or even a marsupial, that potentially had the most deleterious implications for Cuverian correlation.

Such intermediate forms, of course, were posing precisely the same problems for Owen on the other side of the Atlantic, and Hitchcock quickly recognized the magnitude of the "recent discovery of a remarkable animal called by some *Griphosaurus*, and by others, Archæopteryx." Just like the dinornis two decades earlier, this enigmatic new creature, he averred, "throws some light" on the New England footprints, "while they reflect some light upon the feathered fossil."[69] The descriptions of the celebrated fossil in European periodicals that were soon reprinted in the *American Journal of Science* suggested to Hitchcock that its skeleton exhibited the very same features as the perplexing tracks left by members of the *Anomoepus* genus in the Connecticut valley.[70] With Hitchcock able to "ally the Archæopteryx . . . to the Anomœpus," Owen's determination, with the former, to "make it a bird" assumed an even greater importance. His ornithic classification of the feathered fossil now ensconced at the British Museum would, after all, entail that the *Anomoepus*, no matter how incongruous its form, was similarly a bird. And this had profound implications for both the original interpretation of the footprints and the wider validity of the law of correlation. As Hitchcock observed:

> If the Anomœpus were a lizard or a marsupial, we must give up that firmly established law of correlation, which enables us to distinguish different classes of animals by the number and order of phalanges, but if it were a bird, the law can still be reckoned upon among the fossil as well as living animals. . . . If we can presume that Anomœpus was a bird, it lends strong confirmation to another still more important conclusion, which is that all the fourteen species of thick-toed bipeds which I have described . . . were birds. In this case, if we retain the law as to the phalanges, all the characters of the animals as made known by their tracks, belong to birds, with little variation from the existing bird type.

While Hitchcock had feared that the "opinion which I have always maintained was becoming doubtful" and confessed to feeling "tossed on the sea of difficulty" about the footprints, he proclaimed that the "history of the Solenhofen fossil" purchased by the British Museum meant that "we may now with more confidence than ever, maintain the ornithic character of these animals."[71] With Owen's authoritative conclusion that the archaeopteryx was a bird now integral to the fortunes of the law of correlation on both sides of the Atlantic, his lavish expenditure on the Bavarian feathered fossil was beginning to seem a bargain.

When One Part Varies

Akin to the supposition of both Hitchcock and Owen that the New England "*Ornithichnites* are the impressions of the feet of birds, which had the same low grade of organization as the . . . *Dinornis* of New Zealand," Huxley too, in his Friday Evening Discourse at the Royal Institution in February 1868, considered that the "creatures which traversed . . . the sandstones of Connecticut . . . when they were the sandy beaches of a quiet sea" resembled the "but recently extinct (if they be really extinct) birds of New Zealand, *Dinornis*, &c."[72] But, for Huxley, such a "Dinornis of sandstone days," as Hitchcock had termed it, would have been a transitional form that demonstrated the "manner in which Birds may have been evolved from Reptiles."[73] As well as affording valuable evidence for the "doctrine of Evolution," the intermediate bipedal dinosaurian creatures that had been gradually transmuted into giant struthians such as the dinornis, and that combined the "feet of birds" with other distinctly reptilian features, also invalidated the assumption, shared by Cuvier, Owen, and Hitchcock, that a mere footprint could reliably reveal the rest of the animal.[74] Huxley did not go as far as his fellow critic of correlation Hermann von Meyer, who indignantly complained: "Ichnology, or the whole theory of fossil footprints, reposes only upon phenomena of resemblance; and although philosophers of the highest rank are to be found among its defenders . . . it is still destitute of a scientific foundation."[75] In fact, far from impugning the very basis of this new branch of paleontology, Huxley utilized the "numerous figures and descriptions of footprints" in Hitchcock's *Ichnology of New England* (1858) during his own investigations of tracks left by prehistoric crocodiles in the Elgin sandstone of northern Scotland.[76] Huxley's evolutionary speculations on the Connecticut valley footprints, in which he circumvented Owen's ornithic classification of the archaeopteryx and instead invoked that earlier avian icon of the law of correlation the dinornis, nonetheless refuted the conspicuously Cuvierian interpretation of the same tracks that Hitchcock maintained until his death in February 1864.

Huxley's suspicion that the fossil tracks in New England were made by strange intermediate creatures rather than simply large birds was shared by Darwin, who reflected to James Dwight Dana in Connecticut: "How little we know what lived during former times. Oh how I wish a skeleton could be found in your so-called Red Sandstone footstep-beds."[77] It was precisely the absence of such skeletal remains that had induced Hitchcock to place such faith in the heuristic capacity of the footprints, and Darwin's comments to Dana indicate that he did not have the same blithe confidence in their reliability. But Darwin, as has already been seen in previous chapters, was never as scornful as Huxley of the Cuvierian principles on which Hitchcock's trust resided, and, regardless of the strategic alliances forged during the heated paleontological disputes of the mid-1850s, there remained residual differences over correlation between the two men. In fact, in the short interval between the addresses at the Royal Society and the Royal Institution in January and February 1868 in which Huxley not only disputed the ornithic interpretation of the footprints but denied the "existence of any 'law of correlation'" between avian characters, Darwin published a detailed account of his own version of the same law in *The Variation of Animals and Plants under Domestication*. In this two-volume tome, the longest book of Darwin's career, he elaborated on the concept of "correlation of growth" that he had introduced a decade earlier in the *Origin* to denote how an "organization is so tied together during its growth and development, that when slight variations in any one part occur, and are accumulated through natural selection, other parts become modified."[78]

Darwin's interest had first been piqued by the correlation of parts in the late 1830s, when, as seen in chapter 6, he jotted down his initial thoughts on its implications for variation in his secret notebook on transmutation. Three decades later, these clandestine surmises would yield an entire chapter of the expansive new book in which he returned to material originally included in *Natural Selection*, the much longer work he had abandoned in 1858 to hurriedly complete the *Origin*. At the beginning of chapter 25 of *The Variation of Animals and Plants under Domestication*, Darwin announced: "All the parts of the organisation are to a certain extent connected or correlated together. . . . Certain structures always co-exist; for instance, a peculiar form of stomach with teeth of peculiar form, and such structures may . . . be said to be correlated." For Darwin, of course, such correlated structures were perpetually subject to variation, and he observed that "when one part varies, certain other parts always, or nearly always, simultaneously vary." These concurrently modified parts, Darwin proposed, were "subject to the law of correlated variability." While he did not substantively develop the concept of nonadaptational correlative changes being effected by natural selection modifying any one part that he had designated by the "somewhat vague expression of correlation of growth" in the *Origin*, he was now more definite in

according it the status of a veritable scientific law, as well as providing numerous examples of its operation throughout the natural world.[79]

Unlike in the *Origin*, in *Variation* Darwin included footnoted references
to the sources he drew upon. He made no allusions to Cuvier during his
discussion of the structural constraints of correlation, and it has sometimes
been assumed that the French savant's renowned law had no bearing on
Darwin's own considerations. Michael Ghiselin, for instance, has emphatically avowed that the "principle of the correlation of parts . . . is not to be
confused with Darwin's idea of correlated variability."[80] In a portion of the
manuscript of *Natural Selection* that was not subsequently incorporated into
Variation, however, Darwin appears to himself acknowledge precisely such
a connection. Observing how "every modification" is "subjected to the laws
of the correlation of growth & to the direct action of the conditions of existence, as food & climate," he asserted: "This conclusion seems to me to
accord sufficiently well with the famous principle enunciated by Cuvier."[81]
Darwin, as both E. S. Russell and John Reiss have pointed out, seems to have
misunderstood Cuvier's particular conception of the conditions *for* existence,
conflating his own notion of environmental conditions *of* existence to which
organisms become adapted with the specific Cuvierian sense of appropriate
correlations of organs that provide the preconditions that enable creatures
to exist.[82] Darwin was nonetheless alert to the significance of such correlations in understanding animal organization, even if he remained indifferent
to their purported role in allowing organisms to meet the fundamental requirements of existence. After all, he then cited a short section from an 1830
article in the *Annales des sciences naturelles* in which Cuvier insisted on the
"appropriateness of the parts, of their coordination for the role the animal is
to play in nature."[83] Notably, Darwin had actually encountered this pertinent
passage, and annotated it with a marginal "Q," indicating a possible quotation, while reading Étienne Geoffroy Saint-Hilaire's *Principes de philosophie
zoologique* (1830), which only used it in order to fiercely contest Cuvier's particular understanding of animal structure.[84]

Despite Darwin's having apparently come across Cuvier's "famous principle" in the work of his bitterest rival, the Cuvierian law of correlation clearly
served as a stimulus for Darwin's own law of correlated variability. As Timothy
Shanahan has proposed: "Darwin was genuinely pluralistic in his approach
to explaining biological phenomena . . . [and] much impressed by Cuvier's
principle of the 'Correlation of Parts', which treated organisms as integrated
systems."[85] Indeed, for some contemporaries the two men's opinions on the
matter appeared to be virtually interchangeable, with the Unitarian minister
James Martineau insisting in his *A Study of Religion* (1888) that the "same

law which Cuvier announces as 'correlation of organs', appears in Darwin's writings as 'correlation of growth.'"[86] And with the "nictitating membrane a necessary correlation of eyelids," as Darwin jotted in the margins of a book by the German naturalist Gustav Jäger, correlation, in his view, was seemingly no less ineluctable than it was in Cuvier's.[87] Such unvarying and thus potentially perfect correlations could even be taken to imply providential intention, exactly as they had been by the proponents of William Paley's *Natural Theology* (1802) in the opening decades of the century. This was certainly the opinion of the High Church Anglican James Bowling Mozley, who observed in 1869: "The law of correlation of growth has so obviously the look of an arrangement that it figures in Paley's theology as one of the proofs of design. . . . It is correlation in a structure formed for use, and whose use stops half way and waits for correlation to complete it."[88] Patently indebted to Cuvier and operating independently of the randomness and strict utility of natural selection, the concept of correlated variability threatened to reinstate the very Paleyite argument from design that Darwin had otherwise overthrown.

But notwithstanding the parallels discerned by religious commentators such as Martineau and Mozley, Darwin's law of correlated variability diverged decisively from Cuvier's own version of correlation in at least one crucial respect. The simultaneous variations across an entire organism, enacting structurally forced modifications, finally denied the original Cuvierian law its erstwhile antievolutionary potential. Cuvier's argument that the perfect correlation of the animal frame necessitated holistically coordinated alterations that were so complex as to render evolutionary change untenable had, as seen in earlier chapters, done much to check the transformist theories of his Parisian colleague Jean-Baptiste Lamarck, and was subsequently deployed in Britain as the most effective bulwark against domestic versions of transmutationism. By demonstrating that if "one part is modified through continual selection, either by man or under nature," then, without requiring the direct intervention of either domestic breeding or natural selection, "other parts of the organisation will be unavoidably modified," Darwin identified a mechanism that facilitated precisely the intricately coordinated changes across the whole organism that Cuvier and his followers had maintained were impossible.[89]

By the 1890s the French paleontologist Félix Bernard was using Darwin's concept of "variations in *correlation*" and the "simultaneous modifications in various parts of the skeleton" that it entailed to reappraise how Cuvier's "errors . . . were a point of departure for a progressive movement."[90] While Darwin evidently drew upon crucial aspects of the law of correlation, he did it in a way that not only denuded the famous Cuverian axiom of what had long been regarded as its most valuable attribute, but even appropriated it to the new evolutionary paradigm inaugurated by the *Origin*. Although Darwin's surprising but nonetheless keen interest in Cuvier's law of correlation has only rarely been noted by historians, and even then merely in passing, it had hugely significant implications for the continued relevance of the much-contested law in the latter part of the nineteenth century and also complicates conventional assumptions about the triumphant refutation of older, nonevolutionary models of science by Darwin and his scientific naturalist followers.[91]

Variations of Plan

Darwin also departed from Cuvier in contending that those parts, such as jaws and limbs or teeth and hair, that are "identical in form and structure at an early period of embryonic development" and can therefore be considered "homologous" in accordance with the morphological conception of a vertebral archetype, were more liable to "vary in the same manner." This ancillary principle of the "*Correlated Variation of Homologous Parts*," as Darwin termed it in *Variation*, proposed that correlations predominantly occurred between the parallel structures that Geoffroy had earlier claimed indicated that all vertebrates were formed on a single plan, and that Darwin now attributed to a shared evolutionary ancestry.[92] As with these homological parts that had been modified to perform a diverse array of new functions, Cuvier's law of correlation was itself adapted, during the 1860s, to the new scientific priorities of the Darwinian era. In the following decade, the famous paleontological method of reconstruction from only partial fossil remains facilitated by the same Cuvierian law was similarly accommodated with the morphological emphasis on a common plan, although, notably, without any regard to Darwin's evolutionism.

While assisting the elderly Adam Sedgwick with cataloging the fossils held at the Woodwardian Museum in Cambridge, Harry Govier Seeley developed a controversial theory that pterodactyls, which had previously been classified as winged saurians akin to Wagner's feathered reptile *Griphosaurus*, were structurally much closer to birds than reptiles, and even potentially warm-blooded. In *The Ornithosauria* (1870) Seeley suggested that he had been led to this singular conclusion, in part, by his resolute eschewal of the traditional technique of "reasoning from the details of structure," which, he contended, "never makes more than an approximate guess when it endeavours to determine fundamental organization." In his analysis of the Woodwardian's extensive fossil collections, Seeley instead endeavored to "ask, not what the Ptero-

dactyle is like in its several bones, but what common plan it had whereon
its hard structures were necessarily moulded." Seeley elaborated on this apparently "*à priori* method" five years later, when, in a series of lectures at the
Royal Institution on "The Fossil Forms of Flying Animals," he contrasted
his new archetypal approach with that of Cuvier.[93] As the *Illustrated London
News* reported of his opening lecture in April 1875: "Cuvier's doctrine of the
correlation of parts in an animal was examined and explained from his [i.e.,
Seeley's] own point of view, and shown to be inapplicable to the interpretation of fossil animals which did not belong to existing ordinal groups." After
all, the numerous "variations of plan" that were possible in the peculiar fauna
of the ancient past meant that an extinct "animal might no longer be recog-
nised for a bird by the form or proportion of a single bone."[94] This, of course,
was precisely the same problem that transitional "missing links" such as the
"grand *Darwinian* case of the *Archæopteryx*" had posed, only a decade earlier,
to the applicability of the Cuvierian doctrine.

Seeley, though, concurred with his mentor Sedgwick in refusing to make
his "*new Pterodactyle*" a "monstrous form of an old one" and "throw out
species into that 'paradise of fools'—'that limbo large & broad'—now called
transmutation," and he remained defiantly oblivious to the new evolutionary approaches that were transforming the practices of his paleontological
peers.[95] The "plans of animal life," no matter how varied or incongruous they
might appear, were, for Seeley, unchanging "products of divine law."[96] In
fact, far from merely discarding the Cuvierian method of reconstruction with
all its long-standing antievolutionary and natural theological ramifications,
Seeley attempted to reorient it around such fixed transcendental archetypes
rather than the French savant's own prioritization of function. As he declared
in his next Royal Institution lecture:

> Though the Pterodactyle showed how impossible it would be to reconstruct a
> skeleton from a single bone, yet, given enough of the bones to show the plan
> of the ordinal group to which the animal belonged, the limits within which the
> remainder of the skeleton could vary were more or less closely circumscribed.

On the assumption that the "form of one part of the skeleton depends upon
the other parts," just a few representative bones were sufficient to disclose
the archetypal plan—whether avian or reptilian—to which the fragmentary
remains accorded, thereby indicating, with a high degree of certainty, what
the rest of the skeleton must have looked like.[97]

This technique enabled Seeley to illustrate his lectures with hypothetical "drawings of restored pterodactyles" in birdlike postures that he made
by "filling in the outlines of the body according to the indications of the

PTERODACTYLUS LONGIROSTRIS.

FIGURE 9.3. Pterodactylus longirostris. "Royal Institution Lectures," *Illustrated London News* 66 (1875): 447. The illustrations of restored pterodactyls that Seeley used in his Royal Institution lectures depicted them in birdlike postures with hypothetical tails added in accordance with the archetypal plan suggested by the surviving parts of the skeleton. Reproduced by permission of the University of Leicester.

muscles" and adding tails, "of which the fossil gives no indication" but that were "implied by the remainder of the skeleton" (fig. 9.3).[98] As such, this new archetypal technique retained much of the audacious prophetic power and dramatic potential of the original Cuvierian method. In fact, it was sometimes difficult to distinguish between them. Tellingly, a notice of *The Ornithosauria* in the *Popular Science Review* actually mistook the book, in which Seeley had been less explicit about his new method of reconstruction, for an orthodox affirmation of the law of correlation. It observed:

> It is impossible to be perfectly convinced that Mr. Seeley is right in the conclusion he so firmly lays down, for, after all, it is based very much on analogy, and on the supposition that the vague laws of correlation are without exception. But as we know that the law of correlation is occasionally a deceptive guide, it is possible that it may be uncertain even in Mr. Seeley's case, and that, after all, the Pterodactyle may have been a true reptile.[99]

Seeley's ornithic interpretation of the pterodactyl was rendered doubtful by its apparent reliance on the same Cuvierian principle that the *Popular Science Review* had first begun to question during the contretemps over the classification of the archaeopteryx in the early 1860s. Even when the central tenets of

the law of correlation were actually disavowed, it was difficult for new paleontological procedures to escape its lingering shadow.

If There Was a Bias, It Was toward Cuvier

Despite his own keen interest in the archetype and homologies of the vertebrate skeleton, Owen contested Seeley's radical avian interpretation of the
pterodactyl, affirming Cuvier's original reptilian classification so assiduously
that Seeley complained he had let a "feeling of contempt do duty for a knowledge of structure" and arrived at conclusions that were "erroneous, unscientific, and unjust." In 1870 Seeley reiterated Meyer's long-standing concern
that merely a single "cylindrical and shaft-like . . . bone" had, in 1801, been
"regarded by Cuvier . . . as an infallible sign that the Pterodactyle was a saurian reptile, and not a mammal," before adding his own recent proposal that
it was actually a warm-blooded bird.[100] Owen's refutation of this innovative
new theory relied on the same assumption of infallibility that Cuvier had
maintained almost seven decades earlier, and he haughtily pointed out to the
"tyro . . . ambitious of soaring into higher regions of biology" that the "length
and flexibility of the [avian] neck is correlated with the covering necessitated
by the high temperatures of the bird. The cold-blooded flying reptiles have a
comparatively short and rigid neck." In fact, Owen observed, the "constant
correlative structure with hot-bloodedness is a non-conducting covering to
the body. We may with certainty infer that the *Archæopteryx* was hot-blooded
because it had feathers."[101] In retaining the pterodactyl as a true reptile, Owen
once again invoked the law of correlation to demonstrate that the archaeopteryx must equally have been a bird. For Seeley, however, Owen, by not also
considering the larger common plan to which these particular parts had been
molded, was compelled to "substitute personal fancies about an animal's affinities for knowledge."[102]

Notwithstanding his determination to maintain the pterodactyl and the
archaeopteryx as, respectively, well-defined reptiles and birds, Owen, unlike
Seeley, accepted that species underwent progressive modifications and that
the fossil remains of intermediate transitional forms made the reality of this
process indubitable. As he announced in the final volume of *On the Anatomy of Vertebrates* (1866–68), the "progress of Palæontology since 1830 has
brought to light many missing links unknown to the founder of the science."
Such overwhelming evidence of a "tendency to a more generalised, or less
specialised, organisation as species recede in date of existence from the present time" had rendered the same pioneering paleontologist's brusque deni-
als of "every mooted form of transmutation" no longer tenable, and obliged

even Owen to undertake what he conceded was a "reconsideration of Cuvier's conclusions to which I had previously yielded assent." Recalling his youthful reverence for the "great Master" and his remarkable "reconstruction of lost species from fragmentary remains," Owen reflected—notably, in the past tense—of his own ostensibly "unbiased inductive research" that "if there was a bias, it was toward Cuvier."[103]

While Owen's reconsideration of his erstwhile Cuvierian prejudice did not entail relinquishing all "correlative deductions from characters of the petrifiable remains" of extinct animals, it did necessitate combining those same deductions with an appreciation of evolutionary development. Above all, it required a recognition that as "vertebrates rise in the scale . . . the law of correlation, as enunciated by CUVIER, becomes more operative." It was the "structures of the highly-developed Carnivore" and the similarly advanced "herbivorous mammal" that were "so co-ordinated as to justify CUVIER." On the other hand, "as we descend in the scale of life from the grade illustrative of 'Cuvier's Law'" to less specialized and well-adapted forms such as Paleozoic fish the same "law becomes limited in its application accordingly."[104] Cuvier himself, in the "Discours préliminaire," had noted how the reasons for certain structural correlations became "less clear" as one "descends to the orders or subdivisions of the class of hoofed animals," but the trope of descent was here merely figurative and did not carry the same literal phylogenetic connotations that, half a century later, it evidently did for Owen.[105] The successful application of the law of correlation was now dependent on understanding the very process of progressive evolution that its founder had, throughout his own career, resolutely rejected.

In amalgamating Cuvierian correlation with evolution, Owen was careful to make it clear that he was not endorsing what he scornfully dismissed as the "guess-endeavours of LAMARCK, DARWIN . . . and others."[106] Rather, Owen considered Darwin's theory of natural selection mere conjecture, and, since the mid-1840s, had been cultivating an alternative saltational process of evolutionary change by sudden deviations from embryonic types similar to that proposed by the anonymous author of *Vestiges of the Natural History of Creation* (1844). Owen, however, had been reluctant to publish any views that might be anathema to his patrons among the Anglican gentlemen of science at Oxford and Cambridge, and felt it necessary to publicly distance himself from the controversial *Vestiges*.[107] When Darwin broke his own silence on species mutation in the *Origin*, he represented his former friend "as being firmly convinced of the immutability of species."[108] This, as Darwin conceded ten years later, was a "preposterous error," and, only months after the *Origin*'s publication, Owen, in his textbook *Palæontology* (1860), announced his own "axiom of *the continuous operation of the ordained becoming of living things.*"[109] Although, as seen in the previous chapter, even Owen's cannibalistic protégé Charles Carter Blake poured scorn on this rival formula for the origination of species, it soon afforded its author a valuable new means of accounting for the perfect functional correlations of animal structures.

The Higher Marvel of Such a Correlated Organic Machine

Shortly before the *Origin's* publication, Owen received a "keg of spirits" containing the preserved remains of a "most curious and exceptional creature." A rare "living specimen" of the aye-aye, the nocturnal primate native to the island of Madagascar, had been found by sailors trading in the Indian Ocean, but Owen instructed the colonial administrator who procured it to "quietly chloroform the rarity into a better world" and carefully safeguard its cadaver for the perilous sea journey to Britain. While he acknowledged that "if the Zoological Society should know what I have been recommending I should be howled out of the society," Owen considered it a far "greater matter to have its anatomy secured for the scientific world" than to "have one alive." The particular anatomy that Owen directed to be secured in spirits rather than risked, if still alive, to the depredations of boorish sailors, was, he quickly recognized, "wonderfully adapted" to its distinctive habitat and singular lifestyle, and, as he reflected in the late summer of 1859, he was "glad to devote myself this autumn" to explicating its structural peculiarities.[110] By the time Owen finished his painstaking dissection of the aye-aye's pickled remains, Darwin's theory of natural selection was dominating discussion in the elite forums of the scientific community, and, tellingly, Owen announced his conclusions in a paper entitled "On the Character of the Aye-aye as a Test of the Lamarckian and Darwinian Hypothesis of the Transmutation and Origin of Species" at the 1862 meeting of the British Association for the Advancement of Science in Cambridge.

Speaking in the meeting's zoological section, Owen, "referring to the results of a recent dissection," informed his scientific peers that in the structure of the aye-aye

> we have not only obvious, direct and perfect adaptations of particular mechanical instruments to particular functions—of feet to grasp, of teeth to erode, of a digit to feel and extract,—but we discern a correlation of these several modifications with each other, and with modifications of the nervous system and sense-organs—of eyes to catch the least glimmer of light, and ears to detect the feeblest grating of sound,—the whole determining a compound mechanism to the perfect performance of a particular kind of work.

FIGURE 9.4. *Male Aye-aye* (Chiromys madagascariensis, *Cuv.*). Richard Owen, *Monograph on the Aye-aye* (London: Taylor and Francis, 1863), plate I. With its attenuated middle finger enabling it to extract insect larvae from the trunks of hardwood trees, the peculiar structure of the Madagascan aye-aye provided Owen with an opportunity to repulse Darwinian evolution and instead propose that the intricate correlations of its organization were only explicable as the outcome of an ordained law of progressive evolution. © The Trustees of the Natural History Museum, London.

The seamless combination of individually perfect component parts, including an attenuated middle finger that "remains slender as a probe" and that enabled the aye-aye to perform the "delicate manœuvre" of extracting insect larvae hidden deep in the trunks of hardwood trees, could not be accounted for by the "plastic possibilities" in animal structure posited by either Lamarck or, more recently, Darwin (fig. 9.4). After all, there had been "no changes in progress in the Island of Madagascar necessitating a special quest of wood-boring larvæ" that might afford an advantage to particular adaptations in accordance with the "hypothesis of 'natural selection.'" Rather, all the impeccably cohesive and correlated components of the aye-aye, Owen proclaimed, "appear to be structures foreordained—to be predetermined characters."[111]

His habitual health problems prevented Darwin from attending the British Association meeting, and instead he relied on Huxley for jubilant reports of how Owen's "Aye-Aye paper fell flat." It was "meant to raise a discussion of your views," Huxley apprised the dyspeptic Darwin, "but it was all a stale hash, and I only made some half sarcastic remarks which stopped any further attempts at discussion."[112] Owen later recalled the irksome presence of an unnamed "repudiator of final causes" who "objected to the evidences of adaptation" presented in his paper, although he intimated that he had successfully repulsed the fatuous cavils, which he attributed to a "stupid blindness to any meaning in the coadjustment of special modifications."[113] This particular version of events was corroborated by another member of the audience, a

Cambridgeshire clergyman who informed Owen that he had "rejoiced in the complete answer wh. you so ably gave to the views of Prof. Huxley . . . particularly in the case of the Aye Aye."[114] Unsurprisingly, the same appreciative cleric, as he told Owen, also considered "Darwin's book on species . . . unmitigated bosh," adding that its evolutionary suppositions were "entirely inconsistent with a belief in the acc^t. of Creation in the book of Genesis—and I prefer the positive statements of Moses to the speculations of Dr. Darwin."[115]

The clergyman's correspondent, however, was adamant that if faced with the same "alternative—species by miracle or by law?" he would "accept the latter, without misgiving, and recognise such law as continuously operative throughout tertiary time."[116] Indeed, when Owen augmented what he had said at the British Association in his *Monograph on the Aye-aye* (1863), he repeated the passage regarding how "we discern a correlation" between the creature's various "mechanical instruments," before adding:

> Here Paley and the pure teleologist would pause. But I would remark that we further discern the higher marvel of such a correlated organic machine being capable of reproducing itself by the act of generation. That act premised, Aye-aye after Aye-aye becomes what it is, through progressive growth and development, from the condition of a minute pellucid monadiform cell. The whole of its exquisitely adjusted structure is built up according to law.

Although exhibiting precisely the same degree of intricacy and perfection, the mechanical correlations evident in the structure of the aye-aye were not explicable on the conventional Paleyite assumption of a "direct act by the Designer in the formation of an organic and self-reproductive machine." Rather, such "organic mechanism," Owen maintained, was the "necessary, but not the less fore-ordained, result of the nature and adjustment of influences forming part of the general system of our planet, with its varied forces, acting and reacting under certain conditions so as to issue in such a result." Beneath the prolixity of Owen's tortured prose, he was proposing that the "various remarkable correlated structures in the organization of the Aye-aye" were the direct consequences of a "long continued series" of progressive adaptations "stimulated by surrounding changes."[117]

While Owen distinguished his own "derivative hypothesis" from the evolutionary conjectures of Lamarck and Darwin by "invoking a supernatural commencement of organisms" and retaining a quasi-Paleyite "idea of a forecasting, designing Power" operative through secondary laws, the very same harmonious and perfect correlations that Cuvier had first discerned in animal structures seven decades earlier were now attributed to the gradual modifications in response to alterations in the environment that the French savant, when confronted by Lamarck's *Philosophie zoologique* (1809), had explicitly disavowed.[118] As with Darwin and his law of correlated variability, Owen, by the mid-1860s, was able to accommodate the principal elements of Cuvierian correlation, and even the mechanistic metaphors of divine contrivance employed by Paley and conflated with Cuvier's understanding of animal organization by William Buckland, with his own conception of an ordained law of progressive evolution.

Australian Geologists to Class the Australian Fossils

Only three years after announcing his purported "reconsideration of Cuvier's conclusions" in the final volume of *On the Anatomy of Vertebrates*, Owen, in his *Monograph on the Fossil Mammalia of the Mesozoic Formations* (1871), once more proclaimed himself a loyal "disciple of Cuvier, and a believer in his 'Law of Correlation.'" Recent classifications of partial fossil remains had, he insisted, only afforded "further proof of the worth and truth of the principle which Cuvier laid down."[119] This apparently unreconstructed Cuvierianism, regardless of the revisions and modifications necessitated by the new evolutionary paradigm of the post-*Origin* era, was especially evident in Owen's often fractious dealings with collectors and naturalists from across the British Empire. Like Huxley with popular audiences, Owen seems to have adopted a different approach to Cuvier's famous principle when addressing what he condescendingly perceived as colonial subordinates, possibly because such Cuvierian orthodoxy afforded a means of asserting his metropolitan expertise and authority. It was, after all, an underlying assumption of the law of correlation that extinct creatures could be identified from fragmentary remains without the need for knowledge of the local environments in which they were originally found, and that those, such as Cuvier and Owen, who had access to large metropolitan museums that facilitated extensive comparisons were much better placed to make such identifications than those who were closer to where the fossils came from, even if it were thousands of miles away.[120] In fact, Cuvier's paleontological practice, according to Dorinda Outram, prioritized the "immobile *gaze* of the sedentary naturalist," and was predicated on an assumption that "true knowledge of the order of nature comes . . . from the very fact of the observer's *distance* from the actuality of nature. True observation of nature depends on *not* being there."[121] Owen's own just as sedentary inference, made at the Royal College of Surgeons, of the dinornis's erstwhile existence in faraway New Zealand was, of course, the exemplary instance of the distinctly colonialist relationship between center and periphery. By the 1870s, however, such hierarchical arrangements were increasingly resisted, as seen in chapter 3, by colonial naturalists eager to assert their interpretative autonomy.[122]

In Australia, some naturalists were, even by the 1850s, already becoming less inclined to accept Owen's Cuvierian directives in interpreting the rich array of indigenous fossils then being unearthed in the interior of New South Wales. Writing to the *Sydney Morning Herald* in 1858, William Sharp Macleay openly ridiculed Owen's identification of the marsupial *Thylacoleo* as a carnivorous predator akin to a lion, mockingly quoting from *A Midsummer Night's Dream* to equate the purportedly ferocious creature with Shakespeare's decidedly unleonine "very gentle beast." Macleay's scorn for what he called "my friend's notion" was prompted by his reservations, like those of Falconer with the *Plagiaulax* discussed in chapter 6, over the validity of the "famous molar, on which the Professor relies." After all, Macleay insisted, if "'*Ex pede Herculem*' be not always a safe axiom in geology, '*Ex dente Herculem*' is certainly as little to be depended upon."[123] Macleay had been impressed by Huxley when HMS *Rattlesnake's* talented assistant surgeon visited Sydney in 1848, and the young tyro's assault on the law of correlation after returning to Britain in the following decade afforded an opportunity for those in the actual field to dispute the authority of their metropolitan masters. It was certainly seized upon by Macleay, who had long exhorted "Australian geologists to class the Australian fossils" in his role as trustee of Sydney's Australian Museum, with enthusiasm and even a little hauteur.[124]

Macleay's contemptuous rejoinder to Owen over the so-called *Ex dente Herculem* method emboldened one of his successors at the Australian Museum, the German émigré Gerard Krefft, who, notwithstanding his usual animosity toward Macleay, kept a clipping of his letter to the *Sydney Morning Herald* among his private papers.[125] Despite Macleay's caustic criticism, Owen, addressing the fellows of the Royal Society in 1870, reaffirmed his conclusion as to the carnivorous character of the *Thylacoleo* based on "dentition" that he "only knew 'in part,'" taking the opportunity to once more attest that he had been "guided by a principle. It is that laid down by Cuvier" regarding the exact correlation between certain types of teeth and particular feeding habits. This axiom was now more than seven decades old, although Owen insisted that the "aberrations of some contemporary labourers in the field show that it will bear repetition."[126] In reiterating the famous Cuvierian precept with such deliberate condescension, the metropolitan expert was explicitly reasserting his authority over insubordinate colonial collectors. Significantly, Owen repackaged his offprints of the article featuring this necessary repetition, which had originally appeared in the Royal Society's *Philosophical Transactions*, with a new and deliberately provocative title: *A Cuvierian Principle in Palæontology*

Tested by the Evidence of an Extinct Leonine Marsupial (1871). Back in Australia, Krefft reviewed the retitled article, now presented as a pamphlet, in his regular "Natural History" column for the *Sydney Mail*, the weekly edition of the same newspaper to which Macleay had sent his derisive letter.

Seeking to "draw the attention of Australians" to the distinctive peculiarities of their indigenous prehistoric fauna, Krefft proposed that "Cuvier and his principles cannot always be depended on in the classification of Australian fossils." After all, without any local knowledge, the great savant in his Parisian museum would have been unable to "distinguish between the femur of a 'gigantic kangaroo' and that of an elephant." As such, Krefft insisted, "we are justified in discarding Cuvierian principles as far as fossil marsupials are concerned." This jingoistic anti-Cuvierianism, which enabled Krefft to "turn the supposed 'lion' into a leaf-eating phalanger" similar to a wombat, would only really make sense, he suggested, to those who had themselves "experienced the impressions of the teeth of some recent ones [i.e., wombats]," which "crush, but do not tear." The opposite argument, on the other hand, relied upon unrepresentative koala "specimens specially selected by Professor Owen, probably for other than Australian readers," who, never having been "nipped" by a koala, had no means of knowing what their teeth were really like. As such, Krefft proclaimed at the beginning of his review: "Professor Owen spoke boldly when he . . . headed his last treatise . . .—too boldly, in fact—because if the 'Cuvierian Principle in Palæontology' is once found wanting, it must be reduced in value ever afterwards."[127] Antipodean exceptionalism, along with the sometimes painful hands-on experience of colonial settlers, had the potential to derail the entire law of correlation, fatally impugning its claims to universal applicability and unfailing consistency.

Although tailored for a domestic audience, Krefft's *Sydney Mail* review was soon reprinted in Britain in the *Annals and Magazine of Natural History*. Ahead of this, he had already "taken the liberty" of sending a copy to Darwin, explaining, "I am anxious to obtain your opinion on the . . . Thylacoleo," and asking: "Do you believe that it was 'the fellest of Carnivores' as Professor Owen thinks." Darwin read the "newspaper . . . article with *great* interest," telling Krefft it was "lamentable that Prof. Owen . . . shd. adhere in so bigoted a manner" to the conclusions he had reached solely on the basis of the contested method of correlation. The "Cuvierian principle," Darwin observed, "may evidently be extended much too far: Toxodon is a good instance of this, as no one I believe has ventured to surmise from the skull, whether it was an aquatic or terrestrial animal."[128] The toxodon was originally identified, and named, by Owen in 1837 from a single cranium that Darwin had brought back from South America. In the "Fossil Mammalia" volume that

Owen contributed to Darwin's *Zoology of the Voyage of HMS* Beagle (1839), he adumbrated all the "indications of the aquatic habits of the Toxodon" in the classic Cuvierian manner, proposing that there was a "close resemblance between the Toxodon and the Hippopotamus." Like that creature, Owen inferred, its powerful incisors were used to "divide or tear up by the roots the aquatic plants, growing on the banks of the streams, which the Toxodon may have frequented." But Owen, with uncharacteristic equivocation, also acknowledged that the toxodon's peculiar nasal cavities rendered it "extremely improbable that the habits of this species were . . . strictly aquatic."[129] Thirty years later, Darwin now returned to this still enigmatic extinct mammal to endorse Krefft's argument against both Owen's inveterate Cuvierianism and the imperious metropolitan hauteur toward colonial naturalists that it often entailed. Despite having previously acknowledged his erstwhile "bias . . . toward Cuvier," Owen, according to Darwin, still adhered to his law of correlation in a deplorably "bigoted . . . manner."

Conclusion

Krefft, even with Darwin's support, still felt stymied that, as he told the British paleontologist Richard Lydekker in 1880, "here in Australia you must follow the footprints of those ancient gentlemen who still follow Cuvier."[130] Far from instigating a nascent Australian nationalism, Krefft's opposition to Owen and the law of correlation, as well as his fervent advocacy of Darwinism, in fact alienated many of his compatriots, with his column for the *Sydney Mail* regularly being censored by the newspaper's evangelical proprietor even before his dramatic dismissal, on trumped-up charges of drunkenness, from the Australian Museum in 1874.[131] While Owen's unapologetic adherence to an unreconstructed Cuvierianism was evidently perceived as anachronistic by the new breed of refractory colonial naturalists such as Krefft, it did not negate his continuing power and influence, both in Britain and across the empire, at the center of a global network that was now directed from the Natural History Museum in South Kensington.

The consolidation of the British Museum's natural history collections on a separate site had been a long-standing personal project of Owen's, and, after decades of controversy and planning, the new museum finally opened in 1881.[132] Built to Owen's own specifications, the magnificent Romanesque building reflected particular aspects of his distinctive approach to the natural world. In this cathedral of nature, as Nicolaas Rupke has remarked, "Cuvierian functionalism was to stay, as it had become an integral part of the museum, enshrined within the building," not only in the articulated skeletons of the now iconic dinornis and megatherium, but also in the floor plan in which each gallery remained separate from the others, thereby representing their respective flora and fauna as distinct types.[133] Three years after the museum's opening, Cuvier's nephew Frédéric, lamenting that in France the "younger generation of savants . . . speak with disdain of M. Cuvier's work," considered it a "great satisfaction" to know that, across the Channel, there remained a "savant who has especially shown his fidelity to the principles of true scientific observation."[134] Owen's force of personality and prolonged institutional prestige ensured that the Cuvierian law of correlation, even in its most orthodox format, remained a fixture of British science for much longer than it did elsewhere.

Correlation had, in any case, been appropriated into a variety of new approaches to animal structure during the 1860s, which refitted it, if in modified forms, for the new evolutionary and morphological agendas of the period. The profound crisis precipitated by Darwin's *Origin* and the taxonomically anomalous "missing links" that subsequently came to light did not sound a decisive death knell for Cuvier's famous law any more than Huxley's ferocious attacks on necessary correlation a few years earlier. With the exception of Owen's intermittent reversions to the Cuvierian true faith of his youthful prime, however, the law of correlation was subsumed into new scientific axioms such as Darwin's correlated variability that denuded it of many of its earlier attributes, most notably its erstwhile potential to stymie all forms of transformism and evolution. Owen, of course, accepted this and incorporated correlation into his own conception of an ordained law of progressive evolution, and his occasional lurches back to an unreconstructed Cuvierianism have long perplexed historians. Rupke has noted the "enduring nature of Owen's commitment to functionalism," but does not offer any explanation as to why it continued long after the deaths of his influential patrons at the ancient universities, whose remunerative support, as seen in chapter 2, had earlier obliged Owen to maintain a teleological perspective even when his own researches increasingly veered away from those of Cuvier and instead prioritized form over function.[135] For Adrian Desmond, Owen's faithful tending of the "Cuvierian light" into the 1870s and beyond was simply a "gross miscalculation on his part," with "*le baron* fast becoming the whipping boy of Darwinian historiography."[136]

Neither Rupke nor Desmond, however, take account of the opinions of those beyond the scientific community, and the continuing currency of the law of correlation with a variety of other groups, and especially the renewed interest in the predictive powers of paleontologists among both popularizers of science and writers of detective fiction, provides a very different perspective on Owen's ostensible miscalculation. Instead, as the final chapter will show, it was Huxley's opposition to Cuvierian correlation that, following his muddled obfuscation of his actual opinions when addressing popular audiences, appeared most problematic in the final decades of the nineteenth century.

Prophecies of the Past

In the winter of 1876 the publisher John Murray issued a new scientific biography entitled *Life of a Scotch Naturalist: Thomas Edward*, which soon sold out its initial print run. The book's success, however, was not due to the renown of its subject, an obscure and impoverished cobbler from Banff in northern Scotland, and was instead attributable to its author, Samuel Smiles, whose bestselling *Self-Help* (1859) had quickly become one of the most influential works of the nineteenth century. In *Self-Help* Smiles included among his heroes of diligence and perseverance a "profound naturalist . . . discovered in the person of a shoemaker at Banff . . . who . . . has devoted his leisure to the study of natural science." Humble but hardworking autodidacts such as Edward were placed alongside the "cases of strenuous individual application" exhibited by eminent men of science, including "Richard Owen, the Newton of natural history, [who] began life as a midshipman," and "Cuvier, [who] when a youth . . . observed a cuttle fish lying stranded on the beach . . . took it home to dissect, and began the study of the molluscs, which ended in his becoming one of the greatest among natural historians."[1] United by their selfless devotion to nature, all the different ranks of naturalists, whether cobblers or chevaliers, worked in harmony and with an acceptance of the hierarchy of expertise established simply by effort, application, and natural talent. When he subsequently compiled Edward's biography in the mid-1870s, Smiles fashioned him as a moral hero who exemplified the values of stoic endurance and uncomplaining deference toward the established social order. With the Reform Act of 1867 having extended the franchise to much of the urban male working classes, such models of plebeian passivity, as Smiles well knew, were particularly needful.

Unlike most subjects of Victorian biographies, Edward was very much still alive, and, as Anne Secord has shown, his actual behavior following the publication of *Life of a Scotch Naturalist* did not match that of the typical Smilesean hero.[2] As Smiles himself privately conceded: "I fear he will rub the gilt off. I idealized him a bit."[3] This idealization had required, among other things, the suppression of Edward's gloating satisfaction at proving an expert wrong over the identification of a fish.[4] But the discrepancy between ideal and reality was in fact apparent even within Smiles's carefully crafted narrative of respectful humility, especially in an anecdote ostensibly evincing Edward's dogged perseverance. Observing that, while working at the Banff Museum, Edward discerned among "several other fragments of antiquity" a "joint-bone of some extinct animal," Smiles remarked: "The story connected with this bone is rather curious." Edward, as Smiles related, was "struck by its size, thickness, and peculiar shape," and the "idea flashed across his mind that he had somewhere seen something like it in a picture; but he could not remember where." His intuitive conviction that it was a "semi-fossilized bone of some of the pre-Adamite monsters that are dug up now and then" was scornfully dismissed by his superiors at the museum, and "despite all his endeavours, the bone still remained unknown and unnamed." When Roderick Murchison visited the Banff Museum and met Edward, it was the "first thing he put into Sir Roderick's hands," asking, "Can you tell me what it is, sir?" Murchison told the importunate cobbler that he was

> more a *stone* man than a *bone* man. Besides, it is often a difficult matter to distinguish small fragments or single bones of a skeleton . . . and to determine with certainty to what creature it belonged. . . . As regards the bone, I'll tell you what to do. Send the bone to London, to Professor Owen. He's your man. . . . He'll soon tell you all about it.

Notwithstanding the admonitions and advice of such a prestigious metropolitan savant, Smiles divulged that "Edward did not, however, send the bone to London. He knew from experience, that such things, when sent so far away, rarely come back. . . . He therefore kept the bone at home."[5]

As well as not entrusting his unidentified bone to covetous experts such as Owen, Edward remained confident that he could establish the creature from which it came without their assistance. Many years after Murchison's visit, as Smiles recounted,

> when Edward was rummaging over some old books, he came upon the second volume of the *Penny Magazine*. Whilst turning over the pages by chance, he saw a picture of old bones which had much puzzled his brains thirty years before. And now he remembered that it was the picture of the bones here drawn, that had first given him the idea that this bone in the museum was the remnant of some extinct animal.[6]

By means of a woodcut in the *Penny Magazine* from 1833, Edward was finally able to do what Murchison had warned was so prohibitively difficult, and identify the enigmatic bone as the femur of a plesiosaur (fig. 10.1). Significantly, the previous page of the same archaic article offered a "popular exposition" of how "Baron Cuvier . . . has shewn that there reigns such a harmony throughout all the parts of which the skeleton is composed . . . that we are enabled by the inspection of a single bone to say with certainty that it must have belonged to a particular kind of animal."[7] The conspicuous endorsement of the Cuvierian law of correlation in the steam-printed publications with which the Society for the Diffusion of Useful Knowledge (SDUK) flooded the market for cheap print in the 1830s and 1840s was noted in earlier chapters. Such triumphant affirmations of Cuvier's reconstructive abilities, often authored by leading specialists, were intended, in part, to quell proletarian radicalism and supplant the suspect scientific wares touted by working-class publishers. Decades later, however, the very same popularizing accounts, as with Edward's use of the SDUK's *Penny Magazine*, now enabled autodidacts such as the refractory Scottish cobbler to snub the authority of experts and instead make their own identifications from the fragmentary fossils they encountered. With the yellowing pages of the *Penny Magazine* in his hands, Edward could rebuke both Murchison's patrician advice and, with it, the Smilesean model of the docile artisan naturalist.

The story of what Edward called the "auld been" (and Smiles, gauchely deciphering his Scottish pronunciation, rendered as "in other words, the old

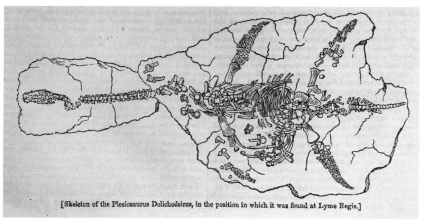

[Skeleton of the Plesiosaurus Dolichodeirus, in the position in which it was found at Lyme Regis.]

FIGURE 10.1. *Skeleton of the* Plesiosaurus dolichodeirus. "Mineral Kingdom. Organic Remains," *Penny Magazine* 2 (1833): 348. It was this woodcut illustration that enabled Edward, in the 1860s, to finally identify his enigmatic "auld been" as the femur of a plesiosaur. Reproduced by permission of the University of Leicester.

bone!") reveals the strange afterlife of the law of correlation in the latter part of the nineteenth century, long after Thomas Henry Huxley's vehement refutation of it in the mid-1850s, and even following the decisive modifications of the Cuvierian axiom in the wake of Darwinian evolution considered in the last chapter.[8] As well as the physical survival of the popular science of the 1830s in dusty copies of the *Penny Magazine*, claims about Cuvier's apparently unerring powers continued to circulate—and potentially even increased—in later formats of popularization aimed at the newly literate mass audience created by the 1870 Elementary Education Act. As with the intractable Banff cobbler, the famous Cuvierian method, which had previously been the preserve of virtuoso experts such as Owen, was defiantly democratized by the newly educated and enfranchised proletarian masses, with even women, as will be seen later, among those who now claimed to be able to identify or actually reconstruct unknown creatures from merely a single part. In such demotic appropriations, the almost preternatural predictive powers that paleontologists had long used to fashion themselves as infallible scientific geniuses often became conflated with the more literal occult beliefs that many plebeian autodidacts embraced as a riposte to the religious and secular orthodoxies of the intellectual elite. Amid the dramatic social, cultural, and technological changes of the final years of the Victorian era, and even into the opening decades of the twentieth century, there was a notable revival of interest in Cuvierian correlation, as well as a no less remarkable recurrence of the very same claims that were first articulated in the opening years of the nineteenth century and that began to reach their apogee when Edward's faded copy of the *Penny Magazine* was first printed.

Having decisively won over most of his fellow scientific practitioners, Huxley, during the 1850s and 1860s, reversed his vaunted condemnation of the law of correlation when addressing plebeian audiences. By the 1880s he was compelled to finally state his original objections to the Cuvierian method in popular form, although, by this time, his insistence that it involved only quotidian and empirical mental processes akin to the procedures used every day by the police was complicated by the emergence of a new genre of fiction whose detective protagonists displayed intuitive and seemingly infallible powers of reasoning. The forensic abilities of both Émile Gaboriau's Père Tabaret and Arthur Conan Doyle's Sherlock Holmes were compared to those of Cuvier reconstructing from a single bone, and the image remained a familiar trope of detective stories into the early twentieth century. As well as perpetuating the perception that the law of correlation was both rational and entirely unerring, the popularity of such stories ensured that Huxley's own empirical arguments were reinterpreted and mistakenly considered congruent with those of fictional sleuths such as Holmes and even Cuvier himself. Faced with such grievous misapprehensions, Huxley, as this chapter will show, continued to reiterate the very same criticisms he had first made in the mid-1850s into the final months of his life forty years later.

Huxley was joined in his disavowal of correlation by a new generation of paleontologists on the other side of the Atlantic, whose unprecedented discoveries in the American West of new and peculiarly configured prehistoric mammals made the Cuvierian method increasingly untenable. When, however, the foremost popular writer on paleontology, Henry Neville Hutchinson, became the very first popularizer to endorse Huxley's anti-Cuvierian arguments in the 1890s, it was already too late, as Hutchinson himself conceded, to stem the tide of popular belief in the law of correlation.

Let None but Working Men In

With urban working-class men brought into the electorate from 1867 and for the first time having political influence in the following year's general election, the popular lectures that Huxley and other leading men of science had been giving since the early 1850s assumed a vital new significance. In the late 1860s the chemist Henry Enfield Roscoe began a scheme to provide the "working men of Manchester" with "science instruction given in a plain but scientific form," with a modest "entry charge of One Penny per lecture." The success of this annual series, which was published as *Science Lectures for the People*, led Roscoe to hope that the "example of Manchester may be followed by other large towns."[9] After all, the popularization of science afforded a means by which new voters, especially those in the industrial cities of the north of England, could become educated and therefore responsible participants in the new era of mass democracy.

As Norman Lockyer, editor of the new commercial science periodical *Nature*, quickly intuited, these recent political transformations also meant that efforts to "extend the knowledge and love of science among the working classes" could have more direct and tangible benefits for the scientific community itself. Lockyer had initially intended *Nature*, which was founded in 1869, to appeal to both scientific practitioners and the general public, although it was already becoming apparent that the increasing specialization of science entailed that only the former could be accommodated by the new journal's emphasis on reports of the latest research.[10] It was nonetheless imperative that the scientific community did not lose touch with the remainder of the populace, even if periodicals such as *Nature* could no longer perform that function. In an editorial in 1871, Lockyer advised *Nature*'s select coterie of specialist readers:

"Success, in fine, will not be attained by a policy of isolation, but by leavening the whole mass of the community with the love of science, and when this is done science will rise to its just place in the councils of the nation." Lockyer accordingly lauded the "Science Lectures for the People, lately delivered in Manchester," as a "great success," which "point to what ought to be the future action of men of science." By leaving their laboratories and "diffusing scientific information among the populace of large towns" that were now the arbiters of political power, scientific practitioners who engaged in popularization would be "forcing the claims of science before the Government."[11]

The new importance of popularization ensured that concerns over precisely what kinds of science were being imbibed by the newly enfranchised audiences for popular lectures loomed larger than ever. In 1868, at a conference on the vexed issue of technical education, Huxley was adamant that the "minds of the young should be imbued with scientific methods and scientific principles" rather than the "branches of science which immediately applied to . . . particular trades and callings," which he scorned as "merely the scum and top surface of science."[12] The Science Lectures for the People in Manchester similarly eschewed applied science, with Roscoe refusing sponsorship from the Society for the Promotion of Scientific Industry.[13] When, after a short hiatus, Roscoe arranged a new series of penny lectures in 1870, he invited Huxley to give the opening one. Exhorting his host to "let none but working men in," Huxley regaled his plebeian auditors on the suitably abstruse topic of the formation of coral reefs, assuring them that Charles Darwin had come to his conclusions on the matter "by way of deduction from simple principles of natural science" that involved the "same sort of reasoning that you would use in the ordinary affairs of life."[14] Huxley, though, would have been considerably less sanguine about other abstract scientific principles that were aligned with the quotidian practices of workingmen in subsequent series of Science Lectures for the People.

Three years after Huxley's own lecture, Samuel Messenger Bradley, assistant surgeon at the Manchester Infirmary, addressed his fellow Mancunians on "Animal Mechanics," an esoteric topic involving anatomy and physiology while neatly circumventing the more practical aspects of his professional activities. There was also a further reason why it was a particularly apposite subject for a lecture in Manchester. Explaining the complicated interdependence of the avian frame, Bradley invoked an experience that would be familiar to those employed in the city's factories and mills:

If we glance for one moment at the skeleton of a bird in its entirety you will find that not only is each part admirably adapted to perform its functions, but that each part is perfectly in harmony with every other part, and in a measure necessitates every other part, just as the smallest segment of a circle gives the whole figure; or as many a man in this room could, I daresay, from the piston of an engine build up the entire machine, so to one who has sufficient knowledge the smallest fragments of an animal will enable him to build up the entire organism, and to infer its habits and nature.

The mechanical proficiency possessed by artisanal engineers in the audience was analogous to the renowned powers of the century's two most celebrated anatomists, one of whom even hailed from the same heavily industrialized county as Bradley and his auditors. He continued:

> In this way Cuvier, from a single tooth, was able to build up the Anoplotherium so perfectly, that when the remains of the animal were discovered somewhat later, there was nothing to alter in the model of the great Frenchman. By following the same law, . . . Owen—of whom we Lancashire men have good reason to be proud— . . . restored those "dragons of the slime", familiar to us in the gardens of the Crystal Palace. By following this same law of induction, you can trace out not only the character of the skeleton, but the shape and character of the internal organs, and the general nature of the animal.[15]

The life-size models of prehistoric creatures, including Cuvier's ungulate anoplotherium, that, since their construction in the mid-1850s, had apparently become familiar even to workers living hundreds of miles from the location of the rebuilt Crystal Palace on the southern outskirts of London, were still considered infallible exemplars of the law of correlation, notwithstanding the suspicions of many specialists noted in chapter 5. The enduring efficacy of the same Cuvierian law, Bradley urged, was in fact a matter of Lancastrian local pride, especially as Owen's vaunted abilities relied on the same mechanical ingenuity characteristic of Manchester's workingmen.

Bradley did admonish his audience, "You must not suppose . . . from my speaking in this light, sketchy way, that it is by any means always easy to follow out these co-relations," and he even warned: "Of course, the application of this [law] needs caution."[16] But few of the engineers and laborers who heard his lecture at the Hulme Town Hall in November 1873 would have suspected that the same scientific principle Bradley had, for the most part, extolled with such apparent certainty, was, in reality, highly contested, and, to other speakers in Science Lectures for the People, even embarrassingly obsolescent. Roscoe's avowal that the penny lectures would afford the "people of Manchester . . . the opportunity of hearing the explanation of scientific truths from the lips of those who have made the extension of natural knowledge their chief occupation in life" certainly implied that all the abstract principles to be imparted were at the cutting edge of their respective disciplines.[17] As the tenor of Bradley's lecture made clear, however, the putative reliability of the law of correlation, as well as its analogies with the mechanical precision of industrial engineering, were more likely to enthuse popular audiences than the uninviting intricacies of the anti-Cuvierian arguments by which Huxley had swayed many of his fellow practitioners.

When Philip Martin Duncan, the vice president of the Geological Society, spoke on "The Great Extinct Quadrupeds" in the Science Lectures for the People series for 1875, he acknowledged that "in introducing the subject of my lecture to you it is absolutely necessary to try and vivify these old bones, because were I to lecture to you only upon the bones of the extinct quadrupeds, in about a quarter of an hour you would all be tired to death." To enliven his potentially soporific lecture, Duncan quickly digressed from prehistoric quadrupeds to more recent extinct bipeds, especially the remains of "huge birds . . . which were called dinornis." Evidently anxious to avoid inducing lethargy, Duncan reminded his audience how "Professor Owen years ago restored these bones, that is to say he drew a figure of what he believed this bird was like from the few bones he then had."[18] Like the brick-and-mortar models at the Sydenham Crystal Palace, anecdotes concerning Owen's Cuvierian prediction of the dinornis's former existence in New Zealand back in 1839 remained a mainstay of science popularization, even when, following the reform legislation of the late 1860s, the significance of popular lectures had been transformed by the new political influence of their proletarian audiences.

The Method of Zadig Pure and Simple

Huxley, of course, had himself been willing to tailor the approach of his self-styled "People's Lectures" to adhere to the mode of lecturing that his artisanal auditors were accustomed to, even if, as seen in chapter 7, this involved contradicting the vehement opposition to the law of correlation that he expressed before more specialist audiences.[19] By the beginning of the 1880s, however, the increased importance of popularization, as well as the continuing endorsement and celebration of the Cuvierian law in popular lectures such as those given in Manchester by Bradley and Duncan, obliged Huxley to change tack, and, for the very first time, detail his long-standing complaints regarding correlation before an audience of workingmen. Rather than the expanding industrial cities of the north, it was at the Working Men's College in Great Ormond Street, central London, that Huxley, in March 1880, delivered his celebrated lecture on "The Method of Zadig: Retrospective Prophecy as a Function of Science."

Noting, at the beginning of the lecture, that "Zadig is cited in one of the most important chapters of Cuvier's greatest work," Huxley seemed to assume that this reference to the "Discours préliminaire" from *Recherches sur les ossemens fossiles* (1812) would be meaningful to his plebeian audience, and he even ventured a dryly urbane witticism regarding the "strict historical accuracy" of the orientalist fantasy by Voltaire that Cuvier had alluded to.[20] The eponymous Babylonian sage in Voltaire's novel could identify the precise characteristics of an escaped dog and horse having seen only their respective footprints, and, as discussed in chapter 1, Cuvier proposed that the "track of a cloven hoof" was, to a naturalist employing the law of correlation, "a more certain mark than all those of Zadig" that it had been made by a ruminant.[21] When Huxley's lecture was published in the *Nineteenth Century* in June 1880, the same statement appeared below the title, in its original French, as a seemingly apposite epigraph. In the "Discours préliminaire," though, Cuvier's tone was actually rather dismissive toward Zadig, whose fictional abilities were inferior to those that the French savant claimed to himself possess in reality. Huxley, on the other hand, was more disposed to esteem what he considered the "foundation of all Zadig's arguments . . . the coarse commonplace assumption . . . that we may conclude from an effect to the pre-existence of a cause competent to produce that effect."[22] Appropriating Zadig from Cuvier's indifferent tenure, Huxley strategically refashioned the ancient philosopher as the archetype of his own empirical approach to paleontology.

As Huxley assured his artisanal hearers:

> The whole fabric of palæontology, in fact, falls to the ground unless we admit the validity of Zadig's great principle, that like effects imply like causes; and that the process of reasoning from a shell, or a tooth, or a bone, to the nature of the animal to which it belonged, rests absolutely on the assumption that the likeness of this shell, or tooth, or bone, to that of some animal with which we are already acquainted, is such that we are justified in inferring a corresponding degree of likeness in the rest of the two organisms. It is on this very simple principle, and not upon imaginary laws of physiological correlation, about which, in most cases, we know nothing whatever, that the so-called restorations of the palæontologist are based.

It was merely the observation of customary correspondences that, according to Huxley, was the basis of both Zadig's astounding capabilities in Voltaire's novel and the achievements of paleontologists in recovering the vanished denizens of the prehistoric past. The "empirical laws of co-ordination of structure" discerned by such prosaic observations could, as Huxley advised, be "confidently trusted, if employed with due caution, to lead to a just interpretation of fossil remains."[23] But these circumspect reconstructions remained only provisional approximations and could never aspire to the claimed precision of more elaborate representations of extinct creatures, such as the brick-and-mortar models of revivified dinosaurs at the Sydenham Crystal Palace, which Huxley disdained as "so-called restorations." After all, the rational laws of correlation on which such ersatz restorations were predicated were themselves an illusory misconception of the empirical principles that Cuvier and subsequent generations of his followers had actually depended on.

This was precisely the same argument that, as seen in chapter 6, Huxley first made in "On the Relation of Physiological Science to Other Branches of Knowledge," which, delivered in 1854, was among the then twenty-nine-year-old's very first public addresses. Now in his midfifties and reaching the height of his celebrity as one of Victorian Britain's foremost public intellectuals, Huxley echoed his earlier lecture's contention that "science is . . . nothing but *trained and organized common sense*" in asserting that scientific reasoning involves the same "carnal common sense . . . upon which every act of our daily lives is based."[24] This reiteration of an argument that was now more than a quarter of a century old was necessary because, as Huxley was finally prepared to acknowledge, his original onslaught against Cuvier's purported powers had, beyond his fellow scientific practitioners, been largely ineffectual.

Huxley conceded to his audience at the Working Men's College:

> But it may be said that the method of Zadig, which is simple reasoning from analogy, does not account for the most striking feats of modern palæontology— the reconstruction of entire animals from a tooth or perhaps a fragment of a bone; and it may be justly urged that Cuvier, the great master of this kind of investigation, gave a very different account of the process which yielded such remarkable results.[25]

When, a year later, Huxley gave a modified version of the same lecture before a more specialist audience at the 1881 meeting of the British Association for the Advancement of Science, he amended this particular line to "that feat of palæontology which has so powerfully impressed the popular imagination, the reconstruction of an extinct animal from a tooth or a bone."[26] As this revision made clear, it was the lingering misconception, perpetuated by popularizers and given credence by a pliable public, that merely a single tooth or bone could, unerringly, yield the remainder of an unknown prehistoric creature that compelled Huxley to recycle a callow twenty-six-year-old lecture, even if he was too tactful to admit this before his plebeian auditors.

Having celebrated Cuvier's deployment of the law of correlation in both

his 1859 workingmen's lecture on "Objects of Interest in the Collection of Fossils" and the popularizing "Explanatory Preface" to *A Catalogue of the Collection of Fossils in the Museum of Practical Geology* (1865), Huxley, by the time of
his lecture on Zadig, was determined to finally expose, before a nonspecialist
audience, the real nature of the putative paleontological accomplishments
that had continued to impress the imagination of the general public. But,
as with the incongruous urbane humor that was noted earlier, Huxley was
seemingly unwilling to do much to adapt his lecture to its particular circumstances, and instead he merely recycled old arguments that he had previously
presented, in the very same terms, to more elevated audiences.

Insisting that Cuvier's descriptions of his own working practices ought not
to be taken at face value, Huxley set his artisanal hearers an improbable test:

> Cuvier is not the first man of ability who has failed to make his own men-
> tal processes clear to himself, and he will not be the last. The matter can be
> easily tested. Search the eight volumes of the *Recherches sur les Ossemens Fos-
> siles* from cover to cover, and no reasoning from physiological necessities—
> nothing but the application of the method of Zadig pure and simple—will
> be found.[27]

Back in 1856 Huxley had similarly contended that a "reader of Cuvier's 'Ossemens Fossiles' might begin at the tenth volume and read onto the second . . .
[but] would find himself very rarely troubled with any remarks upon physiological correlations."[28] This had been in the *Annals and Magazine of Natural History*, though, and at least some of the journal's specialist readership
might have had copies of Cuvier's massively expensive magnum opus to
hand. For even the most prosperous laborers and artisans in the lecture hall
of the Working Men's College, by contrast, Huxley's proposed bibliographic
exercise was entirely unfeasible, especially as the full *Recherches* had never
been translated and was unlikely to be stocked by public libraries or reading rooms. *Lloyd's Weekly Newspaper* reported that Huxley's "interesting lecture . . . was listened to with profound silence and much attention."[29] But
notwithstanding his reputation as an incisive and highly effective popular
lecturer, there were clear reasons why the anti-Cuvierian arguments that
Huxley adumbrated at the Working Men's College might have left his audience somewhat nonplussed, no matter how attentive they seemed.

The Logic of Steno

Huxley appears to have recognized that his lecture on Zadig's method might
have been inappropriate for its initial proletarian audience, and, within days

of delivering it, he offered the manuscript to James Knowles, editor of the *Nineteenth Century*, who gratefully assured him he would "reserve a place . . . for the *promised* paper—Hurrah!"[30] Knowles's monthly review, which cost two shillings and sixpence, was, along with the *Fortnightly Review* and the *Contemporary Review*, an organ of the so-called higher journalism and was aimed primarily at members of the intellectual elite as well as the professional middle classes. In addition to its prohibitive cover price, the *Nineteenth Century*'s austere format of articles that frequently ran to thirty or forty pages, without any illustrations or lighter material, posed considerable problems for those who had only limited leisure time and did their reading in noisy and crowded spaces where it was difficult to maintain sustained concentration.[31] As such, it is unlikely that many of the workingmen who, in March 1880, listened to Huxley's lecture in respectful silence had the opportunity to read it when it was published, without substantive revisions, in Knowles's highbrow review three months later.[32] Whereas Huxley's earlier Lectures to Working Men at the Government School of Mines had been published in the artisanal trade journal the *Builder* or in cheap serial parts, the decision to place "On the Method of Zadig" in the *Nineteenth Century* was an acknowledgment that, all along, it had not been suitable for a mass audience.

This was confirmed at the following year's meeting of the British Association, where Huxley, in an address on "The Rise and Progress of Palæontology," reused the same structure and rhetorical techniques he had employed in his workingmen's lecture on Zadig, albeit with one significant alteration. Rather than Voltaire's ancient Babylonian sage, it was now the seventeenth-century Danish anatomist Nicolas Steno to whom Huxley attributed the "application of the axiom that 'like effects imply like causes', or as Steno puts it . . . 'bodies which are altogether similar have been produced in the same way.'" Cuvier's renowned "reconstruction of an extinct animal from a tooth or a bone," Huxley insisted, was actually "based upon the simplest imaginable application of the logic of Steno," as it was "only when the tooth or bone presents peculiarities which we know by previous experience to be characteristic of a certain group that we can safely predict that the fossil belonged to an animal of the same group."[33] Just as he had with Zadig, Huxley implied that the vaunted achievements of Cuvierian correlation were in fact simple manifestations of the empirical logic of Steno.

"The Rise and Progress of Palæontology" was published in September 1881 in *Nature*, whose readership, as noted earlier, was increasingly confined to scientific practitioners. Even then, Huxley protested to Lockyer at the omission of his address from *Nature*'s "sketch of the proceedings of the British Association," ironically acknowledging the journal's burgeoning influence within the scientific community by asserting: "If it were not that [the] very lecture appears in full in the same number I should be quite prepared to admit that it was not given, so highly do I estimate the authority of Nature."[34] With the fictional method of Zadig supplanted by the more historically reputable method of Steno, Huxley's anti-Cuvierian address was authenticated by its appearance in the prestigious pages of *Nature*, where, after all, it was likely to be considerably more meaningful to its audience than in its original form as a workingmen's lecture.

To both the readers of *Nature*, which had attained its prestige and authority through keeping abreast of the latest developments in scientific research, and the delegates at the British Association's annual celebration of the advancement of science, it would have been inexplicable that Huxley, in "The Rise and Progress of Palæontology," was merely reprising the very same arguments he had first articulated almost thirty years earlier. In "On the Method of Zadig," he ventured a modishly Darwinian forecast that fallacious doctrines such as physiological correlation would be "extinguished in the struggle for survival against their great rival common sense," but otherwise his renewed anti-Cuvierianism of the early 1880s remained remarkably similar to its initial incarnation in the mid-1850s.[35] It was, as Huxley now acknowledged, the law of correlation's continuing hold over the popular imagination that required him to revert to these archaic arguments, although his failure to actually render them suitable for popular audiences ensured that, as had occurred in the 1850s and 1860s, they still barely registered beyond the scientific specialists he had, for the most part, already won over. Unlike in earlier decades, Huxley now at least apprised the plebeian audiences for his workingmen's lectures that anatomical correlations were simply customary correspondences deduced by empirical observation rather than infallible a priori laws. Notably, however, he did so in exactly the same manner in which he addressed the readers of the *Nineteenth Century* or even *Nature*, and it was therefore unlikely that the finer details of his rather recondite distinction between rational and empirical approaches to animal structure would always be appreciated. In such circumstances, errors and misapprehensions were almost inevitable.

Flat-Earth Men and the Like

Working-class demands for political representation, and organizational endeavors to promote plebeian education, had, since the 1850s and 1860s, become increasingly entwined with the promulgation of heterodox beliefs that challenged the established dogmas of intellectual elites, both religious and secular.[36] In his lecture at the Working Men's College, Huxley alluded warily to the "people from whom the circle-squarers, perpetual-motioners, flat-earth men and the like, are recruited, to say nothing of table-turners and spirit-rappers," and he was careful to insist that Zadig's prosaically empirical method manifested only the "unconscious logic of common sense" and, notwithstanding its extraordinary results, involved nothing that was even remotely out of the ordinary. His descriptions of Zadig's earthily quotidian method were nevertheless couched in the idiom of the occult. Self-consciously appropriating the "word 'prophecy'" for a set of empirical principles that facilitated the "seeing of that which to the natural sense of the seer is invisible," Huxley avowed that, for Zadig and nineteenth-century paleontologists alike, this entailed a "process of divination beyond the limits of possible direct knowledge."[37] Just as he had with religious terminology and metaphors in works such as his *Lay Sermons* (1870), Huxley evidently intended to naturalize such signifiers of the supernatural, denuding them of their occult implications and demonstrating that the prophetic divination of the unseen required merely the sagacious application of routine procedures.[38] But with the original auditors of "On the Method of Zadig," if not with those who subsequently read it in the *Nineteenth Century*, Huxley's task was considerably harder than it had been in secularizing the language of the Bible. The renowned capacity of paleontologists to reconstruct strange prehistoric monsters from just fragmentary remnants of their fossilized remains had long had distinctly supernatural connotations that were especially appealing—as well as potentially empowering—to plebeian audiences.

As well as alluding to Zadig, Cuvier, in *Recherches*, had invested his paleontological accomplishments with a prophetic, quasi-scriptural rhetoric of resurrecting the dead, even if he did not have, as he put it, the "almighty trumpet at my disposal."[39] In the mid-1840s, Owen and his close friend William John Broderip, as seen in chapter 4, indulged their shared taste for the narrative pleasures of ghost stories in "Recollections and Reflections of Gideon Shaddoe, Esq.," which they wrote together for *Hood's Magazine*. Broderip, as the pseudonymous antiquary, reflected drolly on "records of prophetic sights and sounds" and occasions when "*second sight* is infallible," while Owen adopted the visionary sobriquet Silas Seer.[40] Shortly afterward, Owen lauded the "prophetic power" of the Cuvierian "principle of palæontological research" in his *History of British Fossil Mammals* (1846), and he and Broderip, in their notorious collaborative notice of Owen's own work for the *Quarterly Review*, commended the "ingenious prevision of contingencies" that enabled Owen's prediction of the dinornis.[41] Despite the eventual debunking of the ghost stories in "Gideon Shaddoe, Esq.," spectral modes of prescience and foresight clearly afforded seductive, if ultimately only figurative, parallels for the prodigious predictive powers of Cuvier and Owen, who used such occult self-fashioning to augment their reputations as infallible scientific geniuses. For other prominent exponents of the law of correlation, however, the same heterodox forms of visionary insight had more serious implications for their paleontological practices.

Cuvier's Swiss protégé Louis Agassiz was "desirous of knowing what to think of mesmerism," the controversial claim that an indiscernible magnetic fluid pervaded all animal bodies and became manifest in states of somnambulism, and in 1839 had himself induced into a "state of mesmeric sleep."[42] In it, Agassiz reported, he "experienced an indescribable sensation of delight, and for an instant saw before me rays of dazzling light which instantly disappeared."[43] Agassiz, as seen in chapter 2, employed the same methods as his mentor for identifying and reconstructing prehistoric fish, often from just single scales, and even before his experiences of mesmeric trance, he had already relied on the divinatory insights of sleep to accomplish his correlative inferences. Having spent "weeks striving to decipher" an "obscure impression" on a "stone slab" in Paris during the early 1830s, Agassiz became convinced that "while asleep he had seen his fish with all the missing features perfectly restored." He put a "pen and pencil beside his bed before going to sleep," and

> toward morning the fish reappeared in his dream . . . with such distinctness that he had no longer any doubt as to its zoölogical characters. Still half dreaming, in perfect darkness, he traced these characters on the sheet of paper at the bedside. In the morning he was surprised to see in his nocturnal sketch features which he thought it impossible the fossil itself should reveal. He hastened to the Jardin des Plantes, and, with his drawing as a guide, succeeded in chiselling away the surface of the stone under which portions of the fish proved to be hidden. When wholly exposed it corresponded with his dream and his drawing.

The verification, in the cold light of day, of what Agassiz's wife called the "track of his vision" was no less spectacular than Cuvier's own prediction, in exactly the same location, of the famous Montmartre opossum, although, notably, the rational triumph of zoological laws was now replaced by uncanny premonition and prophetic reverie (fig. 10.2).[44] For Agassiz, such mystical intuition was a necessary attribute of the scientific study of fossil fish, reflecting his early training, prior to arriving in Paris, in the more speculative methods of German *Naturphilosophie*.[45] After emigrating to America in the mid-1840s, he insisted of an extinct piscine genus: "I can distinguish the

FIGURE 10.2. Antoine Sonrel, *Louis Agassiz (Speaking with Himself)*. Albumen print, circa 1871. Louis Agassiz Papers, bAg 15.70.1, Archives of the Museum of Comparative Zoology, Ernst Mayr Library, Harvard University. This double-exposure photographic portrait, for which Agassiz would have moved into a different position for the second exposure to create a doppelgänger image, suggests his continuing interest in the uncanny effects he had earlier trusted to with the reconstruction of his dream fish. Reproduced by permission of the Ernst Mayr Library, Harvard University.

European species by a single scale; but this not from any definite character, but rather by a kind of instinct."[46] The New World, at this time, was experiencing a frenzy of interest in table-rapping, clairvoyance, and many other manifestations of a putative spirit world, and it was not long before Agassiz's apparent trust in unorthodox modes of vision and augury was seized upon to authorize an explicit attempt to conflate Cuvierian correlation with the occult beliefs embraced by plebeian autodidacts.[47]

I Feel Like a Perfect Monster

During the 1860s, William Denton, a self-taught geologist from Darlington who fled the religious dogmatism of Britain for the intellectual liberty of America, made his living as an itinerant lecturer (fig. 10.3A). In his popular geological lectures, which he published himself, Denton regularly adopted the "language of Cuvier" to explain how the "bony structure of an animal corresponds exactly with the life that it must live; and a man familiar with the relation between them can sometimes, from a small fragment, determine the form of the perfect animal and its habits."[48] Denton's fortitude and per-severance in also pursuing his own researches on fossils collected during his lecture tours meant that, as an admiring biographer remarked, "Smiles would do well in a future edition of 'Self-Help,' to add William Denton to his list of subjects."[49] But Smiles would have found this émigré autodidact even less biddable than the refractory Scottish cobbler Thomas Edward. Having al-ready "entered the mystical realm of Mesmerism" while still in Britain, Den-ton encountered new and even more potent extrasensory forces on the other side of the Atlantic, and, as he later recalled, "being intensely interested in geology and paleontology, it occurred to me that perhaps something might be done by psychometry . . . in these departments of science."[50] Psychometry had been founded in 1842 as a method of extrasensory perception in which physical contact with an object yielded information regarding its history, in the same way that photography could capture the light emitted by different bodies.[51] As an enthusiastic adherent, Denton was soon convinced that the "science of psychometry will shed new light upon many extinct animals" by enabling its clairsentient practitioners—regardless of their lack of scientific training—to evoke more detailed visions of the creatures than were possible even for Cuvier. This was done by having "fossil specimens . . . placed upon the forehead, and held there during the examination."[52]

It was not Denton himself, however, who was capable of this tactile form of paleontological divination. Rather, it was his American wife, Elizabeth (fig. 10.3B), who, when a "small fragment of the enamel of a mastodon's tooth" was pressed against her brow, not only had the "impression . . . that it is a part of some monstrous animal," but herself began to inhabit the crea-ture's colossal frame. During the examination, she declared:

> I feel like a perfect monster, with heavy legs, unwieldy head, and very large
> body. I go down to a shallow stream to drink. (I can hardly speak my jaws
> are so heavy). I feel like getting down on all fours. . . . My ears are very large
> and leathery, and I can almost fancy they flap my face as I move my head. . . .
> (It seems so out of keeping to be talking with these heavy jaws).[53]

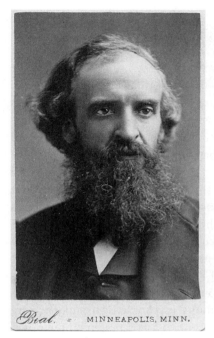

FIGURE 10.3. *A*, William Denton. *B*, Elizabeth Denton. Albumen prints, circa 1860s. Wellesley Historical Society, Wellesley, Massachusetts. These cartes de visites were made at the height of the Dentons' mutual interest in the paleontological applications of psychometry. Courtesy of the Wellesley Historical Society, Wellesley, MA.

Although Cuvier, Agassiz, William Buckland, and Owen had all received invaluable assistance from female family members, Elizabeth's psychometric rapport with this extinct monster was, irrespective of its unconventional nature, the very first time since Cuvier made the claim famous in the opening decades of the nineteenth century that a woman had professed to be able to reconstruct an entire creature from a single tooth or bone.[54] Elizabeth's idiosyncratic process of identification evinced an empathetic connection with the heavy-jawed mastodon, as well as an emphasis on the subjective and internal, that was notably absent from the detached objectivism of her male counterparts.[55]

Denton warned that "when woman shall employ her psychometric power in a scientific direction . . . some of our savans may tremble for their laurels," although across the Atlantic, where imported copies of their jointly authored *The Soul of Things* (1863) soon arrived, such claims provoked derision in the periodicals that deigned to acknowledge the book.[56] The *Athenæum*, in a scathing review, mockingly advised that Elizabeth's "proper place is among the geologists of the British Association . . . for she learns things in a few minutes, by her easy process of holding a pebble to her brow, which the greybeards of Science go pottering about year after year without result."[57] The condescension of such metropolitan weeklies, though, did not prevent the publication of a cheaper British edition of the Dentons' psychometric opus, under the revised title *Nature's Secrets* (1863), and beyond London's professional classes their paleontological clairsentience was often treated with considerably more respect. In Denton's native northeast the *Newcastle Courant* advised purchasers of *Nature's Secrets* that they would be "brought into contact with most singular phenomena . . . as new and extraordinary as they are deeply interesting," and concluded that the "instructive" book was "not unlikely to bring over some readers to opinions which they had regarded as residing on the basis of mere fancy."[58] Those provincial readers who followed the newspaper's promptings and embraced the Dentons' heterodox approach to paleontological reconstruction presumably valued it as a method that helped empower both autodidacts and women. They might also have recognized that, notwithstanding the sardonic scorn of their social superiors in journals such as the *Athenæum*, the Dentons' psychometric techniques were not so very different from how elite paleontologists represented their own discoveries.

Significantly, Elizabeth compared her own "true vision" to the "statement of Professor Agassiz" regarding the "*kind of instinct* . . . which enables him to distinguish" between species of fossil fish "by a single scale." She even urged Agassiz "to analyze the sensations to which he so pleasantly alludes," confident that "he would find himself possessed of a faculty" far beyond mere "instinct, as defined by our lexicographers," and instead partaking of paranormal powers.[59] In both Britain and America, women were considered to be more susceptible to psychic forces than men, and such sensitivities afforded them an authority that was otherwise perceived as inappropriate, permitting Elizabeth to offer unsolicited advice to the most revered naturalist in America.[60] Her audacious intervention certainly drew out the implications of the occult self-fashioning of experts such as Cuvier and Owen, making literal what was initially used only figuratively. It also exposed the more serious supernatural interests of Agassiz, something that historians, regardless of the available evidence, have been reluctant to acknowledge.[61]

In *The Soul of Things*, Denton prompted his wife to also try her psychometric abilities on a "slab containing the impression of two toes of one of the bird-like tracks from the Connecticut Valley," insisting that her apparitions of "rude reptilian birds" would be "appreciated by those who have made the sandstones of the Connecticut Valley and their footprints a matter of study."[62] The fierce controversy over these disputed footprints, involving Huxley, Owen, and Edward Hitchcock, was considered in the previous chapter, although for the Dentons it could be readily resolved by Elizabeth's feminine sensitivity to the psychic traces of prehistoric creatures that still resided in the material remnants of their former existence. Remarkably, some of the paleontological experts involved in the controversy, or at least members of their close families, seemed to agree. Hitchcock's son Charles, himself a prominent geologist who coauthored textbooks on the subject with his father, wrote to Denton, telling him:

> I found a copy of your book . . . and read it with great interest—particularly the sections upon the bird tracks. If you and Mrs Denton ever pass near Amherst my father and myself would be very happy to see you and show you the collection of tracks: I doubt not that Mrs Denton could verify some of the monsters by their feet.[63]

He was still more effusive in a subsequent missive, avowing: "My father . . . I doubt not will be interested in Psychometry—certainly in its practical revelations, if it will discover for him a new locality of bird tracks."[64] While there is no evidence of what Hitchcock père's actual views on the matter might have been, his son's enthusiastic response to *The Soul of Things* demonstrates that interest in the Dentons' heterodox approach to paleontology went well beyond the plebeian autodidacts who were its intended audience, and, as with Agassiz's own supernatural proclivities, extended even into the nascent American scientific community.[65]

By 1881 Denton was sufficiently emboldened to assert that his "investigations in mesmerism, spiritualism, and psychometry, showed me the defectiveness of the theories advanced by Darwin, Huxley, and others."[66] Huxley's own tone of supercilious irony toward the "canons of magician lore" a year earlier in "On the Method of Zadig" might have been suitable for the highbrow audience who read it in the *Nineteenth Century*, but, as the conspicuous success of the Dentons' self-taught paleontological clairsentience made evident, it was unlikely to prevail with the original auditors of his workingmen's lecture.[67]

Reasoning Backwards

Fourteen years after addressing the Working Men's College on Zadig's method, Huxley, now in his sixties and having retired to Eastbourne, reflected that "various sadly comical experiences of the results of my own efforts have led me to entertain a very moderate estimate of the purely intellectual value of lectures." In fact, he conceded ruefully: "I venture to doubt if more than one in ten of an average audience carries away an accurate notion of what the speaker has been driving at." Among all the "grotesque travesties of scientific conceptions" that had been perpetuated in recent years, Huxley was particularly concerned with what he termed "'Science as she is misunderstood' in the sermon, the novel, and the leading article."[68] Both clergymen and journalists, as seen in chapter 7, had long participated in endeavors to make science accessible to new audiences, although apparently with little regard for its accuracy, and Huxley now added a further group to the roster of fallacious popularizers: novelists. His own workingmen's lecture had begun by wryly applauding the imaginative invention of the "biographer of Zadig, one Arouet de Voltaire," but during the 1880s and 1890s it was in fiction that what Huxley actually said about the method of Zadig was most palpably misconstrued.[69]

In addition to naturalizing the supernatural language of "prophecy," Huxley maintained that while the term was in "ordinary use restricted to 'foretelling,'" it could also accommodate the very opposite of the prognostications accomplished by the "foreteller." He proposed a new figure of oracular insight, the "retrospective prophet," who adopts a "backward . . . relation to the course of time" and "strive[s] towards the reconstruction in human imagination of events which have vanished and ceased to be," just as Zadig had done from the footprints of the escaped dog and horse. Such retrospective prophecies, of course, relied on "accurate and long-continued observation" rather than the "imaginary laws of physiological correlation" that, as Huxley brusquely opined, Cuvier and his followers invoked in their fanciful "so-called restorations."[70] The discrepancy between these different manifestations of the imagination—one meticulous and empirical, the other merely illusory—was insisted on throughout "On the Method of Zadig," and demonstrating the divergence was one of Huxley's principal intentions in the lecture. It was nevertheless not long before his distinctive method of retrospective prophecy was confounded with the very paleontological axiom it was actually meant to extirpate.

When the readers of *Beeton's Christmas Annual* for 1887 reached the conclusion of the magazine's seasonal mystery story, they were informed by its inscrutable protagonist:

> In solving a problem . . . the grand thing is to be able to reason backwards. . . . Most people, if you describe a train of events to them, will tell you what the result would be. They can put those events together in their minds, and argue from them that something will come to pass. There are few people, however, who, if you told them a result, would be able to evolve from their own inner consciousness what the steps were which led up to that result. This power is what I mean when I talk of reasoning backwards.

In pursuing precisely this process of retrospective reasoning to solve a strange and gruesome case of murder, the same character also avows: "There is no branch of detective science which is so important and so much neglected as the art of tracing footsteps. Happily, I have always laid great stress upon it, and much practice has made it second nature to me." This proficiency in distinguishing "marks of footsteps upon . . . wet clayey soil" is integral to the resolution of the convoluted plot of *A Study in Scarlet*, the novel, originally published as part of *Beeton's Christmas Annual*, in which Arthur Conan Doyle first introduced his famous detective Sherlock Holmes. The "Science of Deduction and Analysis" that Holmes outlines in *A Study in Scarlet* was, unsurprisingly, almost immediately recognized as exemplifying the very same methods adumbrated by Huxley.[71]

Doyle had trained in medicine at Edinburgh, and his erstwhile tutor Joseph Bell, whose own medical acuity was itself an inspiration for the fictional detective, observed in a review of *The Adventures of Sherlock Holmes* (1892): "Voltaire taught us the method of Zadig. . . . Carried into ordinary life, granted the presence of an insatiable curiosity and fairly acute senses, you have Sherlock Holmes as he astonishes his somewhat dense friend Watson."[72] When, two years later, "On the Method of Zadig" was reprinted in Huxley's *Collected Essays* (1894), the *Derby Mercury* declared: "Had that lecture been delivered during the last year or two it could not have omitted some reference to Dr. Conan Doyle's fascinating detective, for the resemblance alike in theory and practice between the Babylonish philosopher of Voltaire and the English philosopher of Dr. Doyle is worth noting."[73] In the view of such provincial newspapers, the forensic process Holmes hailed as reasoning backward would have afforded an especially effective means of making Huxley's parallel method of retrospective prophecy accessible to his original plebeian audience.

Huxley, after all, had himself already observed in his 1854 lecture "On the Relation of Physiological Science to Other Branches of Knowledge" that a "detective policeman discovers a burglar from the marks made by his shoe, by a mental process identical with that by which Cuvier restored the extinct animals of Montmartre from fragments of their bones."[74] The detectives whose procedures Huxley equated with the paleontological practices of Cuvier were doughty representatives of the official police, who in the 1850s, as noted in chapter 6, had only recently come to prominence through the fiction and journalism of Charles Dickens, where they were presented as shrewd and meticulous rather than inspired or outstandingly gifted.[75] Indeed, it was precisely Huxley's point that Cuvier's reconstructions had required only the exercise of quotidian mental powers that had nothing to do with genius or

unerring rational laws. Over the next three decades, however, rapid transformations in the literary marketplace, including the expansion of reading
audiences eager for excitement and sensation and the reduced cost of book
production, led to the emergence of a new format of short fiction that, following a pattern originally established by Edgar Allan Poe, focused on the career of single detectives as they worked through a number of different cases.[76]
In a departure from the pattern of serialized fiction discussed in chapter 4,
each story was both discrete and yet linked to a larger sequence. In contrast
to Huxley's diligent detective policemen of the 1850s, the protagonists of this
new genre of detective fiction, who were generally private agents rather than
part of the official police force, displayed uncanny and infallible powers of
reasoning, as well as haughtily disparaging the more prosaic efforts of their
colleagues in the employ of the state. Looking back on this literary development from the 1920s, the American detective writer Willard Huntington
Wright contrasted the "plodding, hard working, routine investigator of the
official police" with the "inspired, intuitive, brilliantly logical super-sleuth of
the late nineteenth century."[77] For Huxley, repeating the same arguments he
had initially made in the 1850s in the very different circumstances of the 1880s,
the advent of such supersleuths and the changed conception of the mental
processes employed by detectives had significant implications for the clarity
of his original criticisms of Cuvierian correlation.

At the beginning of A Study in Scarlet, Dr. Watson, on first learning of
his new friend's forensic proclivities, comments: "You remind me of Edgar
Allen [sic] Poe's Dupin. . . . Have you read Gaboriau's works? . . . Does Lecoq
come up to your idea of a detective?" While Holmes responds curtly that
his prospective counterparts are, respectively, a "very inferior fellow" and a
"miserable bungler," Doyle, notwithstanding his metatextual irony, evidently
did draw on the tradition of detective fiction inaugurated by his American
and French precursors.[78] Notably, the two founders of this particular literary
genre had, since its inception, both invoked the same French naturalist as the
principal exemplar of the bold, intuitive, and unfailingly accurate reasoning
exhibited by their heroes. During the 1830s Poe had supplemented his meager
income by translating Cuvier's conchological writings, and in what is generally
regarded as the very first detective story, "The Murders in the Rue Morgue"
(1841), C. Auguste Dupin "understood the full horrors of the murder" of two
Parisian women only after reading a "passage from Cuvier" featuring a "minute anatomical . . . account of the large fulvous Ourang-Outang," which alerts
him that an escaped simian has perpetrated the gruesome crime.[79] Dupin introduces the passage, presumably from Le règne animal (1817), with a theatri-

cal flourish that emphasizes how pivotal Cuvier's unrivaled knowledge of the natural world is to his pioneering style of detection.

In Gaboriau's *L'affaire Lerouge* (1866), the elderly detective Père Tabaret, mentor of Lecoq, boasts of his capacity to "reconstruct all the scenes of an assassination, as a savant who from a single bone reconstructs an antediluvian animal," a claim that is repeated and confirmed by the narrator of *Monsieur Lecoq* (1869).[80] In the same novel, the narrator has already recounted that Tabaret, in resolving an earlier case, had "proved by A. plus B., by a mathematical deduction, so to speak," how a robbery had occurred. In fact, Tabaret claims of the evidence left by criminals that he does not need to "see it with my own eyes," a lowly procedure that he leaves to the official police, and instead he relies on rational and a priori principles akin to the abstract axioms of mathematics.[81] Cuvier had himself regularly compared the necessary correlations in animal anatomy to the invariable relations discerned by mathematicians, and in Gaboriau's detective novels, which were hugely popular in France, his ability to reconstruct from a single bone was, akin to Tabaret's analogous investigations, emphatically affirmed as requiring a mastery of rational laws rather than simple empirical observation.

Across the Channel, the publisher George Routledge issued lurid sixpenny translations of Gaboriau's detective novels in the autumn of 1887 only months ahead of the publication of *A Study in Scarlet* (itself a punning translation of *L'affaire Lerouge*), and Holmes soon augmented the initial explanation of his methods in terms that, even to anglophone readers, would already have been familiar from Tabaret's descriptions of his own procedures. In "The Five Orange Pips," which appeared in the *Strand Magazine* in 1891, Holmes, pondering a series of mysterious missives marked with the letters *K.K.K.*, observes:

> The ideal reasoner . . . would, when he had once been shown a single fact in all its bearings, deduce from it not only all the chain of events which led up to it but also all the results which would follow from it. As Cuvier could correctly describe a whole animal by the contemplation of a single bone, so the observer who has thoroughly understood one link in a series of incidents should be able to accurately state all the other ones, both before and after. We have not yet grasped the results which reason alone can attain to. Problems may be solved in the study which have baffled all those who have sought a solution by the aid of their senses.

Significantly, Holmes has "closed his eyes" when he makes this statement, and, as with Gaboriau's Tabaret, the comparison of the ideal capacities of the detective with the mental processes employed by Cuvier in his renowned feats of paleontological reconstruction is part of an argument in which abstract reasoning from rational principles is privileged over the empirical evidence of the senses.[82]

Holmes's expositions of his methods were, notoriously, often incoherent, and sometimes even contradictory (perhaps evincing a hint of irony at the pretensions of the nascent science of criminology), and in other stories he diverged from this stringently rational outlook, exhibiting the characteristics of an "observing machine" and telling Watson that their sole objective was to "observe and draw inferences from our observations."[83] But, crucially, Holmes's preference for quasi-Cuverian rationalism in "The Five Orange Pips" was, with the exception of actually naming the revered French savant, articulated in almost exactly the same way he had explained his "Science of Deduction and Analysis" in *A Study in Scarlet*, where he claimed that "from a drop of water . . . a logician could infer the possibility of an Atlantic or a Niagara. . . . All life is a great chain, the nature of which is known whenever we are shown a single link of it."[84] The supersleuth's Zadig-like process of backward reasoning through a connected chain of events, which was almost indistinguishable from what Huxley designated retrospective prophecy, was conflated with Cuvier's conviction that the paleontological accomplishments that had made him famous were attributable to his understanding of rational laws. This, of course, was the very thing that Huxley, in "On the Method of Zadig," had endeavored to expose, finally, as a delusive and treacherous fallacy.

The Mere Suspicion of Scientific Thought

In berating "'Science as she is misunderstood' in . . . the novel," Huxley was particularly incensed at the "affliction caused by persons who think that what they have picked up from popular exposition qualifies them for discussing the greatest problems of science."[85] Doyle's own grasp of contemporary science, despite his medical training with Bell, was not always particularly accomplished. He professed unapologetically that when he "read science," he had no interest in the "dust of the pedants" and instead tended to "avoid the text-books, which repel, and cultivate that popular science which attracts." In contrast to Huxley, Doyle was confident that only a "very little reading will give a man such a knowledge of geology," or any other science he chooses, that would enable him to "have a broad idea of general results, and to understand their relations to each other."[86] This cursory but nonetheless adequate understanding of science, for Doyle, often supported heterodox conclusions, and, after becoming a spiritualist, he even acclaimed the "very cogent objective evidence" for the psychometric experiments conducted by "William Denton, the author of 'Nature's Secrets.'"[87] Akin to Holmes's use of Cuvier, meanwhile, the apprehension of a murderer in an American detective novel entitled *A Mystery of New Orleans: Solved by New Methods* (1890) is accomplished by following the same procedures as when a "geological specimen was sent to Professor Denton . . . for psychometric analysis."[88]

Doyle was convinced that the "mere suspicion of scientific thought or scientific methods has a great charm in any branch of literature, however far it may be removed from actual research."[89] Into the early twentieth century, he was still adding self-conscious allure to his journalistic writing with allusions to scientific methods that, by then, were very far removed from current research, asserting, in an account of a real-life murder trial: "As Professor Owen would reconstruct an entire animal out of a single bone, so from this one little letter the man stands flagrantly revealed."[90] In this nonfictional article on crime and detection, Owen's predictive powers were presented as no less unerringly accurate than Cuvier's were assumed to be, by the fictive Holmes, in "The Five Orange Pips." Doyle's famous fictional detective spawned numerous imitations, many of whom adopted the very same scientific methods espoused by Holmes and his creator. In Arthur J. Rees's *The Shrieking Pit* (1919), for instance, the hero, David Colwyn, is commended for the "logical skill and masterly deductive powers with which [he] had reconstructed the hidden events of the night of the murder, like an Owen reconstructing the extinct moa from a single bone."[91] Tellingly, when Doyle created his own successor to Holmes, the irascible Professor Challenger, who was the protagonist of *The Lost World* (1912) and several other stories, he chose to make him a naturalist who could interpret a single "curved bone" as belonging to "a very large, a very strong, and, by all analogy, a very fierce animal."[92] While Doyle acknowledged that what Holmes and his successor practiced were, in reality, only "semi-scientific methods," the enormous influence and popularity of his detective stories, which in the *Strand Magazine* were attracting an unprecedented readership of more than a million, ensured that his pronounced insouciance toward scientific accuracy had important consequences for Huxley's attempts to popularize his hostility toward Cuvierian correlation.[93]

The intuitive and unfailingly accurate powers of detection that helped make Holmes and other heroic supersleuths such lucrative commodities in the late-Victorian and Edwardian literary marketplace were aligned, confusingly, with what Huxley had actually contended was a quotidian and prosaic approach, which could yield only cautiously approximate results. In such circumstances, it is unsurprising that Zadig's simple empirical assumptions regarding cause and effect were subsumed by the grand "feat of palæontology" that, as Huxley had conceded, "so powerfully impressed the popular imagination."[94] By the 1890s even the very title of Huxley's anti-Cuvierian lecture at the Working Men's College had become, paradoxically, a convenient shorthand for how the famous feat of reconstruction was purportedly performed by the French savant and his acolytes. Ironically, it was Owen's discovery of the giant New Zealand struthian from merely a fragment of femur bone—the nineteenth century's most celebrated exemplar of the law of correlation—that was most often attributed to Huxley's nominal method.

A review of *The Life of Richard Owen* (1894) in the *Academy* remarked of Owen's "famous reconstruction of the *Dinornis*," that "by the method of Zadig he reconstructed this bird, which no living man had ever seen."[95] Five years later, the Scottish physician George William Balfour appropriated both the title and elements of the structure of Huxley's workingmen's lecture for his own address to the York Medical Association. Strikingly, though, in "On the Method of Zadig in the Advancement of Medicine," which was delivered in October 1899, Balfour made no reference whatsoever to Huxley himself. Instead, he began by applauding the "tales of that wonderful detective— Sherlock Holmes," who, Balfour averred, was "indissolubly associated" with the approach that was "everywhere known as the method of Zadig." Observing that the "method of Zadig . . . enables us to reconstruct an entire animal from a fragment of bone," Balfour proposed that the most "striking and . . . legitimate instance of this method is told us by the late Sir Richard Owen." Drawing on the same reverential memoir, compiled by Owen's grandson, that had been reviewed in the *Academy*, he regaled his medical auditors with the familiar details of Owen's "reconstruction from so limited a premise" of the "magnificent bird" from New Zealand.[96] This, of course, was the same sanctioned Cuvierian interpretation of the contested discovery that was examined in chapter 3, which was rendered still more spectacular by being discussed in the same vein as the "wonderful" fictive abilities of Sherlock Holmes. If, as Balfour suggested in the closing months of the nineteenth century, the method of Zadig was now known everywhere, its popularity had very little to do with Huxley's original conception of the process.

Huxley's withering disdain for the misapprehension of science "in the sermon, the novel, and the leading article" masked the reality that it was his own muddled obfuscation of his actual opinions on correlation when addressing popular audiences during the 1850s and 1860s, which had then leached into what were considered his more specialist works, that helped perpetuate the continuing acclaim for Cuvier's reconstructive methods. Even his attempt, at the beginning of the 1880s, to finally reveal to a nonspecialist audience the true nature of Cuvier's putative paleontological accomplishments was actually more suited to the professional middle classes who read the *Nineteenth Century*, or even the specialist readership of *Nature*, than to the proletarian auditors for whom it had ostensibly been written. An unexpected opportunity to correct these mistakes and enter the fray one last time occurred in September 1893, when Owen's grandson, seemingly oblivious to their festering antagonism, invited Huxley to contribute a chapter on his late grandfather's scientific work to the forthcoming *Life of Richard Owen*. Huxley, imagining his erstwhile rival's "ghost in Hades . . . grinning over my difficulties," considered it "almost impertinent to trouble the modern world with such antiquarian business," and, as he reflected, "it was not altogether with a light heart that I assented to the proposal."[97] Having accepted the awkward commission, Huxley was adamant that "what I have to say must be addressed not to experts, but to the general public," among whom, he acknowledged, there existed a "popular distaste for anatomical science," particularly those aspects of it regarded as "dry and technical."[98] As Huxley clearly recognized, it was the nonscientific public, rather than the experts he had already won over, who still needed to be apprised of the proper principles of anatomy.

Regardless of his own tribulations with the popularization of science, Huxley remained confident that a "shorter and easier road" might be found by which the public could "obtain a sufficiently accurate general view of the scope of anatomical science." This involved, among other things, assuring them that there is an "empirical law . . . that such and such structures are always found together." On the other hand, as Huxley insisted, "no amount of merely physiological lore would enable" an anatomist "to so much as guess why the one set of characters is thus constantly associated with the other." In fact, Huxley even claimed that "Cuvier . . . advocated hypotheses," including that of "physiological deduction as the basis of palæontology," simply "because they happened to be in favour with the multitude, instructed and uninstructed."[99] Far from tempering the youthful impudence and pugnacity with which his crusade against the law of correlation had begun back in the mid-1850s, the sexagenarian Huxley now, for the first time, charged Cuvier with explicitly recognizing the redundancy of his own avowed principles, and adhering to them for venally pragmatic and political reasons, the very same accusation, as seen in chapter 8, that was made by the disreputable cynic Robert Knox.

The Life of Richard Owen was published in December 1894, and by June of the following year Huxley had joined his adversary in Hades. In the final months of his life, as his health grew increasingly frail, Huxley evidently considered it necessary to continue to reiterate exactly the same arguments concerning the empirical basis of Cuvier's renowned paleontological abilities that he had first begun to make exactly forty years before. The reaction of the *Academy* and George William Balfour to Owen's grandson's memoir, with both employing the book to confusedly acclaim Owen's prediction of the dinornis as a testament to the method of Zadig, nevertheless suggests that Huxley was no more successful in amending popular misconceptions of Cuvierian correlation in death than he was in life.

An Almost Unlimited Number of Combinations

Huxley's continued failure to popularize the same arguments against the law of correlation that had already swayed most of his scientific peers was all the more surprising because, across the Atlantic, an unprecedented profusion of fossil discoveries in America's new western territories was making the Cuvierian law not only increasingly irrelevant but demonstrably erroneous. As Frederick Burritt Peck, professor of geology at Lafayette College, informed an audience in Wyoming in 1903:

> Cuvier, who lived a century ago, used to pride himself on being able to reconstruct with certainty, as he thought, an entire animal from a bone, a tooth or a claw; but his theory of the necessary correlation of organs has been rudely shaken by the remarkable palæontological discoveries of the New World.[100]

Such a refutation had not been immediately apparent, though, and when the potential of the fossiliferous rock formations on the East Coast of America first began to be recognized fifty years earlier, they seemed only to confirm the famous theory. Indeed, Owen, writing in 1851, observed of a "series of remains of the *Mosasaurus Maximiliani*, from a Green-sand formation at New Jersey, United States," that they "gave . . . new proof of the Cuvierian law of correlation of organic structures."[101] During the early 1870s hired teams of bone hunters, funded and directed by the rival East Coast paleontologists Othniel Charles Marsh and Edward Drinker Cope, began excavating the vast badlands of Colorado, Nebraska, and Wyoming.[102] These remote western territories soon yielded extensive remains, and sometimes intact skeletons, of previously unknown extinct mammals whose peculiarly configured organizations could not be accommodated with Cuvier's assumption that there were evident taxonomic connections between particular component parts and the overall structure from which they derived.

Having examined the abundance of mammalian fossils transported back to the Peabody Museum in New Haven, Connecticut, Marsh announced in 1879: "After a lapse of three-quarters of a century, we can now see that Cuvier was wrong on some important points." In particular, Marsh proposed that the law of "Correlation of Structures," as laid down by Cuvier, has been more widely accepted than almost anything else that bears his name; and yet, although founded in truth, and useful within certain limits, it would certainly lead to serious error if applied widely in the way he proposed. . . . We know today that unknown extinct animals cannot be restored from a single tooth or claw, unless they are very similar to forms already known. Had Cuvier himself applied his methods to many forms from the early tertiary or older formations, he would have failed. If, for instance, he had before him the disconnected fragments of an eocene tillodont, he would undoubtedly have referred a molar tooth to one of his pachyderms; an incisor tooth to a rodent; and a claw bone to a carnivore.[103]

When first identifying the tillodont in 1875, Marsh had observed that the "animals are among the most remarkable yet discovered in American strata, and seem to combine characters of several distinct groups, viz: Carnivores, Ungulates, and Rodents."[104] Such an anomalous combination of distinct taxonomic characters, as had initially seemed the case with the archaeopteryx in the previous chapter, cast serious doubts on the viability of Cuvier's most famous contribution to paleontology, and would inevitably have confounded even the revered savant himself.

By the 1890s still stranger prehistoric mammals were arriving from the western badlands, although now they were mostly sent to Henry Fairfield Osborn at the American Museum of Natural History in New York, who, siding with Cope, had contested many of Marsh's interpretations and finally supplanted him as the foremost paleontological collector on the East Coast. Osborn assimilated many of his rival's distinctive working practices, though, and he certainly concurred with Marsh on at least one crucial aspect of paleontological method.[105] As Osborn reflected in 1893:

We recall Cuvier's famous law. . . . No generalization has been more thoroughly routed than that of a law of necessary correlation between tooth and foot structure. Besides the orthodox clawed carnivores and hoofed pachyderms of the great French anatomist, we have discovered hoofed carnivores such as Mesonyx, and clawed pachyderms such as Chalicotherium.

It was the latter incongruously clawed herbivorous mammal, as part of a larger order named by Osborn's mentor, that diverged most dramatically from the Cuvierian understanding of animal organization, with Osborn insisting that the "order Ancylopoda Cope presents the most signal exception to the law of correlation."[106] Out in Wyoming, Peck advised those in the field that they would "find it necessary to modify George Cuvier's theory of correlation of organs and assume an entirely unbiased and unprejudiced position regarding any new form which may be discovered. For though carniverous [*sic*], it may possess horns or even hoofs; and though herbiverous [*sic*], it may be armed with claws."[107] Working in the parched terrain of Wyoming and other equally barren western territories, America's pioneering field paleontologists could no longer rely on the long-standing postulations of the Old World, and instead were compelled by the profusion of peculiar new fossil forms they were exhuming to forsake their prior assumptions regarding the regularity of animal structures.

Osborn went even further in renouncing the scientific shibboleths of the Old World, avowing that "not only is there no correlation of type, but none in the rate of evolution."[108] In fact, the strange new mammalian forms of the American West, for Osborn, were equally "fatal to Darwin's original natural-selection hypothesis," particularly the related law of correlated variability, which, as seen in the last chapter, proposed that the modification of any one part by natural selection necessarily instigates nonadaptational correlative changes to the rest of the organism.[109] Osborn could discern "no correlation in the rate of evolution either of adjoining or of separated parts." There was, however, manifest evidence for what he termed the "independent evolution of parts," indicating, contrary to correlated variability, that modifications could occur in individual organs without requiring simultaneous changes in others (akin to what would later become known as "mosaic evolution").[110] It was, Osborn contended, "through this independent adaptation of different parts to their specific ends [that] there have arisen among vertebrates an almost unlimited number of combinations of foot and tooth structure," as had been revealed so decisively by "herbivores with sloth-like claws, such as *Chalicotherium*." As well as negating the Darwinian concept of correlated variation, Osborn was adamant that the "broad generalization that every part of an animal . . . has its separate and independent basis . . . brings us to the very antithesis of Cuvier's supposed 'law of correlation.'" It confirmed, after all, that there was "no fixed correlation" across the animal frame. Osborn could therefore affirm—finally and incontrovertibly—that it was "impossible for the palaeontologist to predict the anatomy of an unknown animal from one of its parts only."[111]

Both Osborn's and Marsh's anti-Cuvierian sentiments were quickly reprinted in *Nature*, which responded to the requirements of its specialist readers by assuming an increasingly internationalist purview, and the potential of what Osborn called the "revelation of the vast ancient life of the western half of the American continent" to "revolutionize" paleontology was soon no less evident in Britain than on the other side of the Atlantic.[112] Having reached New Haven or New York by railway, some of the plethora of new fossils from the western badlands were then transported by steamship to London, where they were displayed prominently in the Natural History Museum. The South Kensington museum's official *Guide to the Exhibition Galleries of the Department of Geology and Palæontology* (1890) informed visitors of the "skeleton of *Chalicotherium*" that it was "remarkable for the abnormality in the structure of the feet, so much so indeed as to render it for the future unsafe to predict the character of an animal from a single bone, and to invalidate the old maxim, *ex pede Herculem*."[113] Although his name did not appear on its title page, the guidebook was authored by Henry Woodward, who was now the keeper of the museum's geological department.[114] Three decades earlier, as was seen in the last chapter, Woodward, as a junior member of the museum's staff, had been coerced by Owen and his acolytes in the press into retracting his criticism of the same Latin maxim in regard to the archaeopteryx. The transatlantic torrent of equally strangely configured extinct mammals now permitted Woodward to authoritatively reaffirm his original objections without fear of reprisals.

To Correct a Popular Impression

While Alpheus Spring Packard, professor of zoology at Brown University, agreed with compatriots such as Marsh and Osborn on the "falsity of [Cuvier's] assertion that a single facet of a bone was sufficient to reconstruct a skeleton," he recognized the enduring "popular reputation and prestige" of the "famous law of correlation of parts," which, writing in 1901, he attributed to the simple consideration that "it could be easily understood by the layman."[115] The purported clarity of the Cuvierian method, as opposed to the dour caviling of Huxley's empiricism or the complexities of the various neo-Lamarckian mechanisms developed by both Osborn and Packard, had commended it, as noted earlier, to lecturers endeavoring to engage popular audiences into the 1870s and beyond. It was this putative precision that had likewise appealed to novelists suspicious of scientific technicalities such as Doyle. The apparent simplicity and yet unfailing accuracy of the law of correlation was particularly enticing for writers, in Britain, addressing the vastly expanded but often only rudimentarily literate reading audiences drawn from the first generation, reaching adulthood in the 1890s, to undergo the compulsory schooling inaugurated by the 1870 Elementary Education Act.[116]

Correlation's continuing presence in late-nineteenth-century print culture was certainly evident to Henry Neville Hutchinson, who, during the 1890s, established himself as the foremost popularizer of paleontology with a series of best-selling books that, among other things, introduced the new mass reading audience in Britain to the spectacular discoveries of the American West. In *Extinct Monsters* (1892) Hutchinson noted that when "people . . . visit" the "great museum . . . at South Kensington," they often "pass hastily by the cases of bones, teeth, and skeletons," which, he acknowledged, "fail to interest them."[117] As an Anglican priest, Hutchinson adopted the same prophetic language of the book of Ezekiel that Cuvier had first invoked exactly eighty years before, to pledge that, in his popular tome, he would use "reason and imagination" to make "these dry bones live" and "tell to the passer-by their wondrous story." He also, as Ralph O'Connor has pointed out, repeated the "familiar patterns" of many earlier popular writers on paleontology, offering merely "late-Victorian replications" of previously established rhetorical methods of evoking the marvels of the ancient past.[118] There was one particular popularizing trope, however, that Hutchinson, unlike almost all previous popular writers on the subject, and especially those who, like himself, were part of the long-standing tradition of clerical popularizers, resolutely refused to resort to.[119]

Examining the "material from which important conclusions with regard to the structure and habits of an extinct animal may be drawn," Hutchinson abruptly cautioned: "It is, of course, impossible for any one to reconstruct an entire animal from a single bone or a few teeth." While the readers of *Extinct Monsters* would doubtless have encountered this alluring myth in numerous other popular works, or in lectures, journalism, and museum displays, it was no longer tenable, Hutchinson insisted, because the "'Law of Correlation' . . . has been found to be not infallible; as Professor Huxley has shown, it has exceptions. It expresses our experience among living animals, but, when applied to the more ancient types of life, is liable to be misleading."[120] Notably, this was the very first time that the anti-Cuvierian arguments that Huxley first began to articulate in the mid-1850s were embraced unequivocally by a professional popularizer (as opposed to practitioners who occasionally engaged in popularization such as Huxley himself). Notwithstanding his clerical background, Hutchinson evidently perceived himself as a sympathetic publicist for Huxley's particular agenda. In June 1892, only three months before the publication of *Extinct Monsters*, Hutchinson sent Huxley a copy of another of his popular books, recalling, in the accompanying letter, "having a pleasant chat with you in Bristol Cathedral one evening after service in 1884." Huxley, with his customary disdain for popularizers, brusquely jotted in the margin of the missive, "Never was there—that I remember," and he seems not to have otherwise responded to Hutchinson's gift.[121] Despite this apparent rebuff, it was Hutchinson who, following Huxley's death, endeavored to contend with the continuing consequences of his haughtiness toward popularizers and inability to himself shape public opinion.

Like the *Strand Magazine*, which carried Doyle's hugely popular Sherlock Holmes stories, *Pearson's Magazine* was another illustrated sixpenny monthly that regularly sold in excess of a million copies. At the beginning of the twentieth century, its proprietor Arthur Pearson also launched an avowedly populist halfpenny newspaper, the *Daily Express*, which he publicized by funding an audacious "*Express* expedition" to Patagonia to locate a living specimen of the extinct mylodon. The Christmas number of *Pearson's* for 1900 helped promote its new sister title's expedition by commissioning an article on "Prehistoric Monsters" from the leading popular writer on the subject. Rather than immediately entering into the exciting details of how "no expedition of the kind . . . has had such high hopes of success," Hutchinson began his contribution to *Pearson's* by peremptorily announcing:

> Before I proceed further, I should like to correct a popular impression which I find very prevalent. The idea exists in the minds of many that the illustrious Cuvier or Sir Richard Owen, his friend and successor, could, from a single bone, or tooth, "restore" the whole animal! This feat is quite impossible. . . . Of course, if the creature were already known from the discovery of its whole skeleton, the expert could tell you all about it. Possibly this is where the confusion of thought arose.[122]

Eight years after his initial endorsement of Huxley's jeremiads against Cuvier and his law of correlation in *Extinct Monsters*, Hutchinson, addressing a much larger readership than in any of his books, still found the erroneous "popular impression . . . very prevalent," and again resolved, in the opening year of the new century, to finally root it out.

But like Huxley before him, Hutchinson's admonitions, despite his evident proficiency as a popularizer, seem to have done little to amend popular misconceptions regarding Cuvierian correlation. Tellingly, when a new edition of *Extinct Monsters* was published in 1910, Hutchinson considered it necessary to both augment and update the warning he had included in the first edition eighteen years before, which now read: "It is of course impossible for any one to reconstruct an entire animal from a single bone or a few teeth. Not even Owen could do this—in spite of the rather frequent assertions to that effect one sees in newspapers and magazines!"[123] The peevish exclamation mark at the end of the sentence made clear Hutchinson's exasperation that, even into the second decade of the twentieth century, confident avowals of Owen's remarkable reconstructive abilities showed no signs of diminishing in the new era of mass-market magazines and newspapers.[124] If anything, Hutchinson's evident irritation suggests that the prevalence of such hyperbolic claims may actually have increased in the period between 1892 and 1910. Certainly, he felt obliged, later in the new edition of *Extinct Monsters*, to add another new warning of "how little ground there is for the popular belief that a Cuvier or an Owen could 'restore' a whole animal from a jaw, or even a single tooth!"[125] As well as Huxley, Hutchinson also sent advance copies of his popular works to Marsh in New Haven, who "very kindly read many of the proof-sheets," although even the profusion of American fossil specimens explicitly demonstrating the fallacy of Cuvier's understanding of animal structure could do nothing to stem the tide of "popular belief" in the law of correlation.[126]

Conclusion

In the autumn of 1916, with the First World War having descended into a bloody stalemate, a mysterious new weapon was introduced onto the muddy battlefields of the western front. To the *Daily Mail*, another of the new century's populist mass-market newspapers, these weapons appeared to be "fantastic monsters . . . like blind creatures emerging from the primeval slime," while for the *Times* they seemed "like some vast antediluvian brutes which Nature had made and forgotten."[127] The monstrous new weapon was known officially as the tank, and, as Trudi Tate has observed, almost "all the early tank writings compare the machine to some kind of dinosaur emerging out of the primeval mud."[128] It was one dinosaurian genus in particular that most often came to mind, especially on the home front, where few had actually seen the British Army's squat secret weapon, and the powerful yet ponderous tank was soon dubbed the "Diplodocus Galumphant."[129] The diplodocus, a gigantic Jurassic sauropod, was another of the remarkable prehistoric creatures discovered in the badlands of Colorado and Wyoming during the 1870s, and a cast of its colossal skeleton had been one of the most spectacular exhibits at the Natural History Museum in South Kensington since it was donated by the American industrialist Andrew Carnegie in 1905.[130] In the summer of 1917, just as the ostensibly diplodocian tanks were being deployed against German infantry defenses, the mounting of this same cast provoked a heated controversy that ensured that, even amid the more urgent concerns of the ongoing mechanized slaughter, the law of correlation did not disappear entirely from the thinner, more closely printed pages of periodicals necessitated by wartime paper shortages.

It was Hutchinson, now in his sixties and eager to transcend his erstwhile reputation as a popularizer and instead assert his own intellectual credentials, who disputed the "present reconstruction" of the Natural History Museum's iconic diplodocus, which, he complained, had been put together with assistance from American paleontologists, including Osborn, who were "somewhat too anxious to produce something very big and imposing." Writing in the *Geological Magazine* for August 1917, Hutchinson protested that the "American palæontologists wished this great reptile to be as tall as they could make it" and had thus articulated its "huge body high above the ground and limbs erect, as in the elephant." He, on the other hand, proposed "bringing down the vertebral column into a lower position," with the limbs situated laterally to the trunk, so that the creature more closely approximated the squatter posture "we see in lizards and crocodiles," as well as, incidentally, the diplodocian tanks then galumphing across the western front (fig. 10.4). Hutchinson was drawing on earlier complaints made by the paleontologists Oliver Hay and Gustav Tornier, although in reinforcing their interpretation of the diplodocus's distinctly reptilian gait, which was supported by the creature's unprecedentedly long and heavy tail, he employed an unexpected secret weapon of his own.[131] As he proclaimed:

> In all these matters the law of correlation is a useful guide, though we admit by no means an infallible one. In a case like this it seems quite reasonable to make use of this guide, as Cuvier did who first propounded it. Looking at mammalian skeletons generally, we seem to discover that big upright limbs and a proper quadrupedal progression are correlated with small, light tails, as in *Elephas*, *Bos*, etc. Why is this? The answer seems to be . . . that the drag of a heavy tail would be too great. . . . And so our argument is confirmed; *Diplodocus* never had its femur working in a plane parallel to axis of the body as in mammals.[132]

In his prewar popularization, of course, Hutchinson had insisted that the "'Law of Correlation' . . . when applied to the more ancient types of life, is liable to be misleading."[133] While still conceding that it was not infallible, he now seemed to largely disregard such admonitions in his more specialist contribution to the *Geological Magazine*. This, notably, was the complete reverse of Huxley's trajectory of rescinding his expert censure of correlation when addressing more popular audiences. Even more remarkably, the very same scientific law that had first been imported into Britain during the Napoleonic Wars more than a century earlier was still being used to resolve disputes over paleontological reconstruction, and apparently in the same manner as its original propounder, at the height of the next major European conflict.

The burgeoning mass culture of the late nineteenth and early twentieth

FIGURE 10.4. *Photograph of Model of* Diplodocus carnegiei. H. N. Hutchinson, "Observations on the Reconstructed Skeleton of the Dinosaurian Reptile *Diplodocus carnegiei*," *Geological Magazine*, n.s., 4 (1917): plate 23. This photograph shows a model of *Diplodocus carnegiei* as restored by Hutchinson, with the vertebral column brought down into a lower position than the articulated skeleton in the Natural History Museum, and, in accordance with the law of correlation, its tail dragging on the ground. © The Trustees of the Natural History Museum, London.

centuries was inevitably constrained by the rigors of total war, with the prices of cheap books and periodicals rising sharply and publishers focusing on jingoistic propaganda.[134] Even Sherlock Holmes was compelled to forego his usual detective mysteries and, in "His Last Bow" (1917), instead expose German spies. It certainly appears that what Hutchinson termed the "popular belief that a Cuvier or an Owen could 'restore' a whole animal from . . . a single tooth!" echoing Huxley's earlier frustrated acknowledgment of "that feat of palæontology which has so powerfully impressed the popular imagination," began to wane—along with so many other quintessentially Victorian nostrums—during the years of wartime austerity.[135] The famous Cuvierian method, however, had already outlasted Huxley's ferocious onslaught beginning in the mid-1850s, as well as, subsequently, the new era of Darwinian evolution and then Huxley's belated attempts to popularize his criticisms. Hutchinson's paradoxical yet pragmatic recourse to the law of correlation in upholding his own interpretation of the tanklike posture of the diplodocus only demonstrated, once again, its continuing resilience.

Ghosts of Correlation

On the night of 10 May 1941, the German Luftwaffe launched the heaviest raid on London of the entire nine-month bombing campaign known as the Blitz. Hundreds of the city's inhabitants were killed, and several of its unique historic buildings, including the British Museum and the Tower of London, sustained substantial damage. Arthur Keith, the former Hunterian Professor of Comparative Anatomy and Physiology at the Royal College of Surgeons, recorded in his diary: "In clear moonlight, a devilish procession started overhead; presently the sky over London glowed red; it had a hell of a time. All night long I sat up in bed, longing for the morning." But the light of dawn brought little comfort, revealing that, amid the wider carnage, the "Museum of the Royal College of Surgeons had been reduced to a charred mass."[1] The Hunterian Museum had received a direct hit from a "heavy high-explosive bomb," which ignited fires that "raged with terrible ferocity." It was reported that "when the conflagration finally burnt itself out the scene of destruction was indescribable and heartrending," prompting Keith to reflect mournfully: "All that my predecessors and I had laboured to bring about had been wiped out in a night."[2] In particular, it was the Hunterian's renowned comparative anatomy collections that suffered the most severe damage, and the gardens of Lincoln's Inn Fields were strewn with the shattered remnants of strange prehistoric creatures.[3] It was, of course, Keith's mid-nineteenth-century predecessor Richard Owen who had done most to develop the museum's extensive collection of extinct animals, often identifying and reconstructing them from only fragmentary remains using Georges Cuvier's famous law of correlation (see fig. 3.3). The skeletons of those same creatures, including the iconic dinornis and megatherium, had now been returned to an even more fragmentary condition by Adolf Hitler's aerial bombardment (fig. E.1).

FIGURE E.1. Photograph of bomb damage at the Royal College of Surgeons taken for the *Daily Mail*, 1941. A bust of Owen is one of the few items to survive the bombing of the Hunterian Museum on 10 May 1941. Associated Newspapers Ltd. / Solo Syndication.

The diversion of college funds from surgical matters to Owen's paleontological researches was regularly criticized by the *Lancet* in the 1830s and 1840s, and some of the Hunterian's trustees a century later similarly had "no use for the wonders exhibited by the bony framework of the elephant . . . and no veneration for extinct monsters like the Irish elk, the megatherium or the glyptodon."[4] For them, the wartime devastation afforded the perfect "opportunity for reorganisation and for the introduction of new ideas," ensuring that a rebuilt and rationalized "future museum" would no longer be "cluttered with animal skeletons."[5] The method that Owen had used to articulate the fossilized animal bones whose obliteration by German bombs was discreetly commended by their modernizing custodians, however, could not be eradicated so conveniently, even if its presence in the twentieth century became increasingly shadowy if not downright spectral. Indeed, a decade before the London Blitz, perhaps the most celebrated of nineteenth-century revenants had posthumously reasserted the famous claim of Cuvierian paleontologists regarding reconstruction from just a single bone.

In September 1929 Martha Dickinson Bianchi published *Further Poems of Emily Dickinson*, collecting the remaining manuscript verse of the New

England poet that had not been included in the volumes of her writings published in the 1890s. Introducing this new collection, Bianchi declared that "when the little, unexplored package gave these poems . . . it was for one breathless instant as if the bright apparition of Emily had returned to the old house," and she insisted that "Emily may be said to have accomplished death without loss of life."[6] Dickinson, who died almost wholly unheralded in 1886, wrote her verse privately with seemingly no intention of publication, and one of the poems that Bianchi encountered in the unexplored package of handwritten fascicle pages, which she tentatively titled "(With a Daisy)," had been completed at the end of the 1850s.[7] It begins:

> A science—so the savants say,
> "Comparative Anatomy,"
> By which a single bone
> Is made a secret to unfold
> Of some rare tenant of the mold
> Else perished in the stone.

The remainder of the short poem then compares the revelations yielded to the savant by a solitary bone to how, during winter, to the "eye prospective led," the "meekest flower of the mead / . . . stands representative" of the coming spring, with its "rose and lily . . . / And countless butterfly!"[8]

It is likely that Dickinson would have heard effusive affirmations of the power to reconstruct from a single bone in lectures by Edward Hitchcock, New England's inveterate champion of Cuvierian paleontological methods, when she attended Amherst Academy during the 1840s.[9] As with William Makepeace Thackeray's similar comparison in *The Newcomes* (1853–55) of the method by which "Professor Owen . . . takes a fragment of a bone, and builds an enormous forgotten monster out of it" with how the novelist creates the "megatherium of his history," though, Dickinson's poem seems to exhibit a surreptitious skepticism regarding such hyperbolic scientific claims.[10] The chary "so" in the initial line, and the suggestion that the single bone has to be forcibly "made" to unfold its potentially disingenuous mysteries, certainly accords with Dickinson's general misgivings about the possibility of complete disclosure.[11] Yet the poem is also palpably fascinated by the claim it simultaneously mistrusts, as well as the parallels between the powers of the comparative anatomist and the synecdochic process by which the poet can conjure a unified whole representative of new life and hope. Most significantly, the circumstances of its belated publication meant that Dickinson's posthumous poetic evocation of the mid-nineteenth-century enthrallment with Cuvier's vaunted reconstructive abilities first appeared in print only weeks ahead of the Wall Street crash.

The potency of Cuvier's legendary claim evidently extended through the next two decades of global conflict and into the postwar period. The Harvard paleontologist Stephen Jay Gould, who grew up in New York during the 1950s, later recalled:

> Cuvier's principle of correlation lies behind the popular myth that paleontologists can see an entire dinosaur in a single neck bone (I believed this legend as a child and once despaired of entering my chosen profession because I could not imagine how I could ever obtain such arcane and wondrous knowledge).[12]

The despair that temporarily curtailed Gould's resolution to become a paleontologist was a lingering remnant of the misconceptions surrounding that "feat of palæontology which has so powerfully impressed the popular imagination" that Thomas Henry Huxley bemoaned in the 1880s.[13]

Adulatory accounts of this purported feat, as has been seen throughout *Show Me the Bone*, had circulated in various formats of print culture for much of the nineteenth century and into the early twentieth century, often in the process of syndication and abstraction that James Secord has termed "literary replication." Emphasizing the difficulty of maintaining stable textual meanings in such replications, Secord notes that the "problem of stability extends far more widely, for attempts to reproduce the work also extended beyond the original bookseller, printer, or publisher."[14] In this formulation, literary replication takes place almost exclusively at a spatial level (after all, it happens "widely"), with a privileging of the process of extension in space at the expense of its occurrence in time. Gould's youthful despair in 1950s New York, as well as the posthumous publication of Dickinson's 1859 poem "(With a Daisy)," nevertheless indicate that literary replication occurs not only synchronically, but also diachronically. There is no inevitable cutoff point for when textual material can be or is replicated, and often, as in the case of nineteenth-century claims about the arcane and wondrous powers of paleontologists, it continues to reappear in new print contexts for a very long time, contravening convenient—as well as largely arbitrary—historiographic boundaries. As was seen with the sanctioned version of Owen's discovery of the dinornis in chapter 3, moreover, a high degree of textual stability can be maintained even in such temporally distanced literary replications. Indeed, the story had become so established by the final decades of the nineteenth century that the *Pall Mall Gazette* could describe it as how "Professor Owen inferred the proverbial dinornis."[15]

The despair that haunted Gould's childish imagination and Dickinson's "death without loss of life" both suggest that claims about the reputed capacity to reconstruct prehistoric creatures from just a single part of their anatomy persisted only as a ghostly residue. As was seen in the introduction, however, sociologists of science have recently adopted the concept of "undead science" to describe the strange persistence of theories such as cold fusion that are rejected at the closure of controversies, yet that "continue to haunt the house of science" and whose "ghosts produce palpable material effects . . . amongst the scientists with whom they interact."[16] Cuvier's law of correlation and the spectacular paleontological predictions it facilitated had been largely rejected by the mid-Victorian scientific community following the bitter controversy initiated by Huxley in the 1850s, and especially once the profusion of new fossil discoveries in the American West from the 1870s onward made the ostensibly universal law demonstrably erroneous. But claims about paleontologists' unerring and almost prophetic powers of reconstruction continued to circulate in the nonspecialist formats in which science was brought to new audiences even long after the closure of the controversy, and, crucially, this popular perpetuation of the famous Cuvierian method continued to haunt expert practitioners such as Huxley, impacting, as seen in chapter 7, on their specialist activities.

Nor has the spectral afterlife of this particular paleontological principle ended even now. In 2012 the Natural History Museum in South Kensington opened a new permanent exhibition displaying some of the most exceptional objects and specimens in the museum's extensive collections. In pride of place at the entrance of the "Treasures" gallery is the fragment of bone from which Owen ventured his daring inference of the past existence of the dinornis. As the electronic display next to the fragment, as well as the museum's website, explains to visitors under the heading "bold prediction":

> Owen's most dramatic scientific triumph came in 1839 when he studied a short fragment of bone discovered a few years earlier in New Zealand. . . . Owen deduced it must have belonged to an . . . extinct flightless bird. Four years later the world was astonished when more bones revealed he was right. . . . No one in Europe had seen anything like this monstrous bird before. By the time Europeans first arrived in New Zealand in the 1760s, they had already been hunted to extinction. . . . It took Owen's genius to resurrect them.[17]

This, notably, is precisely the same version of the disputed events, with the drama of Owen's predictive genius foregrounded and the contribution of the surgeon who first brought the bone to him entirely expunged, that Owen, his closest supporters, and their acolytes in the press had strategically propagated in the 1840s and 1850s. Their highly partisan rendition of the dinornis's discovery, which quickly became the nineteenth century's most triumphant and potent validation of Cuvier's law of correlation, still persists—as a veritable ghost in the machine—in the digital media of the twenty-first century.

Acknowledgments

This book, which began as a labor of curiosity more than a decade ago, has benefited greatly from the contributions of many people and institutions, and I would like to offer my sincere thanks for the support and generosity shown to me during the time I have been researching and writing it. I am especially grateful to the Leverhulme Trust for the award of a Research Fellowship in 2012–13, during which the bulk of the book was written. The Leverhulme also funded my postdoctoral Research Fellowship between 1999 and 2002, when I first conceived the original idea for the book, so this project is particularly indebted to the trust's liberality (in all senses of the word) and commitment to independent research. The award of a Small Research Grant from the British Academy enabled me to visit archives in the United States, Australia, and New Zealand in 2009 and 2010, while a Research Grant from the British Society for the History of Science helped fund a research trip to Scotland in 2012. I am very thankful to both. Research incentive funds accruing from an Arts and Humanities Research Council Large Grant in the "Science in Culture" theme contributed to the cost of the book's illustrations, and I am grateful both to the AHRC and the University of Leicester for this assistance.

I am greatly obliged to several friends and colleagues whose guidance and advice helped shape the book in crucial ways throughout its gestation. With huge kindness, Bernie Lightman and Sally Shuttleworth both read the manuscript in its entirety, and their enthusiasm, insight, and judicious counsel are very much appreciated. I am also extremely grateful to Geoffrey Cantor, George Levine, Jim Secord, Jon Topham, and Paul White for reading and commenting on manuscript drafts of various chapters. Additionally, the book has benefited greatly from the suggestions, comments, and assistance of: Sam Alberti, David Amigoni, Melinda Baldwin, Ruth Barton, Peter Bowler, Laurel Brake, Janet Browne, Adelene Buckland, Sabine Clarke, Orietta Da Rold, Jamie Elwick, Jim Endersby, Richard England, Aileen Fyfe, Graeme Gooday, Jon Hodge, John Holmes, Sally Horrocks, Frank James, Alice Jenkins, Melanie Keene, Ivan Kreilkamp, Ben Marsden, Margaret Meredith, Andrew Miller, Jim Moore, Ralph O'Connor, Sadiah Quereshi, Michael Reidy, Evelleen Richards, Martin Rudwick, Sharon Ruston, Anne Secord, Peter Shillingsburg, Jonathan Smith, Matt Stanley, Rebecca Stott, the late Frank Turner, and Michael Wolff. My colleagues in the Victorian Studies Centre at the University of Leicester, especially Claire Brock, Felicity James, Gail Marshall, Julian North, and Joanne Shattock, provided a marvelously supportive and stimulating working environment, while the advice of Julie Coleman and Carol Arlett in the School of English has been particularly invaluable. I would also like to thank my colleagues on the "Constructing Scientific Communities: Citizen Science in the 19th and 21st Centuries" project, particularly Geoff Belknap, Sally Frampton, Julie Harvey, Chris Lintott, and Sally Shuttleworth, who, over the last year, have prompted me to think in new ways about public participation in science.

It has been both a pleasure and a privilege to work with the University of Chicago Press, especially my editor Karen Darling, who has been hugely encouraging and supportive since we first began discussing the book in the summer of 2012. Evan White and Mary Corrado have also been very helpful in answering my numerous queries, while my copyeditor, Sue Cohan, did a tremendous job of Americanizing my resolutely British prose, and spotted several errors that I am quite—in the American sense—glad never made it into print. Two anonymous readers for the press provided remarkably detailed, generous, and thoughtful feedback, which, I am sure, has considerably improved the final book. I would like to take this opportunity to thank them for their time and effort.

This book could not have been written, or at least would have been very different, without the assistance of numerous librarians and archivists. I would particularly like to thank Sam Alberti and Sarah Pearson at the Royal College of Surgeons, Anne Barrett at Imperial College Archives, Paul Cooper at the Natural History Museum, Simon Dixon at the University of Leicester Library, Kathleen Fahey at the Wellesley Historical Society, Frank James at the Royal Institution, James Kirwan at Trinity College Library, Alison Pearn of the Darwin Correspondence Project, Cambridge University Library, Magdalene Popp-Grille at the Württembergische Landesbibliothek, and Tom Whitehead at Temple University Library, all of whom were exceptionally generous in helping me to locate and access archival materials in their respective collections. I am very grateful to those libraries and archives that have allowed me to use unpublished material in their possession: Add. 38091, 39954, and 42577 are quoted by permission of the British Library; Add. 5354, the Darwin Manuscript Collection, and the Darwin Library are quoted with the permission of the Syndics of Cambridge University Library; Cod. hist. qt. 413 is quoted by permission of the Württembergische Landesbibliothek; the Denton Family Papers are quoted by permission of the Wellesley Historical Society; the Edward and Orra White Hitchcock Papers are quoted by permission of Archives and Special Collections, Amherst College Library; the Frances Hirtzel Collection of Richard Owen Papers is quoted by permission of Temple University Special Collections Research Center; the Huxley Papers are quoted by permission of the College Archives, Imperial College London; the John Murray Archive is quoted by permission of the National Library of Scotland; the Mantell Family Papers are quoted by permission of the Alexander Turnbull Library; Mss.B.Ow2 and Mss.B.H981 are quoted by permission of the American Philosophical Society; the *Philosophical Transactions* Referees' Reports are quoted by permission of the Royal Society; the Richard Owen Papers are quoted by permission of the Trustees of the Natural History Museum, London; the Richard Owen Papers are quoted by permission of the Royal College of Surgeons of England Archives; RI MS GB2 is quoted by courtesy of the Royal Institution of Great Britain; the Royal Literary Fund Archive is quoted by permission of the Royal Literary Fund; the Samuel Leigh Sotheby Scrap Album is quoted by courtesy of Bromley Local Studies and Archives; the Society for the Diffusion of Useful Knowledge Papers are quoted by permission of University College London Library Services; and the William Whewell Papers are quoted by permission of the Master and Fellows of Trinity College Cambridge.

Earlier versions of portions of chapter 4 appeared as "Literary Megatheriums and Loose Baggy Monsters: Paleontology and the Victorian Novel," *Victorian Studies* 53 (2011): 203–30; and "Paleontology in Parts: Richard Owen, William John Broderip, and the Serialization of Science in Early Victorian Britain," *Isis* 103 (2012): 637–67. An earlier version of parts of chapter 6 appeared as "'The Great O. versus the Jermyn St. Pet': Huxley, Falconer, and Owen on Paleontological Method," in *Victorian Scientific Naturalism: Community, Identity, Continuity*, edited by Gowan Dawson and Bernard Lightman, 27–54 (Chicago: University of Chicago Press, 2014). Permission to reprint is gratefully acknowledged.

Finally, I thank my family and friends for the support, good humor, and patience that have been so helpful in completing a book that, on occasions, seemed as if it might itself become a large, loose, baggy monster of the sort discussed in chapter 4. Helen Wilkinson's love and companionship were, as ever, particularly invaluable. Jon Mortimer made my regular research trips to London much more fun than they really ought to have been. My mum, Stefanie, sister Carla, and brother Joe all patiently endured my newfound enthusiasms for extinct creatures (and long-standing intransigence about proletarian professors!), while Yvonne Wilkinson offered some shrewdly pragmatic advice about the book's subtitle. The love and support of my late grandparents Jan and Mary Bełej has been an enduring presence in my life, and it is to their memory that this book, which I know would have pleased them, is dedicated.

Notes

Introduction

1. Quoted in [Verplanck] 1819, 147. On Mitchill's flamboyance, see Burnett 2007, 44–47.

2. See Gellius 1927, 1:3–5.

3. On the scientific preeminence of Paris in the revolutionary and Napoleonic eras, see Gillispie 2004.

4. [Verplanck] 1819, 80, 132.

5. Francis 1858, 94.

6. Ibid.; and Burnett 2007, 2.

7. See Outram 1980; and J. C. Smith 1993.

8. Cuvier 1818, 321.

9. Knox 1856, 246. On Knox's training in Paris between 1821 and 1822, see Bates 2010, 42–50.

10. Cuvier 1812a, 57, 61 ("des os isolés . . . celui qui posséderoit rationnellement les lois de l'économie organique, pourroit refaire tout l'animal"). Translation from Rudwick 1997, 217, 219. The original wording of translated quotations is given in the notes; published translations are used when available, and my own translations when they are not. By "reconstruct" (or "*refaire*" in the original French), Cuvier presumably meant the reassembly of an animal's skeleton, which is the sense in which the word is used in this book's subtitle, although the term clearly also carries connotations of restoring a creature to what it looked like in its living state. See O'Connor 2007a, 3n.

11. Thiers 1851, 92.

12. Quoted in Daintith and Gjertsen 1999, 136. Archimedes reputedly claimed of his law of leverage: "Give me a firm place to stand on and I will move the Earth." Quoted in Daintith and Gjertsen 1999, 17. As with Cuvier's purported claim, these statements are only attributed to Descartes and Archimedes.

13. On scientific predictions, see Anderson 2005, 15–40; on performativity in the physical sciences, see Morus 2010.

14. On the etymology of *paleontology*, see Rudwick 2008, 47–48.

15. On the role of evolution in biology, see Bowler 2003; on the nebular hypothesis in astronomy, see Schaffer 1989.

16. See Rudwick 1985, 113.

17. Best 1837, 28–29.

18. See O'Connor 2012, 497–98.

19. Rudwick 2005, 364.

20. On the disputes over extinction, see Black 1991; and Rudwick 2005, 243–47 and passim.

21. See Cuvier 1812a, 4.

22. See Corsi 1988a, 23–24; and Fox 2012, 188–89.

23. Knox 1839, 222.

24. On Cuvier's ambiguous attitude toward religion, see Outram 1984, 148–49; and Taquet 2009.

25. "Infidelity in Disguise" 1837, 465.

26. Rudwick 2005, 596. See Appel 1987, 222–30; Brooke 1989, 35–39; and Desmond 1989, 61–66.

27. Reiss 2009, 119. See also Lenoir 1989, 61–65; and Letteney 1999, 401.

28. Reiss 2009, 85. His customary associations with conservative and antievolutionary doctrines, as Joel Black has suggested, "have combined to make Cuvier's achievement the dark discovery of the nineteenth century that few historians of science, especially outside France, have felt inclined to recognize or to investigate in great detail"; Black 1991, 158. For attempts to attach scientific approaches to fixed ideological positions, see Appel 1987; and Desmond 1989.

29. On the blackening of Owen's reputation, see Rupke 1994, 1–10.

30. On attempts to rehabilitate Owen, see E. Richards 1987; Gruber and Thackray 1992; Rupke 1994; Padian 1997; Camardi 2001; and Rupke 2009. On the lingering perception of his obscurantism, see Desmond 1989, 240–54 and passim; and Desmond 2001, 32. Oddly, Owen, despite his paleontological significance, is hardly mentioned in either O'Connor 2007a; or Rudwick 2008.

31. On the overhaul of early nineteenth-century publishing, see Klancher 1987; J. Secord 2000, 24–40; and Fyfe 2012.

32. J. Secord 2014, 241. See also J. Secord 2000, 126–38 and passim.

33. On the new forms of visual spectacle, see Altick 1978; O'Connor 2007a; and Fyfe and Lightman 2007.

34. On science and Victorian consumer culture, see Fyfe and Lightman 2007.

35. See McGill 2003; and Fyfe 2012, 173–252.

36. See Irmscher 2013, 106–7.

37. While endorsing Secord's general argument that textual meanings are made by readers rather than mandated by authors, *Show Me the Bone*, as will be seen in chapter 3, differs from *Victorian Sensation* in acknowledging that authors, and their acolytes, could—and did—exercise some control over the burgeoning range of meanings entailed in the process of "literary replication"; J. Secord 2000, 126.

38. On Darwin's tendency to overshadow studies of nineteenth-century science, see Endersby 2003.

39. J. Secord 2000, 3.

40. On the significance of such geographical factors, see Livingstone 2003; and Livingstone and Withers 2011.

41. On historians' perplexity at Owen's long adherence to the law of correlation, see Desmond 1982, 173; and Rupke 1994, 136–37. On tensions between the colonial center and periphery, see Barton 2000; and Endersby 2008.

42. J. Secord 2004a, 664, 668.

43. On the *long durée* in the history of science, see Holmes 2003.

44. Although such an approach is, of course, facilitated by what has been called the "electronic harvest" of wholesale textual digitization, it is important to recognize the limitations of keyword searching, no matter how sophisticated, in tracking all the manifestations of a particular concept, and, most especially, understanding the varied contexts in which it operated; J. Secord 2005, 463.

45. Simon 2002. The so-called afterlives of a variety of cultural phenomena have also been considered in recent years by literary scholars influenced by postmodern notions of the instability of historical interpretations. See Douglas-Fairhurst 2004; and O'Gorman 2008.

46. On nonspecialist formats and new audiences for science after midcentury, see Lightman 2007b; Fyfe and Lightman 2007; and Bowler 2009.

47. On the scientific naturalists and Huxley's wider campaign, see F. Turner 1974; Lightman 2009; and Dawson and Lightman 2014.

48. Falconer 1856, 493.

49. For such assumptions, see Cohen 2011, 20; and Rudwick 2013, 363.

50. In line with the practice of several recent historians, potentially contentious categories such as "popularizers," "experts," and "practitioners" are retained in *Show Me the Bone* in order to foreground issues of authority and audience and demonstrate where distinctions were perceived to exist, although they are used as neutrally as possible, and should not be seen as implying any evaluative social or intellectual assumptions. See Lightman 2007b, 9–13; O'Connor 2007a, 12–13; and O'Connor 2009b, 339–44.

51. For such assumptions, see F. Turner 1993, 131–228; and Desmond 1997, 254–56.

52. On the constraints on Huxley's approach to popularization, see Lightman 2007b, 368.

53. Cooter and Pumphrey 1994, 250.

54. Topham 2009, 16.

55. O'Connor 2009b, 343.

56. See A. Secord 1994; Winter 1998; J. Secord 2000; Fyfe 2004; Lightman 2007b; Fyfe and Lightman 2007; O'Connor 2007a; and Bowler 2009 for eminent examples of this body of work.

57. Sheets-Pyenson 1985, 549.

58. T. Huxley 1881, 453. On the nineteenth-century meanings of "popular science" and "popularization," see Topham 2007; and O'Connor 2007a, 11–13.

59. On arguments for thinking in terms of "literature *and* science," "science *as* literature," or "literature *as* science," see O'Connor 2007a, 14–15; Dawson and Lightman 2011–12, 1:ix–xi; and A. Buckland 2013, 14–15, 23–27.

60. Beer 1983, 7.

61. O'Connor 2009b, 343. On the potential for conflict in the reciprocal traffic between literature and science, see Dawson 2007, 218–21.

62. On literary forms of popularization, see O'Connor 2007a, 22–23, 104, and passim.

63. T. Huxley 1893–94, 8:viii.

64. Doyle 1891a, 487.

Chapter One

1. Cuvier 1819, 62–63 ("Il est le premier qui ait appliqué la connaissance de l'anatomie comparée à la détermination des espèces de quaprupèdes dont on trouve les dépouilles fossiles. . . . Son tour de force le plus remarquable en ce genre fut la détermination d'un os . . . comme l'os la jambe d'un géant. Il reconnut, par le moyen de l'anatomie comparé, que ce devait être l'os du

rayon d'une giraffe, quoiqu'il n'eût jamais vu cet animal et qu'il n'existât point de figure de son squelette"). Translation from Cuvier 1828, 13.

2. On the strategic purposes of Cuvier's *éloges*, see Outram 1978. The identification of the giraffe was originally announced in Daubenton 1764.

3. [Brewster] 1844, 24−25.

4. See Fox 2012, 9−11.

5. [Grant] 1830, 344.

6. Quoted in Corsi 1988a, 37. The barbed remark was made by Charles Sonnini de Manoncourt.

7. Viénot 1932.

8. See Semmel 2004, 38−71.

9. "Abstract of Lectures" 1837, 380.

10. Knox 1839, 222.

11. Swainson 1834, 86−87.

12. See de Beer 1960, 149 and passim.

13. See Outram 1984, 73−80.

14. On Cuvier's sycophancy to Napoleon, see Fox 2012, 10.

15. Cuvier 1800−1805, 1:xvii ("espèce d'orgueil, utile peut-être en politique," "en méprisant un peu trop les étrangers, en n'estimant et même en ne consultant presque que ses compatriotes," and "sécheresse qui fait le charactère de quelques-uns de ses auteurs en histoire naturelle et en anatomie comparée"). Translation from "Cuvier's Lectures" 1800, 532.

16. "Cuvier's Lectures" 1804, 248.

17. See Rudwick 1985, 133−39; Appel 1987, 222−30; Brooke 1989, 35−39; Desmond 1989, 61−66; and Rudwick 2005, 596−98.

18. See Jacyna 1994 for a notable exception.

19. Rudwick 2005, 596.

20. On the new meanings given to scientific texts in the process of translation, see Rupke 2000.

21. See Topham 2011, 312.

22. R. Burkhardt 1977, 193.

23. See Stevens 1994, 65−72.

24. Cuvier 1817, 1:10 ("principe de la *subordination des caractères*"). My translation.

25. Serres 1842, 51 ("il y avait . . . dans le beau travail de Vicq-d'Azyr le germe du principe de la coexistence et de l'harmonie des parties dont la démonstration constitue un des plus beaux titres de gloire de Cuvier"). My translation.

26. See Coleman 1964, 46−51.

27. Corsi 1988a, 27−28.

28. On Cuvier's reading of Kant, see Letteney 1999, 374−76.

29. Outram 1986, 344−50.

30. Letteney 1999, 421−32. See also Reiss 2009, 107−9.

31. See Outram 1984, 135−37.

32. See R. Richards 2002, 238−39. On Cuvier's relationship with Kielmeyer, see also Taquet 2006, 85−88, 313−23.

33. Lenoir 1989, 61.

34. Reiss 2009, 119.

35. Letteney 1999, 401.

36. See Outram 1986, 324−27.

37. Cuvier 1800–1805, 1:58 ("conditions nécessaires de l'existence"). Translation from Cuvier 1801, 64.

38. Cuvier 1800–1805, 1:57 ("lois de coexistence" and "à la vue d'un seul d'entre eux, conclure jusqu'à un certain point celle de tout le squelette"). Translation from Cuvier 1801, 63.

39. Quoted in Fox 2012, 63.

40. On Cuvier's strategic self-presentation in this period, see Kete 2012, 107–44.

41. Lawrence 1816, 85. Grant similarly noted that the lectures were "attended by numerous assemblies"; [Grant] 1830, 350.

42. [Grant] 1830, 351.

43. On the collaborative nature of *Leçons* and many other of Cuvier's publications, see Gillispie 2004, 657–58.

44. [Grant] 1830, 350. According to Rudwick, however, Cuvier only visited the Montmartre quarries on occasional weekends, and instead paid an assistant to purchase fossils from the local laborers; Rudwick 2000, 54.

45. Rudwick 1997, 286–87, 36 ("Aujourd'hui l'anatomie comparée est parvenu à un tel point de perfection que l'on peut souvent d'après l'inspection d'un seul os, determiner la classe, quelquefois même le genre de l'animal auquel il a appartenu . . . [le] os qui composent chaque partie du corps d'un animal, sont toujours dans un rapport nécessaire, avec toutes les autres parties, de manière qu'on peut conclure, jusqu'à un certain point, de l'une d'elles à l'ensemble," "méthode qu'on employe dans les reserches dont je vais vous entretenir," "reduit à des conjectures plus délicates, & à des conclusions moins certaines," and "degré de probabilité qui lui appartient").

46. Ibid., 289, 40 ("bien connus," "on aurait ainsi l'image, non seulement du squelette qui existe encore mais de l'animal entier, tel qu'il existait autrefois. On pourroit même avec un peu plus de hardiesse deviner une partie de ses habitudes; car les habitudes d'un animal quelconque dépendent de l'organisation & en connoissant celle cy on peut conclure celles là," and "rétabiles").

47. G. Cuvier to J. H. Autenrieth, 29 fructidor VII [15 September 1799], Cod. hist. qt. 413, l. 3, Württembergische Landesbibliothek, Stuttgart ("Ma collection de squelettes est aujourd'hui si complette qu'il me suffit souvent d'un seul fragment d'os, pourvu que les facette articulaire y soient, pour déterminer à quel genre et même à quelle espèce il a appartenu, surtout lorsque c'est un de os des pieds; ou des mâchoires"). My translation.

48. Rudwick 1992, 36. Rudwick makes this claim in relation to Cuvier's similarly unpublished drawings of a conjecturally reconstructed palaeotherium and anoplotherium.

49. See J. Secord 2007.

50. See Csiszar 2010, 83–87.

51. [Brewster] 1844, 37.

52. See Outram 1984, 181.

53. Cuvier 1804, 286 ("Cette operation se fit en presence de quelques personnes à qui j'en avois annoncé d'avance le résultat, dans l'intention de leur prouver par le fait la justesse de nos théories zoologiques, puisque le vrai cachet d'une théorie est sans contredit la faculté qu'elle donne de prévoir les phénomènes"). Translation from Rudwick 1997, 71.

54. See Corsi 1988a, 23–24.

55. On the collaborative role of audiences in scientific performance, see Morus 2010.

56. See Morus 2007, 336.

57. Rudwick 2000, 62.

58. Cuvier 1804, 292 ("degré de . . . probabilité," "sciences exactes," and "il n'est aucune science qui ne puisse devenir presque géométrique: les chimistes l'ont prouvé dans ces derniers

temps pour la leur; et j'espère que le temps n'est pas éloigné où l'on en dira autant des anato-
mistes"). Translation from Rudwick 1997, 71, 72–73.

59. See Corsi 1988a, 24–25.

60. See Anderson 2005, 16–17; and Jenkins 2007, 88–91.

61. Cuvier 1800–1805, 1:47 ("lois qui déterminent les rapports de leurs organes, et qui sont
d'une nécessité égale à celle des lois . . . ou mathématiques"). Translation from Cuvier 1801, 52.

62. R. Owen 1846a, 74.

63. See Topham 2013.

64. Quoted in Topham 2013, 119, 120.

65. See Topham 2011, 320–21.

66. See Watts 2014.

67. "Abstract of a Memoir" 1799, 512, 514.

68. "Buckland's *Bridgewater Essay*" 1836, 312.

69. Notably, some modern commentators on Cuvier still make the same mistaken assumption; see Reiss 2009, 99.

70. See Topham 2011, 318–20.

71. "Notice Concerning the Skeleton" 1796, 638.

72. See Brock and Meadows 1998, 90–95.

73. "Intelligence and Miscellaneous Articles" 1805, 190.

74. Cuvier 1822, 284n ("Je laisse cet article tel qu'il a paru d'abord, dans les *Annales du Muséum*, comme un monument . . . de la force des lois zoologiques"). My translation.

75. Coleman 1964, 119.

76. On the *Philosophical Magazine*'s preempting of other French scientific developments,
see Csiszar 2010, 82.

77. See Topham 2011, 314–15.

78. "Cuvier's Lectures" 1800, 532, 536.

79. "John Allen, Esq." 1843, 96.

80. Cuvier 1801, [iii], [iv], 63.

81. Ibid., 63.

82. On the social background of Edinburgh's students in this period, see Rosner 1991, 25–43.

83. Allen 1849, xxii, xxiii.

84. Quoted in Jacyna 1994, 63.

85. Cuvier 1801, [iv]; and Allen 1849, xxxiii.

86. Jacyna 1994, 73, 75.

87. While the original French edition of *Leçons* did not deal explicitly with the subordination of characters, the first volume contained several tables based on Cuvier's hierarchical model
of classifying animals that Allen's abridged translation did not reproduce; see Coleman 1964, 86.
Jacyna notes that "Allen's reading of Cuvier was, of course, highly selective"; Jacyna 1994, 192.

88. Quoted in Jacyna 1994, 63.

89. Barclay 1822, 321, 330.

90. See Jacyna 1983, 311–29.

91. "Foreign Publications" 1803, 490, 492, 493.

92. "Day in the Country" 1823, 169.

93. "Retrospect of Domestic Literature" 1803, 589. A guinea was equivalent to twenty-one
shillings; thus, Ross and Macartney's version of *Leçons* cost more than ten times as much as Allen's two-shilling translation. On the epistemological significance of a book's cost and format,
see J. Secord 2000, 122–23.

94. "Cuvier's Lectures" 1804, 248.

95. Outram 1986, 362.

96. See Appel 1987, 69–104; and Le Guyader 2004.

97. See Jordanova 1984; and Corsi 1988a.

98. Gillispie contends that it was only in 1812 that Cuvier rescinded his collaborative working relations with Geoffroy and Lamarck and finally "felt obliged to react against what he thought to be the more and more exaggerated views of his colleagues"; Gillispie 2004, 660–61.

99. Cuvier 1800–1805, 1:47 ("fonctions étoit modifiée d'une manière incompatible avec les modifications des autres, cet être ne pourroit pas exister"). Translation from Cuvier 1801, 52. Cuvier 1812a, 58 ("organisé forme un ensemble . . . dont toutes les parties se correspondent mutuellement" and "Aucune de ces parties ne peut changer sans que les autres changent aussi"). Translation from Rudwick 1997, 217.

100. See Corsi 1988a, 122.

101. See Gould 2002, 293–94.

102. Serres 1842, 57–58 ("principe de la corrélation des formes," "Ce principe reposant sur l'idée que les organes d'un même animal forment un tout unique . . . la conséquence de ce principe . . . était . . . d'arreter dans ses développemens la théorie des évolutions qui admet dans les organismes des transformations passagères," "nouveaux principes," and "principe des analogies organiques"). My translation. Serres himself had reservations about some aspects of transformism; see Corsi 1988a, 240–41.

103. E. Russell 1916, 39.

104. [Chalmers] 1814, 271.

105. Corsi 1988a, 181, 185.

106. Cuvier 1812a, 2, 56, 57, 58, 65 ("constance du lecteur," "sentiers pénibles où je suis contraint de l'engager!," "hérissée de difficultés," "faire évanouir tous les embarras," "à la rigueur, être reconnue par chaque fragment . . . de ses parties," "os isolés, et jetés pêle-mêle," and "arrive à des détails faits pour étonner"). Translation from Rudwick 1997, 185, 216, 217, 221.

107. On the rhetorical style of the "Discours préliminaire," see Outram 1984, 151–52; O'Connor 2007a, 61–62; and Lloyd 2011.

108. Cuvier 1812a, 4 ("os fossiles" and "Je développerai les principes sur lesquels repose l'art de déterminer ces os, ou, en d'autres termes, de reconnoître un genre, et de distinguer une espèce par un seul fragment d'os, art de la certitude duquel dépend celle de tout l'ouvrage"). Translation from Rudwick 1997, 186.

109. Cuvier 1812a, 61 ("l'ongle, l'omoplate, le condyle, le fémur, et tous les autres os pris chacun séparément, donnent la dent, ou se donnent réciproquement; et en commençant par chacun d'eux isolément, celui qui posséderoit rationnellement les lois de l'economie organique, pourroit refaire tout l'animal"). Translation from Rudwick 1997, 219.

110. Rudwick 2005, 589n.

111. Gould 2002, 295n.

112. Cuvier 1812a, 63 ("empreinte," "piste d'un pied fourchu," "tout aussi certaine qu'aucune autre en physique," "seule piste," and "une marque plus sûre que toutes celles de Zadig"). Translation from Rudwick 1997, 220.

113. Voltaire 1748, 178 ("hommes avaient tort de juger d'un tout dont ils n'apercevaient que la plus petite partie"). Translation from Voltaire 1964, 95.

114. Cuvier 1812b, 3 ("presque une résurrection en petit, et je n'avois pas à ma disposition la trompette toute puissante," "lois immuables," and "à la voix de l'anatomie comparée, chaque os, chaque portion d'os reprit sa place"). Translation from Rudwick 2005, 413.

115. See Outram 1984, 143, 148–49.

116. See Corsi 1988a, 183–84; and Reiss 2009, 92–93, 103–4.

117. Outram 1984, 78; quoted in Taquet 2009, 133.

118. Outram 1984, 11. See also Rudwick 1997, 259.

119. Balzac 1831, 1:88–91 ("Cuvier n'est-il pas le plus grand poëte de notre siècle?," "Lord Byron a bien reproduit par des mots quelques agitations morales; mais notre immortel naturaliste a reconstruit des mondes avec des os blanchis . . . il fouille une parcelle de gypse, y aperçoit une empreinte, et vous crie: Voyez! Soudain les marbres s'animalisent, la mort se vivifie, le monde se déroule! . . . cette épouvantable résurrection due à la voix d'un seul homme . . . nous fait pitié," and "les œuvres géologiques de Cuvier"). Translation from Balzac 1977, 40–42.

120. Byron 1821, 337, 338. See O'Connor 1999.

121. Balzac 1831, 1:191–92, 2:185 ("prophetesses agitées par un démon," "regards ranimer des vieux ossemens," "nouveau Messie," "une siècle de lumière," and "soumettrait ses miracles à l'Académie des Sciences"). Translation from Balzac 1977, 82, 223. The First Class of the Institut national was reconstituted as the Académie des sciences in 1816 following the restoration of the Bourbon monarchy.

122. Cuvier 1812a, 63, 64, 65 ("moins claires," "il faut que l'observation supplée au défaut de la théorie," "lois empiriques qui deviennent presque aussi certaines que les lois rationelles, quand elles reposent sur des observations suffisamment répétées," "adoptant ainsi la méthode de l'observation comme un moyen supplémentaire," "une constance spécifique . . . entre telle forme de tel organe, et telle autre forme d'un organe différent," "système général de ces rapports," "constance classique," and "raisonnement effectif"). Translation from Rudwick 1997, 220, 221.

123. Cuvier 1812a, 65 ("méthode de l'observation" and "s'aidant avec un peu d'adresse de l'analogie et de la comparaison effective"). Translation from Rudwick 1997, 221.

124. Coleman 1964, 121.

125. See Topham 2011, 314–15.

126. [Playfair?] 1814, 455.

127. See Rudwick 2005, 510–11.

128. See Jacyna 1994, 56–59.

129. R. Jameson 1817, xi.

130. Quoted in Topham 2011, 334.

131. See O'Connor 1999, 31–32.

132. R. Jameson 1813, ix, viii, vi, vii.

133. On Jameson's moderate Presbyterianism, see J. Secord 1991, 13–14.

134. Quoted in Jacyna 1994, 59.

135. R. Jameson 1813, ix.

136. Ibid., vii.

137. Homo 1815, 225.

138. Cuvier 1813, 5.

139. [Chalmers] 1814, 266, 269.

140. Ibid., 270, 271, 272.

141. See Topham 1999.

142. R. Jameson 1813, ix, v.

143. L. Jameson 1854, 30.

144. On Jameson's attendance at Allen's lectures, see Jacyna 1994, 190n.

145. Knox 1839, 222.

146. Cuvier 1826; and Cuvier 1827–34. On the latter, see Cowan 1969; on Grant's translation, see Desmond 1989, 56.

147. Knox 1839, 222.

148. Knox 1856, 247.

149. Quoted in Outram 1984, 240n ("Ce discours a été en Angleterre l'objet d'une faveur particulière; il y a déjà été réimprimé quatre fois en anglais et deux fois en Amérique"). My translation.

150. Cuvier 1818, 321.

151. Quoted in [Verplanck] 1819, 147; Cuvier 1818, 326, 322.

152. R. Jameson 1815, vii. See Rudwick 2005, 529.

153. "Baron Cuvier" 1832, 538.

154. Lee 1833, 229.

155. R. Jameson 1827, viii.

156. See J. Secord 1991, 1–18.

157. See Topham 1999, 165–68.

Chapter Two

1. Spittal and Stevenson 1829, 286, 287.

2. Thomson 1870, 2:28.

3. Ibid.; and Fleming 1829, 279.

4. Fleming 1829, 279.

5. Fleming 1830, 70, 67.

6. Conybeare 1829, 145n, 142.

7. Thomson 1870, 2:28.

8. Fleming 1830, 68.

9. Duns 1859, xxxvii; quoted in Burns 2007, 219.

10. Duns 1859, xxxviii, xxxvii.

11. W. Buckland 1836, 1:94, 1:95n.

12. [Fleming] 1823, 392.

13. Conybeare 1829, 145–46.

14. K. Lyell 1881, 1:259, 1:260.

15. "Infidelity in Disguise" 1837, 465.

16. For such attempts, see Appel 1987; and Desmond 1989.

17. On Cuvier's English, see Lee 1833, 271.

18. Fleming 1822, 2:137, 2:138n.

19. Quoted in Duns 1859, xxxii ("temoignage . . . de estime" and "J'aurais desire toutefois que vous éussiez un peu plus approfondi ma theorie des coexistences d'organisations et les applications nombreuses que j'en ai faites dans mon ouvrage sur les os fossiles, vous auriez probablement reconnu qu'elle s'éloigne moins que vous ne croyez de votre façon de penser, et surtout vous auriez evité de la présenter comme un appui du materialisme"). My translation.

20. Isler 1879, 60 ("nos matérialistes"). My translation.

21. "Fleming's *Zoology*" 1823, 247.

22. Quoted in Burns 2007, 214.

23. Quoted in Duns 1859, xxxiv.

24. Ibid.; and Barclay 1822, 330.

25. On the Lawrence-Abernethy debate, see Jacyna 1983; and Ruston 2005, 38–63, although neither recognizes the centrality of Cuvier's putative materialism to the dispute.

26. Lawrence 1816, 9, 24.

27. Lawrence 1819, 2, 3n.

28. Lawrence 1816, 85, 121–22.

29. Lenoir 1989, 61.

30. On Cuvier's criticisms of phrenology, see Outram 1984, 129–31.

31. Abernethy 1817, 52, 53, 18, 37, 16.

32. Lawrence 1819, 5, 12.

33. See Wallace 1832, 3:146–50.

34. See Hilton 2006, 251–53.

35. Lawrence 1819, 2–3n. On the attractions of Lawrence's materialism to working-class radicals, see Desmond 1989, 117–21.

36. "Lawrence's Lectures" 1822, 959.

37. Lawrence 1823, [iii].

38. See O'Connor 2007a, 104.

39. "To the Editor" 1828, 30.

40. "Cuvier" 1832, 456.

41. On Cuvier's acquiescence to the Ultra-royalists, see Desmond 1989, 47; and Fox 2012, 11–12.

42. "Natural History" 1835, 455.

43. See Hollis 1969.

44. Lascelles 1821, [537], [538].

45. See Rupke 1997.

46. W. Buckland 1820, 4, 3. On the general interest in geology prompted by the Geological Society, see O'Connor 2009a.

47. W. Buckland 1820, 30n, 29, 23, 7, 13.

48. See Ospovat 1981, 33–34.

49. Lawrence 1819, 8; and W. Buckland 1820, 6.

50. W. Buckland 1820, 14.

51. See Brooke 1991b, 192–97.

52. See Eddy and Knight 2006, xviii.

53. Paley 1802, 289, 201.

54. Cuvier 1812a, 58.

55. Paley 1802, 289.

56. See Rupke 1983, 234–39.

57. Kidd 1824, vii, 1, 13, 24, 2.

58. [P. B. Duncan] 1836, vi.

59. Conybeare and Phillips 1822, xlviii.

60. W. Buckland 1820, 2–3.

61. On Buckland's chaotic rooms, see Boylan 1984, 244.

62. W. Buckland 1824, 390, 391.

63. Quoted in Dean 1999, 62.

64. W. Buckland 1835, 426.

65. "Arts and Sciences" 1829, 808.

66. On the restriction of Buckland's humor to his lectures, see O'Connor 2007a, 80; and Sommer 2007, 287.

67. "Literary Criticism" 1831, 25.

68. [Broderip] 1833–43a, 17:189.

69. [Bugg] 1826–27, 1:7–8, 2:330, 1:363, 1:21, 1:22.

70. Ibid., 1:216, 1:217, 1:218.

71. See O'Connor 2007b.

72. Roberts 1837, 242, 243.

73. Ibid., 246, 247.

74. [Bugg] 1826–27, 2:24, 2:25.

75. See Corsi 1988b, 238. Corsi begins his discussion of Nolan's Lamarckianism "Surprisingly," and implies that he should be seen as merely an anomaly. However, when Bugg and Roberts are also taken into account, such transformism can be considered as a distinct tendency among biblical literalists.

76. Ibid., 241.

77. On Jameson and Grant, see J. Secord 1991.

78. Desmond 1989, 61, 47.

79. For this criticism of Desmond, see Topham 1993, 422, 503–4; Rupke 1994, 68–69; and Hilton 2000.

80. See Rupke 1997, 558–61.

81. Newman 1837, 4, 5.

82. See Morrell and Thackray 1981, 227–29.

83. *Report of the First and Second Meetings* 1833, 104–5.

84. Cuvier 1800–1805, 1:18 ("peut être considéré comme une machine partielle"). My translation.

85. Outram 1986, 348.

86. *Report of the First and Second Meetings* 1833, 106; and Gordon 1894, 132.

87. Gordon 1894, 128; and *Report of the First and Second Meetings* 1833, 106, 107.

88. *Report of the First and Second Meetings* 1833, 107.

89. Brooke 1989, 39.

90. Cuvier 1823, 72 ("en contradiction avec les règles de co-existence que nous trouvons établies dans tout le règne animal"). My translation.

91. W. Buckland 1837, 23, 18.

92. Gordon 1894, 129.

93. See "Proceedings of Societies" 1832.

94. "Scientific Meeting" 1832, 753. On Buckland's humorous asides about the megatherium's posterior, see Boylan 1984, 397.

95. *Report of the First and Second Meetings* 1833, 106, 108.

96. Morrell and Thackray 1984, 156.

97. While Martin Rudwick has suggested that Buckland used his British Association lecture to "out-Cuvier Cuvier," he ensured that such an intention was not actually apparent to readers of its printed form; Rudwick 2008, 432.

98. W. Buckland 1836, 1:141.

99. Quoted in Topham 1993, 203.

100. See Topham 1992.

101. W. Buckland 1836, 1:vii–viii, 1:109.

102. Ibid., 1:vii–viii, 1:109, 1:220.

103. See Topham 1993, 209–10.

104. See Appel 1987, 169–70.

105. W. Buckland 1836, 1:162–63, 1:164, 1:222.

106. E. Agassiz 1885, 1:180–81.

107. L. Agassiz 1847, 41–42.

108. W. Buckland 1836, 1:268, 1:269.

109. Quoted in Rupke 1983, 161.

110. On readers' discussions of Buckland's book, see Topham 1998.

111. K. Lyell 1881, 2:7, 2:8, 2:9.

112. J. Secord 2014, 239–40.

113. "Scientific Meeting" 1832, 752.

114. Kirby 1835, 2:242, 2:333, 2:330, 2:208.

115. Topham 1993, 420.

116. Kidd 1833, 328; and C. Lyell 1830–33, 2:14.

117. C. Bell 1833, 65.

118. Ibid., 82, 80; and [C. Bell] 1827–29, 1:16, 1:15.

119. Topham 1992, 414–15.

120. Quoted in Topham 1992, 416.

121. Topham has termed the information conveyed by the *Bridgewater Treatises* "safe science"; ibid., 404.

122. Carlile 1829, 283.

123. See Brooke 1991a; and Yeo 1993, 118–24. On natural theology at Cambridge, see Fyfe 1997.

124. Whewell 1837, 3:476, 3:475, 3:464.

125. W. Whewell to R. Owen, 30 October 1837, Add. 5354 E8, Cambridge University Library.

126. R. Owen to W. Whewell, 31 October 1837, William Whewell Papers, add.Ms.a.2010 (54), Trinity College Library, Cambridge.

127. R. Owen to W. Whewell, 26 March 1840, Whewell Papers, add.Ms.a.2010 (61).

128. R. Owen to W. Buckland, 28 July 1832, Frances Hirtzel Collection of Richard Owen Papers, Temple University Library, Philadelphia.

129. R. Owen to W. Buckland, [1837], Hirtzel Collection.

130. Owen to Whewell, 26 March 1840, Whewell Papers, add.Ms.a.2010 (61).

131. Charlesworth 1846, 23. On Blainville's estrangement from Cuvier, see Appel 1987, 66–67.

132. Blainville 1839b, 49, 53, 54.

133. R. Owen 1842c, 47, 50.

134. "Our Weekly Gossip" 1838, 841. On the debate, see Desmond 1989, 314.

135. Whewell 1842, 89.

136. Owen to Whewell, 31 October 1837, Whewell Papers, add.Ms.a.2010 (54); and Rupke 1994, 163.

137. L. Huxley 1900, 1:161.

138. Quoted in Desmond 1989, 354, 356.

139. See ibid., 354–57.

140. See Rupke 1994, 162–63, 187–88.

141. Blainville 1839a, 143 ("hypothèse purement gratuite"). My translation.

142. R. Owen to W. Whewell, 11 February 1839, Whewell Papers, add.Ms.a.2010 (55). Although Desmond has argued that Owen's principal target in the dispute over the *Didelphis* was Blainville's British supporter, and overt transmutationist, Robert Edmond Grant, Owen's subsequent exasperation supports Rupke's suggestion that "Grant . . . presented less of a threat to

Owen's Cuvierian aspirations than did . . . Blainville," and that transmutation, which Blainville himself opposed, was not the principal issue; Desmond 1989, 313–21; and Rupke 1994, 147.

143. C.-L. Laurillard to R. Owen, 12 October 1843, Richard Owen Papers, 17:205, Natural History Museum, London ("occasion d'exposer de nouveau les principes de la détermination de os, tels que les a formulés Cuiver," "me disiez en conscience, si vous trouvez les attaques de M. de Blainville contra ces principes, fondés en raisons," and "Je ne sais si c'est mon respect pour Cuvier qui fausse mon judgement, mais je troupe si peu de logique dans les raisonnemens de M. de Blainville que je ne puis croire qu'il ait raison"). My translation.

Chapter Three

1. [Duns] 1858, 326.
2. Warren 1853, 59.
3. Mantell 1848a, 225–26.
4. Curwen 1940, 225.
5. Mantell 1848a, 226.
6. Mantell 1851, 94.
7. Rule 1843, 3, 7.
8. Ibid., 7, 8.
9. J. Rule to W. Buckland, 1 July 1843, Richard Owen Papers, 22:444(d), Natural History Museum, London.
10. Pantin 1963, 19.
11. On Murchison's denial of credit to colonial collaborators, see Stafford 1990, 36–37.
12. Desmond 1989, 240, 243.
13. See Gruber 1987; and Rupke 1994, 125–26. Wolfe 2003 provides a popular account of some of the same events.
14. J. Rule to R. Owen, 18 October 1839, Owen Papers, 22:444(a).
15. See Buick 1936, 59.
16. R. Owen 1879a, 1:149n, 1:iii, 1:iv.
17. R. Owen 1840a, 169.
18. Rule to Owen, 18 October 1839, Owen Papers, 22:444(a).
19. R. Owen 1840a, 170–71.
20. See R. Owen 1840b; and R. Owen 1842b.
21. W. J. Broderip to W. Buckland, 20 January 1843, Add. 38091, British Library, London.
22. See J. Smith 1994, 11–44.
23. Blainville 1839–64, 1:34 ("degré de prévision," "qu'un seul os," "recomposer son squelette entier, et par suite le reste de l'organisation de l'animal dont il provient," and "n'est-il, suivant moi, jamais arrivé à personne de mettre à preuve cette prétention"). My translation.
24. R. Owen 1843b, 8, 10.
25. R. Owen 1843a, 269.
26. R. Owen 1846c, 326n. Rule's article in the *Polytechnic Journal* was signed, making Owen's omission of its author's name a clear snub.
27. R. Owen 1879a, 1:iii.
28. Buick 1936, 51, 59.
29. R. Owen 1843a, 235, 262.
30. [Broderip and R. Owen] 1852, 402, 403.

31. Ibid., 404.

32. Spokes 1927, 245.

33. [Broderip and R. Owen] 1852, 396.

34. R. S. Owen 1894, 1:151.

35. Note by C. D. Sherborn, 15 November 1894, Owen Papers, 22:444(b).

36. R. Murchison 1846, 142.

37. Strickland 1845, 214.

38. Mantell 1844, 2:816, 1:xii–xiii.

39. Mantell 1848a, 226.

40. See Klancher 1987.

41. See Topham 2007.

42. [Lhotsky] 1840, 318, 319, 322, 324, 325.

43. See Bennett 1982.

44. See Kinraide 2006; and Fyfe 2012.

45. [Chambers] 1842, 186.

46. On the circulation of *Chambers's*, see Fyfe 2012, 52.

47. See the handwritten attributions for each entry, made by Alexander Ramsay, in the copy of the *Penny Cyclopædia* held at the British Library, 733l.

48. Charles Knight, "The Penny Cyclopædia," 21 June 1832, Society for the Diffusion of Useful Knowledge Papers, Special Topics 53, University College London, Special Collections.

49. See Knight 1864, 2:203.

50. "National Cyclopædia" 1847, 315.

51. [Broderip] 1833–43b, 23:147.

52. W. J. Broderip to R. Owen, 7 February 1842, Owen Papers, 5:114; and W. J. Broderip to R. Owen, 27 February 1842, Owen Papers, 5:116.

53. Rupke 1994, 135–36.

54. "Letter from Richard Owen . . ." 1843, 341.

55. Broderip to Buckland, 20 January 1843, Add. 38091.

56. [Broderip] 1833–43c, 25:508.

57. [Masson] 1862, 367–68.

58. [Broderip] 1833–43c, 25:508.

59. "National Cyclopædia" 1847, 315.

60. J. Secord 2014, 239.

61. R. Murchison 1846, 143n; and Jerdan 1866, 65.

62. "Miscellanea" 1844, 629.

63. [Chambers] 1847, 346; and [Chambers] 1843, 323.

64. W. J. Broderip to W. Buckland, 31 October 1843, Add. 38091.

65. See J. Secord 2000, 83.

66. See Feely 2009.

67. Knight 1854, 241.

68. "Newcastle Paper" 1852, 106.

69. Warren 1853, 59, 55.

70. Ibid., 59, 60. On those other sites, see Altick 1978; O'Connor 2007a; and Fyfe and Lightman 2007.

71. Lydekker 1891, 246.

72. S. Warren to R. Owen, 7 April 1853, Add. 39954, British Library, London.

73. "Wonderful Bone" 1853, 176.

74. [Chambers] 1848, 247, 248.

75. Knight 1854, 288.

76. W. H. Wills to R. Owen, 1 July 1852, Add. 39954.

77. [Morley and Wills] 1852, 383.

78. Ibid.; quoted in Buick 1936, 50.

79. [Morley and Wills] 1852, 383.

80. See Lohrli 1973, 23.

81. T. H. Huxley to J. D. Hooker, 2 December 1858, Thomas Henry Huxley Papers, 2.39, Imperial College of Science, Technology, and Medicine Archives, London.

82. Quoted in Adrian 1966, 134.

83. House, Storey, and Tillotson 1965–2002, 6:780.

84. [R. Owen] 1853, 374.

85. W. J. Broderip to R. Owen, 12 July 1852, Owen Papers, 5:220.

86. "Bird Twenty Feet High" 1852, 4.

87. See House, Storey, and Tillotson 1965–2002, 6:679.

88. B. Silliman to R. Owen, 14 July 1843, Owen Papers, 24:34.

89. [Hitchcock] 1845, 194. On Owen's forwarding the article to Silliman, see R. Owen to E. Hitchcock, 30 August 1844, Edward and Orra White Hitchcock Papers, 3:30, Amherst College Library, Amherst, MA.

90. [Hitchcock] 1845, 194.

91. As well as forwarding his article to Silliman, Owen also sent a "copy by a private opportunity" to Hitchcock, whose abstract of it is initialed "H"; see Owen to Hitchcock, 30 August 1844, Hitchcock Papers, 3:30.

92. J. Secord 2000, 126.

93. Howsam 2006, 3.

94. W. J. Broderip to R. Owen, 23 May 1852, Owen Papers, 5:213.

95. See annotated proof sheets, Owen Papers, MS OC39(4).

96. [Broderip and R. Owen] 1852, 363.

97. R. Owen to G. H. Lewes, 9 August 1853, Mss.B.Ow2, American Philosophical Society, Philadelphia.

98. Quoted in J. Secord 2000, 48.

99. See Desmond 1989, 236–75.

100. "College Conversazioni" 1842, 246.

101. See Chase 2007, 223–24, 271–74.

102. Broderip to Buckland, 31 October 1843, Add. 38091.

103. R. S. Owen 1894, 1:205–6.

104. Ibid., 1:167.

105. Ibid., 1:233, 1:247.

106. R. Owen 1866–68, 3:814.

107. W. J. Broderip to R. Owen, 13 May 1848, Owen Papers, 5:152.

108. R. Owen 1848b, 374, 375.

109. R. S. Owen 1894, 1:321.

110. See Chase 2007, 46–47.

111. R. Ball to R. Owen, 9 November 1843, Owen Papers, 2:113.

112. W. J. Broderip to R. Owen, 9 September 1843, Owen Papers, 5:127.

113. "Meeting of the British Association" 1843, 134.

114. See Topham 1993, 417–20.

115. W. Buckland 1848, 10, 12, 13.

116. Desmond 1989, 331.

117. See Hilton 2000.

118. W. Buckland 1848, 13.

119. Hilton 2006, 334.

120. W. Buckland 1858, 1:84n, 1:xii–xiii.

121. [Chambers] 1843, 323; and [Chambers] 1844, 147.

122. Quoted in J. Secord 2000, 444.

123. Ibid., 109.

124. Sedgwick 1850, clxxi.

125. See E. Richards 1987; and Rupke 1994, 222–23.

126. [Brewster] 1845, 471, 508, 513.

127. [Duns] 1858, 335, 326, 325, 324.

128. Duns 1859, liv.

129. J. Duns to R. Owen, 29 April 1858, Owen Papers, 10:244.

130. J. Duns to R. Owen, 6 May 1858, Owen Papers, 10:246.

131. J. Duns to R. Owen, 17 March 1859, Owen Papers, 10:250.

132. See Buick 1936, 60–61.

133. See Barton 2000.

134. See Endersby 2008, 110.

135. "Eighth Meeting" 1872, 379.

136. Colenso 1892, 468, 469, 478.

137. J. Secord 2014, 241. See also J. Secord 2000, 126–38 and passim.

Chapter Four

1. [Broderip and R. Owen] 1844, 446, 449, 450, 445.

2. Morgan 1973, 578.

3. See Slater 1972; Hughes and Lund 1991; and Law and Patten 2009.

4. [Broderip and R. Owen] 1845, 302.

5. Morgan 1973, 622.

6. The attributions are made in, respectively, Hood and Broderip 1860, 2:148; and R. S. Owen 1894, 1:12.

7. Rupke 1994, 6.

8. See R. S. Owen 1894, 1:12–13, 1:21.

9. "Memorial of Sir Richard Owen" 1893, 308.

10. Broman 1991, 17.

11. Hopwood, Schaffer, and Secord 2010, 251.

12. R. Owen 1849–84, 1:478, 1:476; R. Owen 1848a, 123; and R. Owen 1846b, 101.

13. Law and Patten 2009, 144.

14. "Miscellanies" 1839, 370.

15. Marcou 1896, 1:29.

16. E. Agassiz 1885, 1:256.

17. L. Agassiz 1847, 41.

18. R. Owen 1840a, 169–71.

19. R. Owen 1879b, 263, 260.

20. R. Owen 1843b, 8.

21. Ibid., 146.

22. R. Owen 1879a, 1:vi.

23. R. Owen 1849–84, 1:ii–iii.

24. Advertising supplement in *Penny Magazine* 1 (1832): i.

25. Ibid.

26. Johns 1998, 630.

27. Arnold 1841, 39.

28. On the SDUK's prohibition of fiction, see Gray 2006, 44, 57.

29. See Yeo 2001, 25–27.

30. Knight 1864, 2:201.

31. "Penny Cyclopædia" 1834, 123.

32. "Charles Knight's English Cyclopædia" 1854, 10.

33. "Penny Cyclopædia" 1834, 123.

34. R. Owen 1848c, 198.

35. Knight 1864, 2:229–30.

36. [Broderip] 1833–43b, 23:147.

37. Todd 1835–59, 1:v–vi.

38. K. Lyell 1881, 2:93.

39. [Broderip] 1833–43c, 25:506, 25:508.

40. Dickens 1836–37, 1:vii.

41. Pycroft 1844, 91–92.

42. *English Cyclopædia* 1856, 469; and "Charles Knight's English Cyclopædia" 1854, 10.

43. "William John Broderip" 1859, 357.

44. [Broderip] 1833–43c, 25:508; and "Dinornis" 1854–55, 344.

45. Quoted in Patten 1978, 185.

46. R. S. Owen 1894, 1:290.

47. See W. J. Broderip to R. Owen, 4 January 1853, Richard Owen Papers, 5:225, Natural History Museum, London.

48. "Sir Richard Owen" 1895, 347, 362.

49. R. S. Owen 1894, 1:104.

50. Ibid., 1:201.

51. Hayward 1997, 35.

52. R .S. Owen 1894, 1:233–34, 1:296.

53. Ibid., 1:318.

54. See Altick 1980, 83–84.

55. R. S. Owen 1894, 1:318.

56. Royal Literary Fund Annual Reports 6 (1844), 23–24, Royal Literary Fund Archive, Loan 96, British Library, London.

57. R. S. Owen 1894, 1:104.

58. Ibid., 2:97.

59. [Cleghorn] 1845, 85.

60. [Mitford] 1891, 277.

61. Hughes and Lund 1991, 11.

62. Dickens 1838–39, 2:477–78.

63. R. Owen 1860, 297.

64. R. Owen 1846a, 220.

65. R. Owen 1879a, 1:iii.

66. R. S. Owen 1894, 1:164.

67. Dickens 1838–39, 1:x.

68. J. Rule to R. Owen, 18 October 1839, Owen Papers, 22:444(a).

69. R. S. Owen 1894, 1:319.

70. "Humourists" 1848, 271, 273–74.

71. R. S. Owen 1894, 1:233. Caroline's diary records that Owen sat for Pickersgill on 26 April and 17 May 1844, and was reading the latest installment of *Martin Chuzzlewit* on 3 May.

72. R. Owen 1879b, 268.

73. R. Owen 1870, 520.

74. R. S. Owen 1894, 1:318.

75. Ibid., 1:239.

76. "David Copperfield" 1851, 158.

77. [Simms] 1844, 322.

78. Bissell 1851, 278; and [Simms] 1844, 323.

79. "Great Expectations" 1861, 218.

80. F. Burkhardt et al. 1985–, 1:299; and Darwin 1845, 83.

81. "Megatherium" 1849, 106.

82. "Penny Cyclopædia" 1834, 124.

83. See O'Connor 2007a, 328–29.

84. [Thackeray] 1847b, 166. For Thackeray's first invocation of the Megatherium Club, see [Thackeray] 1847a, 153.

85. [Thackeray] 1848, 95.

86. [Thackeray] 1846, 19.

87. [Lowell] 1847, 208. On Lowell's trope of literary megatheria, see Dawson 2011.

88. R. S. Owen 1894, 1:358, 2:147, 2:23, 2:81–82.

89. Thackeray 1853–55, 2:81.

90. Levine 1981, 164.

91. Thackeray 1853–55, 1:225–26.

92. Ibid., 1:226.

93. Altick 1991, 782.

94. Ray 1946, 3:239.

95. Harden 1979, 13.

96. R. Owen 1839, 1:106.

97. W. Buckland 1836, 1:144–45.

98. R. Owen 1842a, 149. The lines are from the first epistle of Alexander Pope's *An Essay on Man* (1734).

99. Pope 1734, 21–22.

100. R. Owen 1861, 41.

101. "Thackeray's Pendennis" 1850, 1213.

102. R. Owen 1842a, 162.

103. Thackeray 1853–55, 1:225.

104. Ibid., 2:373, 2:292.

105. R. Owen 1857, 7.

106. Dickens 1852–53, 1.

107. "Landor's Last Fruit" 1854, 118.

108. Norton 1894, 1:375.

109. James 1869, 56; and James 1886, 186.

110. James 1908, 1:ix, 1:x.

111. Ibid., 1:x, 1:xi.

112. "Penny Cyclopædia" 1834, 124.

113. Cope 1872, 319–21.

114. Sternberg 1905, 127.

115. Lubbock 1920, 2:247.

116. McMaster 1978.

117. Pound 1918, 20.

118. James 1914, 21.

119. James 1907, v.

120. Brownell 1905, 498, 497.

121. "Miscellanies" 1836, 289–90.

122. R. Owen 1846a, 76.

123. Patten 1978, 60.

Chapter Five

1. Logan 1851, 248.

2. R. Owen 1851, 251.

3. C. Lyell 1851, lxxv.

4. Logan 1851, 248; and Harrington 1883, 273.

5. Ellis 1851, 2:960, 2:961.

6. On the role of science in the Great Exhibition, see Bellon 2007.

7. Warren 1851, 9, 10.

8. On the reporting of the Great Exhibition in the *Illustrated London News*, see Sinnema 1998, 147, 180–81.

9. Warren 1851, 60–61, 79, 80, 81, 82, 83, 213–14, 215.

10. R. Owen 1852a, 219, 224.

11. Harrington 1883, 275.

12. Warren 1854, v, 58, 22.

13. "Literature" 1851, 707, 708. On Blackwood's regrets, see Oliphant 1897, 2:456.

14. See J. Secord 2000, 152.

15. J. Secord 2004b, 146.

16. Bohn 1856, 3; and "Crystal Palace" 1855, 6.

17. "Literature" 1851, 707.

18. Warren 1854, ii.

19. S. Warren to R. Owen, 12 September 1851, Add. 39954, British Library, London.

20. Warren 1851, i.

21. Warren to Owen, 12 September 1851, Add. 39954.

22. Ibid.; and S. Warren to R. Owen, 22 September 1851, Add 39954.

23. Warren to Owen, 22 September 1851, Add. 39954.

24. R. Owen 1852c, 78. On the Great Exhibition's embodiment of such progressive values, see Auerbach 1999; and Young 2009.

25. Warren to Owen, 12 September 1851, Add. 39954.

26. For such portrayals of Owen, see, for instance, Desmond 1989, 338–39; and Desmond 2001, 32.

27. Warren 1853, 65–66, 55.

28. R. Owen to the editor of *Nature*, n.d., Richard Owen Papers, 21:25, Natural History Museum, London.

29. "One Hundred and First Anniversary Dinner" 1855, 587; and Owen to the editor of *Nature*, n.d., Owen Papers, 21:25.

30. Crystal Palace Company: Extract from Minutes of a Meeting of the Board of Directors, 10 August 1852, Mantell Family Papers, MS-0305-0083-032, Alexander Turnbull Library, Wellington, New Zealand.

31. Atmore 2004, 192.

32. Crystal Palace Company: Extract from Minutes of a Meeting of the Board of Directors, 10 August 1852, Mantell Family Papers, MS-0305-0083-032.

33. L. Huxley 1900, 1:93–94.

34. Mantell 1825, 179, 181, 185.

35. Mantell 1848b, 1:447, 1:420, 1:421.

36. See Dell 1983, 87–88.

37. Curwen 1940, 273.

38. G. Mantell to W. Mantell, 11 August 1852, Mantell Family Papers, MS-0305-0083-107.

39. Dell 1983, 90.

40. J. Secord 2004b, 155; and Dean 1999, 261.

41. Mantell 1851, 461, 477–79.

42. Quoted in Dean 1999, 62.

43. Mantell 1848a, 226.

44. Mantell 1846, 106.

45. Mantell 1850a, 386.

46. Mantell 1850b, 921; and Mantell 1850a, 379.

47. Mantell to Mantell, 11 August 1852, Mantell Family Papers, MS-0305-0083-107.

48. Mantell 1844, 2:849–50.

49. Mantell 1854, 804.

50. Owen's later claim that it was he who "recommended to . . . the Directors of the Crystal Palace the employment of Mr. B. W. Hawkins to realize & carry out my ideas" (Owen to the editor of *Nature*, n.d., Owen Papers, 21:25) is, again, patently false, as the Crystal Palace Company had already advised Mantell that Hawkins was "likely to be of great use in modelling the intended restorations of extinct animals" (W. Thomson to G. Mantell, 12 August 1852, Mantell Family Papers, MS-0305-0083-032).

51. [R. Owen] 1852b, 842. On the fierce rivalry between Owen and Mantell, see Dean 1999, 278–82 and passim; and Torrens 2012.

52. "Dinner to Professor Owen" 1854, 7.

53. [Wynter] 1853, 612.

54. Hales 2006, 118.

55. On Pestalozzi's theories, see Keene 2008, 17–18 and passim.

56. Jones 1854, 11, 23, 20.

57. "Crystal Palace—Visit of Her Majesty" 1854, 267.

58. [Wills and Sala] 1854, 317.

59. On the "virtual tourism" of the Crystal Palace, see O'Connor 2007a, 279.

60. Rossetti 1867, 53, 51–52.

61. Sotheby 1855a, 21, 25, 33.

62. B. W. Hawkins to S. L. Sotheby, 18 December 1854, vol. 1, Samuel Leigh Sotheby Scrap

Album "Papers and Memorandum of the Crystal Palace Company 1852–56," 2 vols., Bromley Local Studies Library, Kent, UK.

63. Sotheby 1855a, 5.

64. [H. Martineau] 1854, 542. On Sotheby's shareholding, see Bohn 1856, iv.

65. J. E. Gray to S. L. Sotheby, 28 December 1854, vol. 1, Sotheby Scrap Album.

66. Sotheby 1855a, 33.

67. S. L. Sotheby to R. Owen, 6 January 1855, Add. 42577, British Library, London.

68. Sotheby 1855a, 5. On Sotheby's perpetual disagreements with Laing, see Atmore 2004, 200–208.

69. R. S. Owen 1894, 1:398.

70. Hawkins 1854, 447.

71. *Routledge's Guide* 1854, 18, 20.

72. Sotheby 1855a, 32.

73. B. W. Hawkins to R. Owen, 6 March 1855, Owen Papers, 14:525.

74. B. W. Hawkins to R. Owen, 16 May 1855, Owen Papers, 14:527.

75. Sotheby 1855b, 15.

76. Hawkins to Owen, 6 March 1855, Owen Papers, 14:525.

77. J. Paxton to R. Owen, 22 May 1854, Add. 5354/65, Cambridge University Library.

78. Phillips 1854, 24.

79. B. W. Hawkins to R. Owen, 11 March 1854, Owen Papers, 14:520.

80. "Geology of the Crystal Palace" 1854, 279.

81. Mantell 1851, 8.

82. Dell 1983, 90.

83. Hawkins to Owen, 11 March 1854, Owen Papers, 14:520.

84. Hawkins 1854, 444.

85. Rossetti 1867, 52.

86. [H. Martineau] 1854, 540.

87. Owen to the editor of *Nature*, n.d., Owen Papers, 21:25.

88. Hawkins 1854, 447.

89. Hawkins to Owen, 6 March 1855, Owen Papers, 14:525.

90. Gray to Sotheby, 28 December 1854, vol. 1, Sotheby Scrap Album; and J. E. Gray to S. L. Sotheby, 24 February 1855, vol. 1, Sotheby Scrap Album.

91. Owen to the editor of *Nature*, n.d., Owen Papers, 21:25.

92. R. Owen 1854, 17, 38.

93. R. Owen 1860, 194.

94. R. Owen 1854, 16.

95. Ibid., 10.

96. See J. Secord 2004b, 157; and McCarthy and Gilbert 1994, 76–77.

97. B. W. Hawkins to R. Owen, 19 May 1854, Add. 5354/64.

98. "People's Palace" 1853, 49.

99. My argument on this point develops that of J. Secord 2004b, 153–55.

100. Gray to Sotheby, 28 December 1854, vol. 1, Sotheby Scrap Album.

101. "To the Public" 1853, 1.

102. House, Storey, and Tillotson 1965–2002, 7:370.

103. [Wills and Sala] 1854, 317.

104. On the coverage of the Great Exhibition, see Sinnema 1998, 147, 180–81.

105. On Ingram's friendship with Paxton, see Bailey 1996, 89; on his relations with Mackay, see Mackay 1877, 2:67, where the editor reflects: "Mr. Ingram was believed by many of the good people, who subscribed to the *Illustrated London News*—to write the whole paper—every line of it, except the advertisements."

106. "Crystal Palace at Sydenham" 1854, 22.

107. "Gigantic Bird of New Zealand" 1854, 22.

108. Hawkins 1854, 444.

109. Adams 1857, 29–30.

110. "Opening of the Crystal Palace" 1854, 245.

111. *Routledge's Guide* 1854, 186.

112. [Arthur] 1854, 237.

113. [Williamson] 1855, 363, 364.

114. "Crystal Palace" 1855, 6. In nineteenth-century British money, the letter *l* (for livres) indicated pounds; *s* indicated shillings; and *d* (derived from denarius, a small Roman coin) indicated pence.

115. Gray to Sotheby, 28 December 1854, vol. 1, Sotheby Scrap Album.

116. "Geology of the Crystal Palace" 1854, 282.

117. Desmond 1979, 228–29. For other criticisms of Desmond's argument, see Rupke 1994, 134; and Torrens 2012, 27–28.

118. On the importance of the models being arranged in a "*temporal* panorama," see Rudwick 1992, 144–45.

119. "Your Company Is Requested" 1854, 52–53.

120. For such satires, see, for instance, [Leech] 1855.

121. [H. Martineau] 1854, 540, 541.

122. Marshall 2007, 292.

123. [H. Martineau] 1854, 545.

124. *Routledge's Guide* 1854, 185–86.

125. "Meeting of the British Association" 1853, 5.

126. "Scientific" 1854, 626.

127. See Hopwood and De Chadarevian 2004.

128. "People's Palace" 1854, 231. On the inadequacies of the Pestalozzian agenda, see J. Secord 2004b, 159.

129. Gray to Sotheby, 28 December 1854, vol. 1, Sotheby Scrap Album.

130. F. Burkhardt et al. 1985–, 5:351, 5:224.

131. Ibid., 5:194, 5:224; and K. Lyell 1890, 2:357–58.

132. Ruskin 1857, 24.

133. Ruskin 1854, 5, 6.

134. "Excursion to the Crystal Palace" 1876–78, 377.

135. Qureshi 2011, 193–211. On the Crystal Palace as an "ethnographic training site," see also Sera-Shriar 2013, 95–99.

136. Ray 1946, 3:62.

137. Harden 1994, 1:643.

138. Ray 1946, 3:386.

139. Thackeray 1853–55, 2:374, 2:81.

140. See "Literature" 1853, 17.

141. See McMaster 1991, 1–7 and passim.

142. "David Copperfield" 1851, 158.

143. See Fyfe 2012, 97–171.

144. Phillips 1854, 15, 193.

145. *Railway Readings* 1847, 80.

146. "Waiting Room" 1857, 60.

147. [Wynter] 1853, 608.

148. "Crystal Palace" 1854, 378.

149. "Crystal Palace Handbooks" 1854, 166.

150. [Arthur] 1854, 235.

151. H. Tennyson 1897, 1:277.

152. "Manners, Traditions and Superstitions" 1846, 636.

153. W. Buckland 1836, 1:160n.

154. Paley 1802, 71.

155. W. Buckland 1836, 1:159; and R. Owen 1861, 35.

156. See Piggott 2004, 148–53.

157. *Routledge's Guide* 1854, 196, 198.

158. W. Buckland 1858, 2:33.

159. W. Buckland 1836, 1:419, 2:24.

160. Hawkins 1876, 18.

161. "Nature's Hints to Inventors" 1868, 38.

162. Dickens 1852–53, 1.

163. W. Buckland 1836, 1:920.

164. W. Buckland 1858, 2:33–34.

165. Hawkins 1860, 26, 27.

166. See Bramwell and Peck 2008, 82–85.

167. Hawkins 1854, 447, 444.

168. Hawkins 1858, 207.

169. Buckland 1858, 1:xi. On the resilience and flexibility of nineteenth-century natural theology, see Brooke 1991b, 192–225.

170. Cantor 2011, 133–37.

171. Flyer for "Crystal Palace Company's News Room," vol. 1, Sotheby Scrap Album.

172. Sotheby 1855a, 15.

173. Hawkins 1858, 207.

174. "Crystal Palace News" 1855, 27.

Chapter Six

1. "Centenary Dinner" 1854, 564–65.

2. "Twenty-Second Ordinary Meeting" 1854, 447.

3. "Centenary Dinner" 1854, 563.

4. Quoted in Cantor 2011, 151.

5. T. Huxley 1870b, 86; and "Educational Exhibition" 1854, 622.

6. T. Huxley 1870b, 86–87.

7. Hawkins 1854, 444, 445, 447.

8. See Shpayer-Makov 2011, 199–200.

9. Dickens 1851, 265.

10. See Bibby 1959, 125.

11. [T. Huxley] 1856d, 2–3.

12. See C. Murchison 1868, xli.

13. "Delenda est Carthago" 1854, 1202; and [Reeve] 1855, 575.

14. Falconer 1856, 477–78.

15. Ibid., 476; and Falconer 1868b, 314.

16. See Markovits 2009, 12–62.

17. T. Huxley 1856b, 192; and "Express from Paris" 1856, 10.

18. Falconer 1856, 476.

19. [T. Huxley] 1855a, 562–63.

20. [T. Huxley] 1854b, 517, 491–92.

21. Quoted in Desmond 1994, 196.

22. On the military metaphor, see Moore 1979, 19–49 and passim; Barton 1983; Lightman 1987, 133–34; and Desmond 1997, 14–15, 250–53.

23. [W. Russell] 1855, 6.

24. Falconer 1856, 492.

25. [T. Huxley] 1854b, 511.

26. Desmond 1997, 252.

27. Reid 2000, 255.

28. See Piggott 2004, 61, 64.

29. Phillips 1855, 45.

30. On the dispatches to the War Department, see "Despatches from the Crimea" 1856, 10; on the Royal Institution's Friday Evening Discourses, see Morus 1998, 26–29.

31. T. Huxley 1856b, 188, 189–90.

32. Ibid., 190–91.

33. Ibid., 191.

34. Ibid., 192.

35. While the Athenaeum's gentlemanly and aristocratic Ordinary Members were elected by a ballot of all members, Hooker and Darwin had intended to get Huxley elected under Rule II, which gave the club's committee the authority to elect, each year, nine members "of distinguished eminence in Science, Literature, or the Arts, or for Public Service"; Collini 1991, 16.

36. F. Burkhardt et al. 1985–, 6:106, 6:112, 4:172. Darwin and Hooker first met Huxley, respectively, only in 1853 and 1851.

37. Ibid., 7:496.

38. Falconer 1859, 7, 2, 4.

39. Prestwich 1901, 6.

40. Falconer 1856, 476, 485, 483, 486.

41. Ibid., 488, 483, 487.

42. Ibid., 492.

43. Collins 1857, 38; and Falconer 1856, 492.

44. F. Burkhardt et al. 1985–, 6:176.

45. Ibid., 6:147.

46. L. Huxley 1918, 1:427.

47. "On Systems and Methods" 1829, 313.

48. L. Huxley 1918, 1:427.

49. F. Burkhardt et al. 1985–, 6:176.

50. Ibid., 6:178.

51. Darwin 1849, 168.

52. R. Owen 1839, 1:64.

53. F. Burkhardt et al. 1985–, 1:368. See Rachootin 1985.

54. Darwin 1851–54, 2:599.

55. See Gould 2002, 333–39.

56. Barrett et al. 1987, 410; and Darwin 1859, 5.

57. Wilson 1970, 226.

58. T. Huxley 1853, 62.

59. Rudwick 2013, 363. Cohen likewise proposes that Huxley opposed the "loi de corréla-tion" because of its "fixiste" implications; Cohen 2011, 20.

60. F. Burkhardt et al. 1985–, 7:161, 7:368; and Falconer 1863, 77.

61. Falconer 1863, 77.

62. F. Burkhardt et al. 1985–, 6:100.

63. Ibid., 5:133, 5:213; and T. Huxley 1856a, 483.

64. See White 2003, 109–10; and Endersby 2008, 332.

65. F. Burkhardt et al. 1985–, 7:246, 6:176, 17:26.

66. L. Huxley 1918, 2:322.

67. On *scientific naturalism* as a mid-nineteenth-century actors' category for the outlook of those who employed exclusively secular, naturalistic methods and epistemologies in under-standing the earth, life, and humanity, see Dawson and Lightman 2014, 1–24. See also F. Turner 1974; and Lightman 2009.

68. [Spencer] 1857, 345. Having previously published an unsigned article on "The Develop-ment Hypothesis" in the *Leader* in 1852, Spencer had outlined his conception of progressive evolution, without the veil of anonymity, in *The Principles of Psychology* (1855).

69. Ibid., 351, 350, 348.

70. D. Duncan 1911, 83.

71. T. Huxley 1856c, 50.

72. Ibid., 50, 54.

73. Cohen 2002, 132.

74. T. Huxley 1856c, 50, 54.

75. F. Burkhardt et al. 1985–, 4:172, 4:267.

76. C. Lyell to T. H. Huxley, 18 July 1856, Thomas Henry Huxley Papers, 6.11, Imperial Col-lege of Science, Technology, and Medicine Archives, London.

77. F. Burkhardt et al. 1985–, 6:175, 6:176, 6:178.

78. G. Dickie to T. H. Huxley, 16 April 1856, Huxley Papers, 13.144.

79. McCosh and Dickie 1857, 445n.

80. T. Huxley 1856c, 48.

81. Falconer 1868a, 373.

82. C. Lyell 1838, 508–9.

83. Falconer and Cautley 1846, 13, 20.

84. T. Huxley 1856c, 44.

85. Falconer 1856, 493.

86. Prestwich 1901, 6.

87. Ibid., 2.

88. Boylan 1977, 13.

89. Falconer 1862, 350, 351, 352.

90. Falconer 1856, 492.

91. R. Owen 1871a, 104, 111, 108, 96.

92. F. Burkhardt et al. 1985–, 10:524.

93. H. Falconer to T. H. Huxley, 8 April [1864], Huxley Papers, 16.1.

94. T. Huxley 1864, 4, 5.

95. F. Burkhardt et al. 1985–, 11:14, 11:29, 13:48.

96. T. Huxley 1856b, 194.

97. Falconer 1856, 488n.

98. T. H. Huxley to F. Dyster, 3 November 1856, Huxley Papers, 15.78.

99. F. Burkhardt et al. 1985–, 6:260.

100. Huxley to Dyster, 3 November 1856, Huxley Papers, 15.78. See also Dickie to Huxley, 16 April 1856, Huxley Papers, 13.144, which states, "I am concerned to hear that you have been suffering from illness."

101. Thackray 1999, vii.

102. R. Owen 1857, 4–5, 6, 5.

103. van der Klaauw 1932, 341.

104. R. Owen 1857, 8, 9. On the insinuations of onanism, see Dawson 2007, 1–2.

105. [Spencer] 1857, 347.

106. R. Owen 1857, 6.

107. A. Tennyson 1842, 2:125.

108. Ibid., 2:135.

109. Ibid., 2:129.

110. L. Huxley 1900, 1:151.

111. T. H. Huxley to F. Dyster, December 1856, Huxley Papers, 15.80.

112. T. H. Huxley to J. Churchill, 22 January 1857, Huxley Papers, 12.194.

113. Huxley to Dyster, December 1856, Huxley Papers, 15.80.

114. Ibid. See Desmond 1994, 230; and White 2003, 52.

115. Huxley to Dyster, December 1856, Huxley Papers, 15.80.

116. R. S. Owen 1894, 2:61.

117. R. I. Murchison to ?, 27 February 1857, Richard Owen Papers, 20:67, Natural History Museum, London.

118. Synopsis of a Course of Lectures on the Osteology and Palæontology, or the Framework and Fossils, of the Class Mammalia (1857), Mss.B.Ow2.1, American Philosophical Society, Philadelphia.

119. R. Owen 1858, 160. This entry was "among the spin-offs" from the lecture series, according to Nicolaas Rupke; Rupke 1994, 95.

120. R. Owen 1858, 160n.

121. "Professor Owen's Lectures at the Museum of Practical Geology" 1857, 220.

122. T. Huxley 1856c, 44.

123. "Professor Owen's Lectures at the Museum of Practical Geology" 1857, 220.

124. R. S. Owen 1894, 2:61.

125. "Professor Owen's Lectures" 1857, 251–52.

126. F. Burkhardt et al. 1985–, 11:26, 11:29.

127. [Spencer] 1857, 352.

128. L. Huxley 1900, 1:161.

129. Quoted in White 2003, 52; see also 51–58.

130. Gill, Scott, and Osborn 1897, 26.

Chapter Seven

1. R. I. Murchison to ?, 27 February 1857, Richard Owen Papers, 20:67, Natural History Museum, London.

2. Four Lectures on the Principles of Natural History (1857), RI MS GB2 f. 97, Royal Institution Archives, London.

3. See Shteir 1996, 83.

4. Four Lectures on the Principles of Natural History, RI MS GB2 f. 97.

5. Ibid.

6. L. Huxley 1900, 1:149.

7. Synopsis of a Course of Lectures on the Osteology and Palæontology, or the Framework and Fossils, of the Class Mammalia (1857), Mss.B.Ow2.1, American Philosophical Society, Philadelphia; and R. Owen 1846a, 74.

8. See Cooter and Pumphrey 1994; Topham 2009; and O'Connor 2009b.

9. Lightman 2007b, 357.

10. F. Turner 1980, 589.

11. For recent reappraisals of such assumptions, see Desmond 2001; and Dawson and Lightman 2014.

12. [T. Huxley] 1854a, 256, 255.

13. T. Huxley 1856b, 190.

14. J. Secord 2000, 499; and [T. Huxley] 1854c, 438. Grub Street, of course, was the London street that, since the eighteenth century, had become synonymous with indigent hack writers producing shoddy literature for money.

15. L. Huxley 1900, 1:132; and Collie 1991, 82.

16. Falconer 1856, 488.

17. F. Burkhardt et al. 1985–, 6:147.

18. Lightman 2007b, 359.

19. [T. Huxley] 1855b, 246.

20. Topham 2004, 63–66.

21. Timbs 1858, v.

22. Ibid., v, 141.

23. See Topham 2004, 64–65.

24. Mackie 1858a, 1. On the proposed *Scientific Review*, see L. Huxley 1900, 1:139.

25. Mackie 1858c, i.

26. Ibid.; and Mackie 1858a, 2.

27. Mackie 1858b, 75. On Kingsley's passionate interest in popular geology, see A. Buckland 2013, 198–208.

28. See Margaret Green 2004.

29. Quoted in Bonython and Burton 2003, 199.

30. Brodie 1858, 373.

31. Wood 1859–63, 1:10.

32. Lightman 2007b, 37–43.

33. Gloag 1859, 38, 57, 61, 62.

34. Falconer 1856, 481–82; see also 484, 487.

35. L. Huxley 1900, 1:158, 1:159.

36. Ibid., 1:139.

37. W. Elwin to T. H. Huxley, 27 January 1859, Thomas Henry Huxley Papers, 15.193, Imperial College of Science, Technology, and Medicine Archives, London.

38. [Elwin] 1849, 316, 322.

39. Elwin to Huxley, 27 January 1859, Huxley Papers, 15.193.

40. Quoted in Desmond 1994, 185. On the *Quarterly*'s rates of pay, see Spencer 1904, 2:19.

41. L. Huxley 1900, 1:154.

42. Elwin 1902, 1:171.

43. Whewell 1858, 2:251–52.

44. W. Elwin to J. Murray, 19 February 1859, John Murray Archive, MS.42197, National Library of Scotland, Edinburgh.

45. L. Huxley 1900, 1:424.

46. Spencer 1904, 2:19–20, 2:21.

47. W. Elwin to J. Murray, 3 May 1859, John Murray Archive, MS.42197.

48. Elwin 1902, 1:350.

49. W. Elwin to J. Murray, 3 December 1859, John Murray Archive, MS.42197.

50. W. Elwin to J. Murray, 29 December 1859, John Murray Archive, MS.42197.

51. L. Huxley 1900, 1:247.

52. W. B. Carpenter to J. Chapman, 1 September 1855, Huxley Papers, 12.80.

53. T. Huxley 1864, 5; and T. Huxley 1869a, 4–5.

54. On Cuvier's natural system of classification, see Coleman 1964, 74–106; Outram 1986, 333–40; and Appel 1987, 42–46.

55. T. Huxley 1864, 4, 5, 85, 86.

56. "Professor Huxley on Classification" 1869, 285.

57. T. Huxley 1869a, iii.

58. F. Burkhardt et al. 1985–, 12:399, 13:6–7.

59. On the posture of being disenchanted with writing, see Salmon 2013, 74–82; on writing as part of scientific practice, see A. Buckland 2013, 15–21.

60. L. Huxley 1900, 1:138.

61. Ibid.; quoted in Desmond 1994, 231.

62. L. Huxley 1900, 1:138.

63. "Science and Religion" 1859, 35.

64. On the tendency of successful popular science lecturers to avoid scripts, see Lightman 2007a, 102, 107–8.

65. "How to Become an Orator" 1884, 1.

66. "Science and Religion" 1859, 35.

67. F. Burkhardt et al. 1985–, 7:246.

68. "Science and Religion" 1859, 35.

69. Ibid., 36.

70. T. Huxley 1856c, 44; and "Science and Religion" 1859, 35.

71. T. Huxley 1863, 4.

72. Report on Lecture to Working Men on "Objects of Interest in the Collection of Fossils," Huxley Papers, 44.4.

73. L. Huxley 1900, 1:207. On Huxley's later concerns with being misrepresented, see Dawson 2004, 178–81.

74. "How to Become an Orator" 1884, 2.

75. T. Huxley 1861, 84.

76. M. Ward 1859, 158, 142, 148.

77. See Lightman 2007b, 103–4.

78. Timbs 1858, 240.

79. See Yanni 1999, 53–56.

80. *Third Report of the Department of Science and Art* 1856, 20.

81. See Alberti 2007.

82. Forgan and Gooday 1996, 439.

83. See Fyfe 2007.

84. *Third Report of the Department of Science and Art* 1856, 18.

85. F. Burkhardt et al. 1985–, 6:505, 6:516.

86. R. Hunt 1856, 4.

87. L. Huxley 1900, 1:136–37.

88. White 2003, 102; and T. Huxley 1865, x.

89. T. Huxley 1865, lx, xiii.

90. Ibid., lxii, lxi.

91. T. Huxley 1864, 4.

92. R. Owen 1846a, 74, 75–76; and T. Huxley 1865, lx.

93. See F. Burkhardt et al. 1985–, 6:505–6.

94. Ibid., 13:134.

95. R. Murchison 1865, i.

96. Hamilton 1866, xli.

97. T. Huxley 1898–1902, 1:v.

98. T. Huxley 1893–94, 8:v.

99. Collie 1991, 23, 24.

100. Ibid., 24.

101. Cooter and Pumphrey 1994, 250; and O'Connor 2009b, 343.

102. J. Secord 2004a, 661.

103. Carey 1995, 139. See also Lightman 2007b, 355–56.

104. *Report from the Select Committee on Scientific Instruction* 1868, 401.

105. Hatton 1882, 248.

106. Forgan and Gooday 1996, 438–46.

107. Proctor 1870, 316.

108. See Lightman 2007b, 302–3.

Chapter Eight

1. Knox 1856, 245.

2. T. Huxley 1856c, 48.

3. See Desmond 1994, 25–26. There is no evidence to support Efram Sera-Shriar's claim that Huxley was taught by Knox himself; Sera-Shriar 2013, 114.

4. See Richardson 1987, 132–40; and McCracken-Flesher 2012.

5. "Noctes Ambrosianæ" 1829, 388.

6. "Pencillings of Eminent Medical Men" 1844, 246.

7. Lonsdale 1870, xi.

8. See Bates 2010, 105–15.

9. "Late Dr. Knox" 1862, 684.

10. E. Richards 1989, 409.

11. See Dawson 2007.

12. Blake 1870, 333; and "Pencillings of Eminent Medical Men" 1844, 245. On the *Lancet's* irascible style, see Desmond 1989, 15 and passim; and Brown 2014.

13. "Lectures on the Skeleton" 1839, 4.

14. Blainville 1839–64, 1:34.

15. Knox 1839, 217.

16. Ibid., 222.

17. R. Owen 1840a, 170–71.

18. "Mr. Warburton's Bill" 1829, 787.

19. "Lectures on the Skeleton" 1839, 5.

20. Knox 1839, 222.

21. Knox 1831, 483, 484, 479.

22. Lonsdale 1870, 281.

23. T. Huxley 1856b, 191.

24. Falconer 1856, 481, 482.

25. Rehbock 1983, 55.

26. Falconer 1856, 482.

27. E. Richards 1989, 386; and T. Huxley 1892, 35.

28. Knox 1831, 483; and T. Huxley 1856b, 190.

29. T. Huxley 1856b, 190; and [T. Huxley] 1854c, 438.

30. Milne-Edwards 1856, vi.

31. "*Great Artists and Great Anatomists*" 1852, 935; and H. Falconer to T. H. Huxley, 8 April [1864], Thomas Henry Huxley Papers, 16.1, Imperial College of Science, Technology, and Medicine Archives, London.

32. van der Klaauw 1932, 341.

33. R. Owen to Council of the Royal Society (draft), [1860s], Richard Owen Papers, MS0025/1/6/1/15/Q (xvii), Royal College of Surgeons, London.

34. T. Huxley 1856b, 190–91, 193, 195.

35. T. Huxley 1866, 632, 637.

36. "Noctes Ambrosianæ" 1829, 388; quoted in McCracken-Flesher 2012, 57.

37. "Pencillings of Eminent Medical Men" 1844, 246.

38. See E. Richards 1989, 383–86.

39. "Late Dr. Knox" 1862, 684.

40. "*Great Artists and Great Anatomists*" 1852, 935.

41. Knox 1852, 48, 49.

42. [Forbes] 1852, 478.

43. See Forbes 1855, 141–44.

44. Desmond 1994, 206.

45. Forbes 1855, iv.

46. Falconer 1856, 478, 492, 477, 488.

47. Milne-Edwards 1856, vi, vii.

48. Milne-Edwards 1841–42, 278 ("obvious signs of a principle of coordination"). My translation.

49. Milne-Edwards 1856, 189, xxii.

50. Ibid., v; and "Batch of Scientific Books" 1856, 258.

51. Knox 1856, 245.

52. E. Richards 1989, 377.

53. Knox 1856, 298.

54. "Bibliographic Notices" 1841, 323.

55. Knox 1856, 246, 247, 270-71, 299.

56. Ibid., 247, 297.

57. T. Huxley 1856b, 189.

58. Knox 1856, 247.

59. Quoted in Blake 1870, 334. For respectful obituaries, see, for instance, "Late Rev. Dr. Buckland" 1856, 12.

60. T. Huxley 1856c, 54.

61. See Bartrip 2009; and Dawson 2009.

62. Milne-Edwards 1863, xi, 199-200.

63. Ibid., xi, xii; Blake 1870, 334; and Blake 1875, v.

64. On this "anthropological schism," see Sera-Shriar 2013, 106-13.

65. Knox 1850, v.

66. Ibid., vi, 169, 445.

67. Knox 1852, 18, 19.

68. Knox 1850, 169-70.

69. E. Richards 1989, 407.

70. Van Evrie 1861, v, 92-93.

71. Ibid., 92, 44.

72. See Irmscher 2013, 219-69; and Desmond and Moore 2009, 228-66.

73. J. Hunt 1863, 48.

74. J. Hunt 1864, 4.

75. J. Hunt 1866, 336.

76. See E. Richards 1989, 410-13.

77. J. Hunt 1863, v, vi.

78. [Blake] 1863, 564, 568.

79. "Mr. Huxley on the Negro-Question" 1864, 288.

80. See Sera-Shriar 2013, 126-27.

81. Blake 1870, 332. On Knox's teaching career in the 1840s and 1850s, see Bates 2010, 140-42.

82. Broca 1864, v.

83. C. C. Blake to R. Owen, 18 August 1857, Richard Owen Papers, 4:196, Natural History Museum, London.

84. F. Burkhardt et al. 1985-, 6:260.

85. R. Owen 1857, 8; and T. Huxley 1861, 83.

86. T. Huxley 1861, 84.

87. R. Owen 1839, 1:35, 1:41.

88. See Rachootin 1985.

89. L. Huxley 1900, 1:149.

90. F. Burkhardt et al. 1985-, 11:26.

91. Blake 1861a, 442-43.

92. R. Owen to Council of the Royal Society (draft), [1860s], Owen Papers, MS0025/1/6/1/15/Q (xvii).

93. F. Burkhardt et al. 1985-, 12:65.

94. Ibid., 10:13.

95. Rachootin 1985, 170.

96. See Blake 1861b.

97. Mackie 1858c, i.

98. Mackie 1859, 39.

99. Blake 1865, 308–9.

100. See Jefferson 1799; and Rudwick 2005, 373–75.

101. Quoted in Desmond and Moore 2009, 338.

102. Blake 1862b, 323, 325n.

103. [Blake] 1863, 563, 564.

104. Blake 1862a, 81, 84.

105. Ibid., 86.

106. F. Burkhardt et al. 1985–, 11:517.

107. "Scientific" 1865, 252.

108. See Sera-Shriar 2013, 113; and Wallen 2013. On the X Club, see Barton 1998a; and Desmond 2001.

109. Pike 1868, 254, 246; and Knox 1855, 358.

110. Pike 1868, 256.

111. Pike 1866, 5, 6.

112. E. Richards 1989, 413.

113. Sera-Shriar 2013, 109–45.

114. Beddoe 1910, 212–13.

115. [Blake] 1863, 563; and F. Burkhardt et al. 1985–, 19:78.

116. [Blake] 1871, 298, 307.

117. C. C. Blake to C. Darwin, 10 July 1879, Darwin Manuscript Collection, 160:200, Cambridge University Library.

118. My thanks to Alison Pearn of the Darwin Correspondence Project for this information.

119. Desmond 1994, 320–21.

120. See ibid., 326.

Chapter Nine

1. R. S. Owen 1894, 2:131–32.

2. "Feathered Enigma" 1862, 9.

3. "Our Weekly Gossip" 1862, 699.

4. Ruskin 1865, 80, 81.

5. Wagner 1862, 261, 266, 267.

6. "Geology and Palæontology" 1862, 524.

7. Meyer 1862, 369.

8. Meyer 1832, 197 ("Schlüsse aus einem Theil des Skelettes auf das ganze Thier irrig ausfielen"); and T. Huxley 1870a, xxxvi. The translation is Huxley's.

9. Rupke 1994, 72.

10. "Feathered Enigma" 1862, 9.

11. Wagner 1862, 266.

12. Mackie 1863a, 2.

13. See Barton 1998b.

14. F. Burkhardt et al. 1985–, 11:23; and Mackie 1863b, 359.

15. Woodward 1862, 317.

16. [Woodward] 1893, 52.

17. Woodward 1862, 317, 319.

18. Ibid., 313.

19. Ibid., 317.

20. "Proceedings of Geological Societies" 1863, 32.

21. R. Owen 1863b, 39, 40, 46.

22. F. Burkhardt et al. 1985–, 11:5.

23. Mackie 1863a, 3–4.

24. Woodward 1875, 347.

25. Mackie 1863b, 360.

26. H. Falconer, Referees' Report on "*Archeopteryx*," 20 December 1862, RR/5/162, Royal Society, London.

27. Falconer 1862, 350; and Falconer, Referees' Report on "*Archeopteryx*," 20 December 1862, RR/5/162.

28. Falconer, Referees' Report on "*Archeopteryx*," 20 December 1862, RR/5/162.

29. On their friendship, see Boylan 2008, 61–62.

30. Quoted in Joan Evans 1943, 115–16.

31. John Evans 1865, 419–20, 421.

32. Quoted in Joan Evans 1943, 115, 116.

33. F. Burkhardt et al. 1985–, 11:5.

34. Mackie 1863a, 7.

35. F. Burkhardt et al. 1985–, 11:26.

36. John Evans 1865, 418.

37. Mackie 1863a, 7, 8.

38. F. Burkhardt et al. 1985–, 11:517.

39. Mackie 1863c, 65.

40. F. Burkhardt et al. 1985–, 11:55.

41. Mackie 1863b, 362–63.

42. Quoted in Joan Evans 1943, 116.

43. John Evans 1865, 421.

44. F. Burkhardt et al. 1985–, 11:11.

45. Mackie 1863a, 4.

46. Desmond 1982, 126.

47. T. Huxley 1868, 243.

48. Ibid., 244, 248.

49. See Desmond 1982, 128–29.

50. T. Huxley 1877b, 58–59.

51. T. Huxley 1868, 248.

52. T. Huxley 1869b, 282, 278, 285, 281, 283.

53. See Desmond 1982, 131.

54. T. Huxley 1869b, 285.

55. See Desmond 1975, 29–31; and Bramwell and Peck 2008, 77–80.

56. T. Huxley 1869b, 285.

57. Ibid., 286.

58. F. Burkhardt et al. 1985–, 11:5.

59. See Dean 1969.

60. Hitchcock 1836, 312, 316, 332, 340.

61. Hitchcock 1844, 292. Certainly it was the same translation of the "Discours préliminaire" that Mitchill had edited for American readers that Hitchcock quoted from.

62. Ibid., 292.

63. Cohen 2011, 36 ("dans le sillage de la grande tradition cuviérienne"). My translation.

64. Gerstner 1970, 137.

65. Hitchcock 1844, 311.

66. R. Owen to E. Hitchcock, 30 August 1844, Edward and Orra White Hitchcock Papers, 3:30, Amherst College Library, Amherst, MA.

67. Quoted in Silliman 1843, 185.

68. Hitchcock 1865, viii, 25, 26.

69. Ibid., vii, 28.

70. See "Discovery of Remains of Vertebrated Animals" 1863.

71. Hitchcock 1865, 29, 28, 32, 33.

72. Silliman 1843, 187; and T. Huxley 1869b, 286, 281.

73. Hitchcock 1844, 311; and T. Huxley 1869b, 287.

74. T. Huxley 1869b, 287, 286.

75. Meyer 1862, 367.

76. T. Huxley 1877a, 49.

77. F. Burkhardt et al. 1985–, 11:23.

78. Darwin 1859, 143.

79. Darwin 1868, 2:319, 2:320.

80. Ghiselin 1969, 239.

81. Stauffer 1975, 383.

82. E. Russell 1916, 239; and Reiss 2009, 136–37.

83. Stauffer 1975, 383–84 ("convenance des parties, de leur coordination pour le role que animal doit jouer dans la nature"). My translation.

84. Di Gregorio 1990, 301.

85. Shanahan 2004, 111.

86. J. Martineau 1888, 1:285.

87. Annotation on p. 114 and back flyleaf of Gustav Jäger, *In Sachen Darwin's insbesondere contra Wigand* (Stuttgart: E. Schweizerbart, 1874), Darwin Library, Cambridge University Library. "Correlation" is incorrectly transcribed as "constituent" in Di Gregorio 1990, 430–31.

88. [Mozley] 1869, 174.

89. Darwin 1868, 2:320.

90. Bernard 1895, 24–25, 8 ("variations en *corrélation*," "modifications simultanées dans diverses parties du squelette," and "erreurs . . . ont été le point de départ d'un mouvement progressif"). My translation.

91. The fullest account of Darwin's concern with correlation is in Gould 2002, 332–39, although hardly any consideration is given to Cuvier's role in prompting Darwin's interest. For an earlier corrective of such conventional assumptions, see Ospovat 1981, 60–86, which argues for the importance of natural theological conceptions of perfect adaptation in Darwin's formulation of natural selection.

92. Darwin 1868, 2:322.

93. Seeley 1870a, 25.

94. "Royal Institution Lectures" 1875, 375.

95. Quoted in Desmond 1982, 189.

96. Seeley 1872, 42–43.

97. "Royal Institution Lectures" 1875, 402.

98. Ibid., 447.

99. "Pterodactyles" 1870, 294.

100. Seeley 1870b, 139–40, 129, 133–34.

101. R. Owen 1849–84, 1:495, 1:489.

102. Seeley 1870b, 147.

103. R. Owen 1866–68, 3:790, 3:789, 3:788, 3:786.

104. Ibid., 1:xxvii, 1:xxviii, 1:xxx.

105. Cuvier 1812a, 62 ("moins claires" and "descendensuite aux ordres ou subdivisions de la classe des animaux à sabots"). Translation from Rudwick 1997, 219–20.

106. R. Owen 1866–68, 1:xxxvi.

107. See E. Richards 1987; and Rupke 1994, 222–23. On Owen's evolutionism, see also Camardi 2001.

108. Darwin 1869, xx.

109. Ibid., xx; and R. Owen 1860, 3.

110. T. Ward 1884, 189, 190.

111. R. Owen 1863c, 114–16.

112. F. Burkhardt et al. 1985–, 10:450.

113. R. Owen 1863a, 59–60n.

114. C. H. Bingham to R. Owen, 13 January 1863, Richard Owen Papers, 4:150, Natural History Museum, London.

115. C. H. Bingham to R. Owen, 20 January 1863, Owen Papers, 4:153.

116. R. Owen 1866–68, 3:793.

117. R. Owen 1863a, 59, 60, 63, 61.

118. Ibid., 63, 65, 61.

119. R. Owen 1871a, 110, 111.

120. On the relation between museum-based comparative and field-based environmental modes of paleontology, see Rachootin 1985.

121. Outram 1996, 262.

122. See Barton 2000; and Endersby 2008, 110 and passim.

123. Macleay 1859, 5.

124. Quoted in Holland 1996, 131.

125. Gerard Krefft Papers, A 267, State Library of New South Wales, Sydney.

126. R. Owen 1871b, 227, 228.

127. Krefft 1872, 626–27.

128. F. Burkhardt et al. 1985–, 20:204, 20:311, 20:312.

129. R. Owen 1839, 1:23, 1:24, 1:21, 1:28.

130. Quoted in Butcher 1988, 147.

131. See Finney 1993, 108–13.

132. See Yanni 1999, 111–46.

133. Rupke 1994, 219. On the intellectual implications of the museum's floor plan, which, of course, conflicted with Owen's incipient evolutionism, see Stearn 1981, 57–59.

134. F. Cuvier to R. Owen, 25 January 1884, Owen Papers, 9:178 ("jeune génération de savants . . . parle avec dedain des travaux de M. Cuvier," "vive satisfaction," and "savant qu'ont précisement illustré sa fidelité aux principes de la veritable observations scientifique"). My translation.

135. Rupke 1994, 137.

136. Desmond 1982, 173.

Chapter Ten

1. Smiles 1859, 9–10, 15, 79–80.

2. See A. Secord 2003.

3. Quoted in A. Secord 2003, 170.

4. Ibid., 169.

5. Smiles 1876, 365, 366, 369, 370, 371.

6. Ibid., 371.

7. "Mineral Kingdom" 1833, 347.

8. Smiles 1876, 368.

9. Roscoe 1871a, [iii].

10. See Baldwin 2012.

11. Lockyer 1871, 81.

12. "Conference on Technical Education" 1868, 199.

13. See Riley 2003, 138–40.

14. Quoted in Roscoe 1906, 127; and T. Huxley 1871, 13, 17.

15. Bradley 1873, 80, 81.

16. Ibid., 81.

17. Roscoe 1871b, [v].

18. P. M. Duncan 1875, 71, 73.

19. L. Huxley 1900, 1:138.

20. T. Huxley 1880, 929.

21. Cuvier 1812a, 63 ("piste d'un pied fourchu" and "une marque plus sûre que toutes celles de Zadig"). Translation from Rudwick 1997, 220.

22. T. Huxley 1880, 932.

23. Ibid., 935, 939.

24. T. Huxley 1870b, 86; and T. Huxley 1880, 932.

25. T. Huxley 1880, 937.

26. T. Huxley 1881, 453.

27. T. Huxley 1880, 937.

28. T. Huxley 1856c, 47.

29. "Working Men's College" 1880, 1.

30. J. Knowles to T. H. Huxley, 17 March 1880, Thomas Henry Huxley Papers, 20.48, Imperial College of Science, Technology, and Medicine Archives, London.

31. See Dawson 2004, 183–89.

32. A summary of the lecture in *Lloyd's Weekly Newspaper* ("Working Men's College" 1880) suggests that its content was virtually identical to the version published in the *Nineteenth Century*.

33. T. Huxley 1881, 453.

34. T. H. Huxley to N. Lockyer, 12 September 1881, Huxley Papers, 21.277.

35. T. Huxley 1880, 940.

36. See Barrow 1986.

37. T. Huxley 1880, 934, 933, 931.

38. See Lightman 1987, 119–23.

39. Cuvier 1812b, 3 ("à ma disposition la trompette toute puissante"). Translation from Rudwick 2005, 413.

40. [Broderip and R. Owen] 1845, 116, 117.

41. R. Owen 1846a, 76; and [Broderip and R. Owen] 1852, 397.

42. Townshend 1840, 567, 571. On mesmerism, see Winter 1998.

43. Townshend 1840, 570.

44. E. Agassiz 1885, 1:181, 1:182.

45. See Lurie 1988, 50–52.

46. L. Agassiz 1850, 25.

47. See Cox 2003, 69–107.

48. W. Denton 1868, 212, 213.

49. Powell 1870, 33.

50. Ibid., 16; and W. Denton and E. Denton 1863, 35–36.

51. See Cox 2003, 109–10.

52. W. Denton and E. Denton 1863, 44, 36.

53. Ibid., 55.

54. On female assistance to Cuvier, see Orr 2007; to Agassiz, see Irmscher 2013, 270–310; to Buckland, see Gordon 1894, 90–93; to Owen, see chapter 4.

55. On the gendering of nineteenth-century notions of objectivity, see Levine 2002, 126–29.

56. W. Denton and E. Denton 1863, 286.

57. "Literature" 1863, 296.

58. "Literary Notices" 1863, 5. There were also positive reviews in *Weldon's Register of Facts* and, more predictably, the *Spiritual Magazine*.

59. W. Denton and E. Denton 1863, 54, 55, 328–29.

60. See A. Owen 1989; and Braude 1989.

61. Agassiz's early experiences with mesmerism and continuing interest in vision and augury are not mentioned in either Lurie 1988 or Irmscher 2013, despite the latter's attempt to broach aspects of Agassiz's career, especially his attitudes to race, that have previously been occluded.

62. W. Denton and E. Denton 1863, 57, 62.

63. C. H. Hitchcock to W. Denton, 17 August 1863, Denton Family Papers, Wellesley Historical Society, Wellesley, MA.

64. C. H. Hitchcock to W. Denton, 7 September 1863, Denton Family Papers.

65. The Dentons seem even to have lent fossils to Agassiz's Museum of Comparative Zoology at Harvard; see A. Agassiz to W. Denton, 7 April 1865, Denton Family Papers.

66. W. Denton 1881, 3.

67. T. Huxley 1880, 932.

68. T. Huxley 1893–94, 8:vi, 8:viii.

69. T. Huxley 1880, 929.

70. Ibid., 931, 933, 935.

71. Doyle 1887, 92, 93, 18, 12.

72. J. Bell 1892, 79.

73. "Sherlock Holmes's Predecessor" 1894, 6.

74. T. Huxley 1870b, 86.

75. See Shpayer-Makov 2011, 199–200.

76. See ibid., 238–44.

77. Wright 1927, 27.

78. Doyle 1887, 14.

79. Poe 1841, 176. On Poe's translation of Cuvier, see Gould 1995.

80. Gaboriau 1866, 21 ("reconstruire toutes les scènes d'un assassinat, comme ce savant qui sur un os rebâtissait les animaux perdus"). Translation from Gaboriau 1887b, 10. See also Gaboriau 1869, 1:414.

81. Gaboriau 1869, 1:399, 1:412 ("prouva par A plus B, par une déduction mathématique, pour ainsi dire" and "juger par mes yeux"). Translation from Gaboriau 1887a, 1:112, 1:115. On Tabaret's disinterest in visual information, and its relation to Cuvier, see Goulet 2006, 95–100.

82. Doyle 1891a, 487.

83. Doyle 1891b, 61; and Doyle 1893, 67. On the incoherence of Holmes's methods, see McDonald 1997, 167–68.

84. Doyle 1887, 12.

85. T. Huxley 1893–94, 8:viii.

86. Doyle 1907, 248, 249.

87. Doyle 1926, 2:162.

88. Holcombe 1890, 44.

89. Doyle 1907, 253.

90. Doyle 1901, 365.

91. Rees 1919, 282.

92. Doyle 1912, 40, 46.

93. Doyle 1924, 112. On the *Strand*'s readership, see McDonald 1997, 142.

94. T. Huxley 1881, 453. Notably, while several literary critics and historians have noted the resemblances between Huxley's method of backward reasoning and Holmes's forensic techniques, they, like Doyle, mistakenly conflate Huxley's and Cuvier's approaches to paleontological reconstruction, and so do not recognize that Doyle misinterpreted Huxley's intention in "On the Method of Zadig"; see, for instance, J. Smith 1994, 225–29; Frank 2003, 23–24; and Snyder 2004, 104–5.

95. Benn 1895, 73.

96. Balfour 1900, 209, 212–13.

97. L. Huxley 1900, 2:373; and T. Huxley 1894, 2:274.

98. T. Huxley 1894, 2:274, 2:275.

99. Ibid., 2:284, 2:295–96.

100. Peck 1904, 25–26.

101. R. Owen 1849–84, 1:189.

102. See Jaffe 2000; and Brinkman 2010, 7–16.

103. Marsh 1879, 24–25.

104. Marsh 1875, 221.

105. See Brinkman 2010, 22–26.

106. Osborn 1893, 6, 43.

107. Peck 1904, 26.

108. Osborn 1893, 7.

109. Osborn 1917, 240.

110. Osborn 1911, 20:588, 20:587. On mosaic evolution, see Gould 2002, 293–94.

111. Osborn 1911, 20:587.

112. Ibid., 20:582.

113. [Woodward] 1890, 34.

114. Woodward's authorship is acknowledged in the preface; ibid., ix.

115. Packard 1901, 144.

116. On the new mass audiences for science, see Bowler 2009; on the limitations of the standard of literacy achieved by the Elementary Education Act, see Vincent 1989, 92–93, 270–73.

117. Hutchinson 1892, 2.

118. O'Connor 2007a, 435, 436.

119. On Hutchinson as a clerical popularizer, see Lightman 2007b, 450–60.

120. Hutchinson 1892, 81–82, 6.

121. H. N. Hutchinson to T. H. Huxley, 7 June 1892, Mss.B.H981, American Philosophical Society, Philadelphia.

122. Hutchinson 1900, 586, 579.

123. Hutchinson 1910, 141.

124. On the treatment of science in such mass-market publications, see Broks 1996; and Bowler 2009, 190–202.

125. Hutchinson 1910, 302. This statement had originally appeared in 1894 in Hutchinson's *Creatures of Other Days*, but was now transferred, sixteen years later, to the new edition of *Extinct Monsters*.

126. Hutchinson 1894, xiv.

127. "From W. Beach Thomas" 1916, 5; and "Mysterious Tanks" 1916, 10.

128. Tate 1997, 72.

129. Williams-Ellis and Williams-Ellis 1919, 30. See also Nieuwland 2010, 61.

130. See Rea 2001; and Nieuwland 2010.

131. On the complaints of Hay and Tornier, see Nieuwland 2010, 65–66; and Rieppel 2012, 472–74.

132. Hutchinson 1917, 356, 366, 365.

133. Hutchinson 1892, 6.

134. See Potter 2007.

135. Hutchinson 1910, 302; and T. Huxley 1881, 453.

Epilogue

1. Keith 1950, 673.

2. G. Turner 1946, 58; and Keith 1950, 673.

3. See Marr 2009, 386.

4. G. Turner 1946, 66. On the *Lancet*'s criticisms, see Rupke 1994, 28.

5. G. Turner 1946, 64, 73.

6. Bianchi and Hampson 1929, v, vi.

7. The poem, without the title Bianchi gave it, split into two stanzas and with minor editorial alterations, is dated "around 1859" in Johnson 1955, 1:77.

8. Bianchi and Hampson 1929, 26.

9. On Dickinson's connections to Hitchcock, see Baym 2002, 138–43.

10. Thackeray 1853–55, 2:81.

11. See Socarides 2007, 39–43.

12. Gould 1983, 100–101.

13. T. Huxley 1881, 453. Gould had resolved to become a paleontologist at the age of five after visiting the American Museum of Natural History; see Michelle Green 1986.

14. J. Secord 2000, 126.

15. "Names of Novels" 1867, 10.

16. Simon 2002, 220.

17. "Moa Bone Fragment," Natural History Museum, London, accessed 21 March 2014, www.nhm.ac.uk/nature-online/collections-at-the-museum/museum-treasures/moa-bone-fragment/.

References

Archives

Add. 5354, Cambridge University Library.

Add. 38091, 39954, and 42577, British Library, London.

Caroline Owen Commonplace Book, Royal College of Surgeons, London.

Codices historici, Württembergische Landesbibliothek, Stuttgart, Germany.

Darwin Library, Cambridge University Library.

Darwin Manuscript Collection, Cambridge University Library.

Denton Family Papers, Wellesley Historical Society, Wellesley, MA.

Edward and Orra White Hitchcock Papers, Amherst College Library, Amherst, MA.

Frances Hirtzel Collection of Richard Owen Papers, Temple University Library, Philadelphia.

Gerard Krefft Papers, State Library of New South Wales, Sydney, Australia.

John Murray Archive, National Library of Scotland, Edinburgh.

Louis Agassiz Papers, Ernst Mayr Library of the Museum of Comparative Zoology, Harvard University, Cambridge, MA.

Mantell Family Papers, Alexander Turnbull Library, Wellington, New Zealand.

Marked copy of the *Penny Cyclopædia*, British Library, London.

Mss.B.H981 and Mss.B.Ow2, American Philosophical Society, Philadelphia.

Philosophical Transactions Referees' Reports, Royal Society, London.

Richard Owen Papers, Natural History Museum, London.

Richard Owen Papers, Royal College of Surgeons, London.

RI MS GB2, Royal Institution Archives, London.

Royal Literary Fund Archive, Loan 96, British Library, London.

Samuel Leigh Sotheby Scrap Album "Papers and Memorandum of the Crystal Palace Company 1852–56," Bromley Local Studies Library, Kent, UK.

Society for the Diffusion of Useful Knowledge Papers, University College London, Special Collections.

Thomas Henry Huxley Papers, Imperial College of Science, Technology, and Medicine Archives, London.

William Whewell Papers, Trinity College Library, Cambridge.

Printed Primary Sources

Abernethy, John. 1817. *Physiological Lectures*. London: Longman.

"An Abstract of a Memoir upon the Fossil Bones of Animals." 1799. *Journal of Natural Philosophy, Chemistry and the Arts* 2:512–14.

"Abstract of Lectures Delivered before the College of Surgeons by Professor Stanley." 1837. *London Medical Gazette* 20:379–86.

Adams, H. G. 1857. "Hours with the Antediluvians." *Ladies' Cabinet* 11:29–32.

Agassiz, Elizabeth Cary, ed. 1885. *Louis Agassiz: His Life and Correspondence*. 2 vols. Boston: Houghton, Mifflin.

Agassiz, Louis. 1833–43. *Recherches sur les poissons fossiles*. Neuchâtel, Switzerland: H. Nicolet.

———. 1847. *An Introduction to the Study of Natural History*. New York: Greeley and McElrath.

———. 1850. *Lake Superior*. Boston: Gould, Kendall, and Lincoln.

Allen, John. 1849. *Inquiry into the Rise and Growth of the Royal Prerogative in England*. 2nd ed. London: Longman.

Arnold, Thomas. 1841. *Christian Life: Its Course, Its Hindrances and Its Helps*. London: B. Fellowes.

[Arthur, William]. 1854. "Guidebooks to the Crystal Palace." *London Quarterly Review* 3:232–79.

"Arts and Sciences." 1829. *Literary Gazette*, no. 673, 808.

Balfour, George W. 1900. "On the Method of Zadig in the Advancement of Medicine." *Edinburgh Medical Journal*, n.s., 7:209–30.

Balzac, Honoré de. 1831. *La peau de chagrin*. 2 vols. Paris: Charles Gosselin.

———. 1977. *The Wild Ass's Skin*. Translated by Herbert J. Hunt. Harmondsworth, UK: Penguin.

Barclay, John. 1822. *An Inquiry into the Opinions, Ancient and Modern, Concerning Life and Organization*. Edinburgh: Bell and Bradfute.

"Baron Cuvier." 1832. *London Medical and Surgical Journal* 1:537–38.

"A Batch of Scientific Books." 1856. *Leader* 7:257–58.

Beddoe, John. 1910. *Memories of Eighty Years*. London: Simpkin, Marshall.

[Bell, Charles]. 1827–29. *Animal Mechanics*. 2 vols. London: Baldwin and Craddock.

———. 1833. *The Hand*. London: William Pickering.

Bell, Joseph. 1892. "The Adventures of Sherlock Holmes." *Bookman* 3:79–81.

Benn, Alfred W. 1895. "Literature." *Academy* 47:73–74.

Bernard, Félix. 1895. *Éléments de paléontologie*. Paris: J. B. Baillière.

Best, Samuel. 1837. *After Thoughts on Reading Dr. Buckland's Bridgewater Treatise*. London: J. Hatchard.

Bianchi, Martha Dickinson, and Alfred Leete Hampson, eds. 1929. *Further Poems of Emily Dickinson*. Boston: Little, Brown.

"Bibliographic Notices." 1841. *Annals and Magazine of Natural History* 7:322–26.

"A Bird Twenty Feet High." 1852. *Times*, 12 July, 4.

Bissell, Champion. 1851. "Serials and Continuations." *Sartain's Union Magazine* 8:278–79.

Blainville, H. M. Ducrotay de. 1839a. "Mémoire sur les traces qu'ont laissées à la surface de la terre, les Édentés terrestes." *Comptes rendus hebdomadaires des séances de l'Académie des sciences* 8:139–46.

———. 1839b. "New Doubts Relating to the Supposed Didelphis of Stonesfield." *Magazine of Natural History*, n.s., 3:49–57.

———. 1839–64. *Ostéographie*. 4 vols. Paris: Arthus Bertrand.

Blake, Charles Carter. 1861a. "On the Discovery of *Macrauchenia* in Bolivia." *Annals and Maga-zine of Natural History*, n.s., 7:441–43.

———. 1861b. "On the Discovery of *Macrauchenia* in Bolivia." *Geologist* 4:354–55.

———. 1862a. "Fossil Monkeys." *Geologist* 5:81–86.

———. 1862b. "Past Life in South America." *Geologist* 5:323–30.

[———]. 1863. "Professor Huxley on Man's Place in Nature." *Edinburgh Review* 117:541–69.

———. 1865. "On the Distribution of the Fossils of South America." *Proceedings of the Geologists' Association* 1:308–9.

———. 1870. "The Life of Dr. Knox." *Journal of Anthropology* 1:332–38.

[———]. 1871. "The Evolution of Species." *British and Foreign Medico-Chirurgical Review* 48:285–309.

———. 1875. *Zoology for Students.* London: Dalby, Isbister.

Bohn, Henry G. 1856. *The Crystal Palace Company: Deed of Settlement, Royal Charters and List of Shareholders.* London: Henry G. Bohn.

Bradley, S. M. 1873. "Animal Mechanics." In *Science Lectures for the People: Fifth Series,* 77–96. Manchester, UK: John Heywood.

[Brewster, David]. 1844. "*Eloge Historique de Baron Cuvier.*" *North British Review* 1:1–41.

[———]. 1845. "Vestiges of the Natural History of Creation." *North British Review* 3:470–515.

Broca, Paul. 1864. *On the Phenomena of Hybridity in the Genus Homo.* Translated by C. Carter Blake. London: Longman, Green.

[Broderip, William John]. 1833–43a. "Pangolins." In *Penny Cyclopædia,* 17:186–89. 27 vols. Lon-don: Charles Knight.

———. 1833–43b. "Struthionidæ." In *Penny Cyclopædia,* 23:133–47. 27 vols. London: Charles Knight.

———. 1833–43c. "Unau." In *Penny Cyclopædia,* 25:501–8. 27 vols. London: Charles Knight.

[———. and Richard Owen]. 1844. "Recollections and Reflections of Gideon Shaddoe, Esq." *Hood's Magazine* 2:442–50.

[———. and Richard Owen]. 1845. "Recollections and Reflections of Gideon Shaddoe, Esq." *Hood's Magazine* 3:114–19, 3:294–303.

[———. and Richard Owen]. 1852. "Progress of Comparative Anatomy." *Quarterly Review* 90:362–413.

Brodie, P. B. 1858. "Contributions to the Geology of Gloucestershire." *Geologist* 1:369–77.

Brownell, W. C. 1905. "Henry James." *Atlantic Monthly* 95:496–519.

Buckland, William. 1820. *Vindiciæ Geologicæ.* Oxford: Oxford University Press.

———. 1824. "Notice on the Megalosaurus or Great Lizard of Stonesfield." *Transactions of the Geological Society,* n.s., 1:390–96.

———. 1835. "On the Discovery of Fossil Bones of the Iguanodon." *Transactions of the Geologi-cal Society,* n.s., 3:425–32.

———. 1836. *Geology and Mineralogy Considered with Reference to Natural Theology.* 2 vols. London: William Pickering.

———. 1837. "On the Adaptation of the Structure of the Sloths to Their Peculiar Mode of Life." *Transactions of the Linnean Society* 17:17–27.

———. 1848. *A Sermon Preached in Westminster Abbey.* London: John Murray.

———. 1858. *Geology and Mineralogy Considered with Reference to Natural Theology.* 2 vols. 3rd ed. Edited by Francis T. Buckland. London: George Routledge.

"Buckland's *Bridgewater Essay on Geology.*" 1836. *British Critic* 20:295–328.

[Bugg, George]. 1826–27. *Scriptural Geology.* 2 vols. London: Hatchard.

Byron, Lord. 1821. *Sardanapalus, A Tragedy; The Two Foscari, A Tragedy; Cain, A Mystery.* London: John Murray.

Carlile, Richard. 1829. "Design." *Lion* 3:281–84.

"The Centenary Dinner." 1854. *Journal of the Society of Arts* 2:563–69.

[Chalmers, Thomas]. 1814. "Review of Cuvier's Theory of the Earth." *Edinburgh Christian Instructor* 8:261–74.

[Chambers, Robert]. 1842. "The Easily Convinced." *Chambers's Journal* 11:185–86.

[———]. 1843. "Popular Information on Science." *Chambers's Journal* 12:322–23.

[———]. 1844. *Vestiges of the Natural History of Creation.* London: John Churchill.

[———]. 1847. "The Ancient World." *Chambers's Journal,* n.s., 7:344–47.

[———]. 1848. "The Dinornis." *Chambers's Journal,* n.s., 9:247–49.

"Charles Knight's English Cyclopædia." 1854. *Times,* 12 October, 10.

Charlesworth, Edward. 1846. "On the Occurrence of a Species of Mosasaurus in the Chalk of England." *London Geological Journal* 1:23–32.

[Cleghorn, Thomas]. 1845. "Writings of Charles Dickens." *North British Review* 3:65–87.

Colenso, W. 1892. "Status Quo: A Retrospect." *Transactions and Proceedings of the New Zealand Institute,* n.s., 24:468–78.

"The College Conversazioni." 1842. *Lancet:* 245–47.

Collins, Charles. 1857. "Spiritual Despotism." *Methodist Quarterly Review* 39:34–47.

"Conference on Technical Education." 1868. *Journal of the Society of Arts* 14:183–209.

Conybeare, W. D. 1829. "Answer to Dr Fleming's View of the Evidence from the Animal Kingdom, as to the Former Temperature of the Northern Regions." *Edinburgh New Philosophical Journal,* n.s., 7:142–52.

Conybeare, W. D., and William Phillips. 1822. *Outlines of the Geology of England and Wales.* London: William Phillips.

Cope, Edward D. 1872. "On the Geology and Paleontology of the Cretaceous Strata of Kansas." In *Preliminary Report of the United States Geological Survey of Montana and Portions of Adjacent Territories,* edited by Ferdinand Vandeveer Hayden, 318–49. Washington, DC: US Government Printing Office.

"The Crystal Palace." 1854. *John Bull* 34:378.

"Crystal Palace." 1855. *Daily News,* 20 September, 6.

"The Crystal Palace at Sydenham." 1854. *Illustrated London News* 24:22.

"The Crystal Palace Handbooks." 1854. *Illustrated Crystal Palace Gazette* 1:166.

"Crystal Palace News." 1855. *Crystal Palace Herald* 2:27–28.

"The Crystal Palace—Visit of Her Majesty." 1854. *John Bull* 34:267.

"Cuvier." 1832. *Isis* 1:455–56.

Cuvier, Georges. 1800–1805. *Leçons d'anatomie comparée.* 5 vols. Paris: Baudouin.

———. 1801. *An Introduction to the Study of Animal Economy.* Translated by John Allen. Edinburgh: Ross and Blackwood.

———. 1802. *Lectures on Comparative Anatomy.* Translated by William Ross and James Macartney. 2 vols. London: Longman, Rees.

———. 1804. "Mémoire sur le squelette presque entier d'un petit quadrupède du genre des Sariques." *Annales du Muséum national d'histoire naturelle* 5:277–92.

———. 1812a. "Discours préliminaire." In Cuvier 1812c, *Recherches sur les ossemens fossiles,* 1:1–116.

———. 1812b. "Introduction." In Cuvier 1812c, *Recherches sur les ossemens fossiles,* 3:1–8.

———. 1812c. *Recherches sur les ossemens fossiles.* 4 vols. Paris: Deterville.

———. 1813. *Essay on the Theory of the Earth.* Translated by Robert Kerr. Edited by Robert Jameson. Edinburgh: William Blackwood.

———. 1817. *Le règne animal.* 4 vols. Paris: Deterville.

———. 1818. *Essay on the Theory of the Earth.* Translated by Robert Kerr. Edited by Robert Jameson and Samuel Latham Mitchill. New York: Kirk and Mercein.

———. 1819. "Éloge historique de Daubenton." [1800.] In *Recueil des éloges historiques,* 1:37–80. 3 vols. Paris: F. G. Levrault.

———. 1821–24. *Recherches sur les ossemens fossiles.* 2nd ed. 5 vols. Paris: G. Dufour et E. d'Ocagne.

———. 1822. "D'une petite espèce de Sarigue." In Cuvier 1821–24, *Recherches sur les ossemens fossiles,* 3:284–97.

———. 1823. "Sur l'osteologie des Paresseux." In Cuvier 1821–24, *Recherches sur les ossemens fossiles,* 5:71–95.

———. 1826. *Researches into Fossil Osteology.* [Translated by Edward Pidgeon et al.] London: George B. Whittaker.

———. 1827–34. *The Animal Kingdom.* [Translated by Edward Pidgeon et al.] Edited by Edward Griffith. 16 vols. London: G. and W. B. Whittaker.

———. 1828. "Biographical Memoir of M. Daubenton." [Translated by Robert Jameson.] *Edinburgh New Philosophical Journal,* n.s., 5:1–22.

"Cuvier's Lectures on Comparative Anatomy." 1800. *Critical Review* 29:529–36.

"Cuvier's Lectures on Comparative Anatomy." 1804. *Monthly Review* 43:247–56.

Darwin, Charles. 1845. *Journal of Researches into the Natural History and Geology of the Countries Visited during the Voyage of H.M.S. Beagle Round the World.* London: John Murray.

———. 1849. "Geology." In *A Manual of Scientific Enquiry,* edited by John Herschel, 156–95. London: John Murray.

———. 1851–54. *A Monograph on the Sub-class Cirripedia.* 2 vols. London: Ray Society.

———. 1859. *On the Origin of Species.* London: John Murray.

———. 1868. *The Variation of Animals and Plants under Domestication.* 2 vols. London: John Murray.

———. 1869. *On the Origin of Species.* 5th ed. London: John Murray.

Daubenton, Louis-Jean-Marie. 1764. "Mémoire sur des os et des dents remarquables par leur grandeur." *Histoire de l'Académie royale des sciences, année 1762*: 206–29.

"David Copperfield and Pendennis." 1851. *Prospective Review* 7:157–91.

"A Day in the Country." 1823. *Edinburgh Magazine* 12:166–71.

"Delenda est Carthago." 1854. *Spectator* 27:1202.

Denton, William. 1868. *Our Planet, Its Past and Present.* Boston: William Denton.

———. 1881. *Is Darwin Right?* Wellesley, MA: Denton Publishing.

Denton, William, and Elizabeth M. F. Denton. 1863. *The Soul of Things.* Boston: Walker, Wise.

"Despatches from the Crimea." 1856. *Times,* 16 February, 10.

Dickens, Charles. 1836–37. *The Posthumous Papers of the Pickwick Club.* 2 vols. London: Chapman and Hall.

———. 1838–39. *The Life and Adventures of Nicholas Nickleby.* 2 vols. London: Chapman and Hall.

———. 1851. "On Duty with Inspector Field." *Household Words* 3:265–70.

———. 1852–53. *Bleak House.* London: Bradbury and Evans.

"Dinner to Professor Owen in the Iguanodon." 1854. *Leader* 5:6–7.

"Dinornis." 1854–55. In *English Cyclopædia*, Natural History, 2:343–47. 4 vols. London: Bradbury and Evans.

"Discovery of Remains of Vertebrated Animals Provided with Feathers." 1863. *American Journal of Science*, n.s., 35:129–33.

Doyle, Arthur Conan. 1887. "A Study in Scarlet." *Beeton's Christmas Annual* 28:1–95.

———. 1891a. "The Five Orange Pips." *Strand Magazine* 2:481–91.

———. 1891b. "A Scandal in Bohemia." *Strand Magazine* 2:61–75.

———. 1893. "The Adventure of the Cardboard Box." *Strand Magazine* 5:61–73.

———. 1901. "Strange Studies from Life." *Strand Magazine* 21:363–70.

———. 1907. *Through the Magic Door.* London: Smith, Elder.

———. 1912. *The Lost World.* London: Hodder and Stoughton.

———. 1917. "His Last Bow: The War Service of Sherlock Holmes." *Strand Magazine* 54:227–36.

———. 1924. *Memories and Adventures.* London: John Murray.

———. 1926. *The History of Spiritualism.* 2 vols. London: Cassell.

Duncan, David, 1911. *The Life and Letters of Herbert Spencer.* London: Williams and Norgate.

Duncan, P. Martin. 1875. "The Great Extinct Quadrupeds." In *Science Lectures for the People: Seventh Series,* 69–84. Manchester, UK: John Heywood.

[Duncan, Philip Bury]. 1836. *A Catalogue of the Ashmolean Museum.* Oxford: S. Collingwood.

[Duns, John]. 1858. "Professor Owen's Works." *North British Review* 28:313–45.

———. 1859. "Memoir." In John Fleming, *The Lithology of Edinburgh,* i–civ. Edinburgh: William P. Kennedy.

"Educational Exhibition." 1854. *Journal of the Society of Arts* 2:621–30.

"Eighth Meeting. *28th October,* 1871." 1872. *Transactions of the New Zealand Institute* 4:376–79.

Ellis, Robert, ed. 1851. *Official Descriptive and Illustrated Catalogue of the Great Exhibition.* 3 vols. London: Spicer Brothers.

[Elwin, Whitwell]. 1849. "Popular Science." *Quarterly Review* 84:307–44.

———. 1902. *Some XVIII Century Men of Letters.* Edited by Warwick Elwin. 2 vols. London: John Murray.

"The *English Cyclopædia.*" 1856. *Medical Times and Gazette,* n.s., 12:468–69.

Evans, John. 1865. "On Portions of a Cranium and of a Jaw, in the Slab Containing the Fossil Remains of the Archæopteryx." *Natural History Review,* n.s., 5:415–21.

"Excursion to the Crystal Palace." 1876–78. *Proceedings of the Geologists' Association* 5:377.

"Express from Paris." 1856. *Times,* 23 February, 10.

Falconer, Hugh. 1856. "On Prof. Huxley's Attempted Refutation of Cuvier's Laws of Correlation." *Annals and Magazine of Natural History,* n.s., 17:476–93.

———. 1859. *Descriptive Catalogue of the Fossil Remains of Vertebrata . . . in the Museum of the Asiatic Society of Bengal.* Calcutta: Baptist Mission Press.

———. 1862. "On the Disputed Affinity of the Mammalian Genus Plagiaulax." *Quarterly Journal of the Geological Society* 18:348–69.

———. 1863. "On the American Fossil Elephant." *Natural History Review,* n.s., 3:43–114.

———. 1868a. "Abstract of an Extempore Discourse on the Colossochelys atlas" [1844]. In Falconer 1868c, *Palæontological Memoirs,* 1:372–77.

———. 1868b. "Note on a Correction of Published Statements Respecting Fossil Quadrumana" [1862]. In Falconer 1868c, *Palæontological Memoirs,* 1:309–14.

———. 1868c. *Palæontological Memoirs and Notes of the Late Hugh Falconer.* Edited by Charles Murchison. 2 vols. London: Robert Hardwicke.

Falconer, Hugh, and Proby T. Cautley. 1846. *Fauna Antiqua Sivalensis.* London: Smith, Elder.

"A Feathered Enigma." 1862. *Times*, 12 November, 9.

Fleming, John. 1822. *The Philosophy of Zoology*. 2 vols. Edinburgh: Archibald Constable.

[————]. 1823. "Cuvier's *Theory of the Earth*, by Jameson." *New Edinburgh Review* 4:381–98.

————. 1829. "On the Value of the Evidence from the Animal Kingdom, Tending to Prove That the Arctic Regions Formerly Enjoyed a Milder Climate Than at Present." *Edinburgh New Philosophical Journal*, n.s., 6:277–86.

————. 1830. "Additional Remarks on the Climate of the Arctic Regions." *Edinburgh New Philosophical Journal*, n.s., 8:65–74.

"Fleming's *Zoology*." 1823. *New Edinburgh Review* 4:229–53.

[Forbes, Edward]. 1852. "*Great Artists and Great Anatomists* by R. Knox." *Literary Gazette*, no. 1847, 478.

————. 1855. *Literary Papers by the Late Professor Edward Forbes*. London: Lovell Reeve.

"Foreign Publications." 1803. *Anti-Jacobin Review* 15:490–93.

Francis, John W. 1858. *Old New York*. New York: Charles Roe.

"From W. Beach Thomas." 1916. *Daily Mail*, 18 September, 5.

Gaboriau, Émile. 1866. *L'affaire Lerouge*. Paris: E. Dentu.

————. 1869. *Monsieur Lecoq*. 2 vols. Paris: E. Dentu.

————. 1887a. *Monsieur Lecoq*. 2 vols. London: George Routledge.

————. 1887b. *The Widow Lerouge*. London: George Routledge.

Gellius, Aulus. 1927. *The Attic Nights*. Translated by John C. Rolfe. 3 vols. London: W. Heinemann.

"Geology and Palæontology." 1862. *Popular Science Review* 1:523–26.

"The Geology of the Crystal Palace." 1854. *Hogg's Instructor* 2:279–86.

"Gigantic Bird of New Zealand." 1854. *Illustrated London News* 24:22.

Gill, Theodore, William Berryman Scott, and Henry Fairfield Osborn. 1897. *Addresses in Memory of Edward Drinker Cope*. Philadelphia: American Philosophical Society.

Gloag, Paton J. 1859. *The Primeval World*. Edinburgh: T. & T. Clark.

Gordon, Elizabeth Oke. 1894. *The Life and Correspondence of William Buckland*. London: John Murray.

[Grant, Robert Edmond]. 1830. "Baron Cuvier." *Foreign Review* 5:342–80.

"*Great Artists and Great Anatomists* by R. Knox." 1852. *Athenæum*, no. 1297, 935–37.

"Great Expectations." 1861. *Sharpe's London Magazine*, n.s., 20:218–20.

Hamilton, William John. 1866. "Anniversary Address Delivered by the President." *Proceedings of the Geological Society* 22:xxx–cii.

Harrington, Bernard J. 1883. *Life of William E. Logan*. London: Sampson Low.

Hatton, Joseph. 1882. *Journalistic London: A Series of Sketches of Famous Pens and Papers of the Day*. London: Sampson Low.

Hawkins, B. Waterhouse. 1854. "On Visual Education as Applied to Geology." *Journal of the Society of Arts* 2:444–47.

————. 1858. "Geology, and the Extinct Animals of the Ancient World." *Popular Lecturer*, n.s., 3:193–207.

————. 1860. *A Comparative View of the Human and Animal Frame*. London: Chapman and Hall.

————. 1876. *The Artistic Anatomy of the Dog and Deer*. London: Winsor and Newton.

Haworth, Erasmus. 1896–1908. *The University Geological Survey of Kansas*. 9 vols. Topeka, KS: W. Y. Morgan.

Hitchcock, Edward. 1836. "Ornithichnology." *American Journal of Science* 29:307–40.

————. 1844. "Report on Ichnolithology, or Fossil Footmarks." *American Journal of Science* 47:292–322.

[————]. 1845. "Bibliographical Notices: On Dinornis (I)." *American Journal of Science* 48:194–201.

————. 1858. *Ichnology of New England*. Boston: William White.

————. 1865. *Supplement to the Ichnology of New England*. Boston: Wright and Potter.

Holcombe, William H. 1890. *A Mystery of New Orleans: Solved by New Methods*. Philadelphia: J. B. Lippincott.

Homo. 1815. "On Jameson's Preface to Cuvier's Theory of the Earth." *Philosophical Magazine* 46:225–29.

Hood, Tom, and Frances Freeling Broderip, eds. 1860. *Memorials of Thomas Hood*. 2 vols. London: Edward Moxon.

"How to Become an Orator." 1884. *Pall Mall Gazette*, 24 October, 1–2.

"Humourists: Dickens and Thackeray." 1848. *English Review* 10:257–75.

Hunt, James. 1863. *On the Negro's Place in Nature*. London: Trübner.

————. 1864. *The Negro's Place in Nature*. New York: Van Evrie, Horton.

————. 1866. "On the Application of the Principle of Natural Selection to Anthropology." *Anthropological Review* 4:320–40.

Hunt, Robert. 1856. *A Descriptive Guide to the Museum of Practical Geology*. London: Her Majesty's Stationery Office.

Hutchinson, H. N. 1892. *Extinct Monsters*. London: Chapman and Hall.

————. 1894. *Creatures of Other Days*. London: Chapman and Hall.

————. 1900. "Prehistoric Monsters." *Pearson's Magazine* 10:578–87.

————. 1910. *Extinct Monsters*. 6th ed. London: Chapman and Hall.

————. 1917. "Observations on the Reconstructed Skeleton of the Dinosaurian Reptile *Diplodocus carnegiei*." *Geological Magazine*, n.s., 4:356–70.

Huxley, Leonard. 1900. *Life and Letters of Thomas Henry Huxley*. 2 vols. London: Macmillan.

————. 1918. *Life and Letters of Sir Joseph Dalton Hooker*. 2 vols. London: John Murray.

Huxley, Thomas Henry. 1853. "On the Morphology of the Cephalous Mollusca." *Philosophical Transactions of the Royal Society* 143:29–65.

[————]. 1854a. "Contemporary Literature: Science." *Westminster Review*, n.s., 5:254–70.

[————]. 1854b. "Schamyl, the Prophet-Warrior of the Caucasus." *Westminster Review*, n.s., 5:480–519.

[————]. 1854c. "The Vestiges of Creation." *British and Foreign Medico-Chirurgical Review* 13:425–39.

[————]. 1855a. "Contemporary Literature: Science." *Westminster Review*, n.s., 7:551–63.

[————]. 1855b. "Contemporary Literature: Science." *Westminster Review*, n.s., 8:240–63.

————. 1856a. "Lectures on General Natural History." *Medical Times and Gazette* 12:481–84.

————. 1856b. "On Natural History, as Knowledge, Discipline, and Power." *Notices of the Proceedings of the Royal Institution* 2:187–95.

————. 1856c. "On the Method of Palæontology." *Annals and Magazine of Natural History*, n.s., 18:42–54.

[————]. 1856d. "Owen and Rymer Jones on Comparative Anatomy." *British and Foreign Medico-Chirurgical Review* 35:1–27.

————. 1861. "On a New Species of *Macrauchenia*." *Quarterly Journal of the Geological Society* 17:73–84.

————. 1863. *On Our Knowledge of the Causes of the Phenomena of Organic Nature*. London: Robert Hardwicke.

————. 1864. *Lectures on the Elements of Comparative Anatomy*. London: John Churchill.

———. 1865. "Explanatory Preface." In Thomas Henry Huxley and Robert Etheridge, *A Catalogue of the Collection of Fossils in the Museum of Practical Geology*, v–lxxix. London: Her Majesty's Stationery Office.

———. 1866. "On the Advisableness of Improving Natural Knowledge." *Fortnightly Review* 3:626–37.

———. 1868. "Remarks upon *Archæopteryx lithographica*." *Proceedings of the Royal Society* 16:243–48.

———. 1869a. *An Introduction to the Classification of Animals*. London: John Churchill.

———. 1869b. "On the Animals Which are Most Nearly Intermediate between Birds and Reptiles" [1868]. *Notices of the Proceedings of the Royal Institution* 5:278–87.

———. 1870a. "The Anniversary Address of the President." *Quarterly Journal of the Geological Society* 26: xxix–lxiv.

———. 1870b. "On the Educational Value of the Natural Historical Sciences" [1854]. In *Lay Sermons, Addresses and Reviews*, 81–103. London: Macmillan.

———. 1871. "Coral and Coral Reefs." In *Science Lectures for the People: First and Second Series*, 3–17. Manchester, UK: John Heywood.

———. 1877a. *The Crocodilian Remains Found in the Elgin Sandstone*. Memoirs of the Geological Survey Monograph 3. London: Her Majesty's Stationery Office.

———. 1877b. "The Hypothesis of Evolution." In *American Addresses*, 31–70. London: Macmillan.

———. 1880. "On the Method of Zadig: Retrospective Prophecy as a Function of Science." *Nineteenth Century* 7:929–40.

———. 1881. "The Rise and Progress of Palæontology." *Nature* 24:452–55.

———. 1892. *Essays upon Some Controverted Questions*. London: Macmillan.

———. 1893–94. *Collected Essays*. 9 vols. London: Macmillan.

———. 1894. "Owen's Position in the History of Anatomical Science." In Richard S. Owen, *The Life of Richard Owen*, 2:273–332. 2 vols. London: John Murray.

———. 1898–1902. *The Scientific Memoirs of Thomas Henry Huxley*. Edited by Michael Foster and E. Ray Lankester. 4 vols. London: Macmillan.

"Infidelity in Disguise—Geology." 1837. *Church of England Quarterly Review* 2:450–91.

"Intelligence and Miscellaneous Articles." 1805. *Philosophical Magazine* 21:183–200.

Isler, M. 1879. *Briefe von Benj. Constant . . . und vielen Anderen. Auswahl aus dem handschriftlichen Nachlasse des Ch. de Villers*. Hamburg: Otto Meissner.

James, Henry. 1869. "Gabrielle De Bergerac." *Atlantic Monthly* 24:55–71.

———. 1886. *The Bostonians*. London: Macmillan.

———. 1907. *The American*. New York ed. New York: Charles Scribner, [1877].

———. 1908. *The Tragic Muse*. 2 vols. New York ed. New York: Charles Scribner, [1890].

———. 1914. *Notes of a Son and Brother*. London: Macmillan.

Jameson, Laurence. 1854. "Biographical Memoir of the Late Professor Jameson." *Edinburgh New Philosophical Journal* 57:1–49.

Jameson, Robert. 1813. Preface to Georges Cuvier, *Essay on the Theory of the Earth*, translated by Robert Kerr, edited by Robert Jameson, v–ix. Edinburgh: William Blackwood.

———. 1815. Preface to Georges Cuvier, *Essay on the Theory of the Earth*, translated by Robert Kerr, edited by Robert Jameson, v–x. 2nd ed. Edinburgh: William Blackwood.

———. 1817. Preface to Georges Cuvier, *Essay on the Theory of the Earth*, translated by Robert Kerr, edited by Robert Jameson, v–xv. 3rd ed. Edinburgh: William Blackwood.

———. 1827. Preface to Georges Cuvier, *Essay on the Theory of the Earth*, translated by Robert Kerr, edited by Robert Jameson, v–viii. 5th ed. Edinburgh: William Blackwood.

Jefferson, Thomas. 1799. "A Memoir on the Discovery of Certain Bones of a Quadruped of the Clawed Kind in the Western Parts of Virginia." *Transactions of the American Philosophical Society* 4:246–60.

Jerdan, William. 1866. *Men I Have Known.* London: George Routledge.

"John Allen, Esq." 1843. *Gentleman's Magazine*, n.s., 20:96–97.

Jones, Owen. 1854. *An Apology for the Colouring of the Greek Court in the Crystal Palace.* London: Bradbury and Evans.

Keith, Arthur. 1950. *Autobiography.* London: Watts.

Kidd, John. 1824. *An Introductory Lecture to a Course in Comparative Anatomy Illustrative of Paley's Natural Theology.* Oxford: Oxford University Press.

———. 1833. *On the Adaptation of External Nature to the Physical Condition of Man.* London: William Pickering.

Kirby, William. 1835. *On the History, Habits and Instincts of Animals.* 2 vols. London: William Pickering.

Knight, Charles. 1854. *The Old Printer and the Modern Press.* London: John Murray.

———. 1864. *Passages of a Working Life during Half a Century.* 3 vols. London: Bradbury and Evans.

Knox, Robert. 1831. "Observations on the Stomach of the Peruvian Lama." *Transactions of the Royal Society of Edinburgh* 11:479–98.

———. 1839. "Lectures by M. De Blainville on Comparative Osteology." *Lancet*: 217–22.

———. 1850. *The Races of Men.* London: Henry Renshaw.

———. 1852. *Great Artists and Great Anatomists.* London: John Van Voorst.

———. 1855. "Some Remarks on the Aztecque and Bosjieman Children." *Lancet*: 357–60.

———. 1856. "On Organic Harmonies: Anatomical Co-relations, and Methods of Zoology and Paleontology." *Lancet*: 245–47, 270–71, and 297–300.

Krefft, Gerard. 1872. "Natural History: *A Cuvierian Principle in Palaeontology* . . . by Professor Owen." *Sydney Mail*, 18 May, 626–27.

"Landor's Last Fruit off an Old Tree." 1854. *New Monthly Magazine* 100:116–20.

Lascelles, Rowley. 1821. *The University and City of Oxford.* London: Sherwood, Neely and Jones.

"The Late Dr. Knox." 1862. *Medical Times and Gazette*, 683–85.

"The Late Rev. Dr. Buckland." 1856. *Times*, 16 August, 12.

Lawrence, William. 1816. *An Introduction to Comparative Anatomy and Physiology.* London: J. Callow.

———. 1819. *Lectures on Physiology, Zoology, and the Natural History of Man.* London: J. Callow.

———. 1823. *Lectures on Comparative Anatomy, Physiology, Zoology, and the Natural History of Man.* London: R. Carlile.

"Lawrence's Lectures." 1822. *Republican* 6:959.

"Lectures on the Skeleton." 1839. *Lancet*: 4–6.

Lee, Mrs R. 1833. *Memoirs of Baron Cuvier.* London: Longman.

[Leech, John]. 1855. "A Visit to the Antediluvian Reptiles at Sydenham." *Punch* 28:xvii.

"Letter from Richard Owen. . . ." 1843. *American Journal of Science* 44:341–45.

[Lhotsky, Jan]. 1840. "Fossil Osteography, Zoology and Geology." *Foreign Quarterly Review* 25:318–36.

"Literary Criticism." 1831. *Edinburgh Literary Journal* 5:19–28.

"Literary Notices." 1863. *Newcastle Courant*, 20 November, 5.

"Literature." 1851. *Tait's Edinburgh Magazine* 18:707–11.

"Literature." 1853. *Crystal Palace Herald* 1:17.

"Literature." 1863. *Athenæum*, no. 1871, 295–97.

Lockyer, Norman. 1871. "Science Lectures for the People." *Nature* 4:81.

Logan, W. E. 1851. "Occurrence of a Track and Foot-prints of an Animal in the Potsdam Sandstone of Lower Canada." *Quarterly Journal of the Geological Society* 7:247–50.

Lonsdale, Henry. 1870. *A Sketch of the Life and Writings of Robert Knox*. London: Macmillan.

[Lowell, James Russell]. 1847. "Disraeli's *Tancred*." *North American Review* 65:201–24.

Lydekker, Richard. 1891. *Catalogue of the Fossil Birds in the British Museum (Natural History)*. London: British Museum.

Lyell, Charles. 1830–33. *Principles of Geology*. 3 vols. London: John Murray.

———. 1838. "Anniversary Address Delivered by the President" [1837]. *Proceedings of the Geological Society* 2:479–523.

———. 1851. "Anniversary Address of the President." *Quarterly Journal of the Geological Society* 7:xxv–lxxvi.

Lyell, Katharine M., ed. 1881. *Life, Letters and Journals of Sir Charles Lyell, Bart*. 2 vols. London: John Murray.

———, ed. 1890. *Memoir of Leonard Horner*. 2 vols. London: Women's Printing Society.

Mackay, Charles. 1877. *Forty Years' Recollections of Life, Literature, and Public Affairs*. 2 vols. London: Chapman and Hall.

Mackie, S. J. 1858a. "The Geologist." *Geologist* 1:1–5.

———. 1858b. "Letter from the Rev. C. Kingsley." *Geologist* 1:75–77.

———. 1858c. "Preface." *Geologist* 1:i.

———. 1859. "Notes and Queries." *Geologist* 2:37–43.

———. 1863a. "The Aeronauts of the Solenhofen Age." *Geologist* 6:1–8.

———. 1863b. "Fossil Birds." *Popular Science Review* 2:354–64.

———. 1863c. "Notes and Queries." *Geologist* 6:64–70.

Macleay, W. S. 1859. "The Native 'Lion' of Australia." *Sydney Morning Herald*, 1 January, 5.

"Manners, Traditions and Superstitions of the Shetlanders." 1846. *Fraser's Magazine* 33:631–48.

Mantell, Gideon. 1825. "Notice on the Iguanodon, a Newly Discovered Fossil Reptile." *Philosophical Transactions of the Royal Society* 115:179–86.

———. 1844. *The Medals of Creation*. 2 vols. London: Henry G. Bohn.

———. 1846. "On the Fossil Remains of Birds in the Wealden Strata of the South-East of England." *Quarterly Journal of the Geological Society* 2:104–6.

———. 1848a. "On the Fossil Remains of Birds Collected in Various Parts of New Zealand." *Quarterly Journal of the Geological Society* 4:225–38.

———. 1848b. *The Wonders of Geology*. 6th ed. 2 vols. London: Henry G. Bohn.

———. 1850a. "On the Pelorosaurus; an Undescribed Gigantic Terrestrial Reptile." *Philosophical Transactions of the Royal Society* 140:379–90.

———. 1850b. "On the Pelorosaurus; an Undescribed Gigantic Terrestrial Reptile." *Abstracts of the Papers Communicated to the Royal Society* 5:921–22.

———. 1851. *Petrifactions and Their Teachings*. London: Henry G. Bohn.

———. 1854. *The Medals of Creation*. 2nd ed. London: Henry G. Bohn.

Marcou, Jules. 1896. *Life, Letters, and Works of Louis Agassiz*. 2 vols. New York: Macmillan.

Marsh, O. C. 1875. "New Order of Eocene Mammals." *American Journal of Science*, n.s., 9:221.

———. 1879. *History and Methods of Palæontological Discovery*. New Haven, CT: Tuttle, Morehouse, and Taylor.

[Martineau, Harriet]. 1854. "The Crystal Palace." *Westminster Review*, n.s., 6:534–50.

Martineau, James. 1888. *A Study of Religion*. 2 vols. Oxford: Clarendon Press.

[Masson, David]. 1862. "Universal Information and *The English Cyclopædia*." *Macmillan's Magazine* 5:357–70.

McCosh, James, and George Dickie. 1857. *Typical Forms and Special Ends in Creation*. 2nd ed. Edinburgh: Thomas Constable.

"Meeting of the British Association at Cork." 1843. *Illustrated London News* 3:132–34.

"Meeting of the British Association for the Advancement of Science, at Hull." 1853. *Illustrated Crystal Palace Gazette* 1:5.

"The Megatherium." 1849. *Church of England Magazine* 26:106–7.

"Memorial of Sir Richard Owen." 1893. *Nature* 47:307–9.

Meyer, Hermann von. 1832. *Palaeologica*. Frankfurt: Siegmund Schmerber.

———. 1862. "On the *Archæopteryx lithographica*." *Annals and Magazine of Natural History*, n.s., 9:366–70.

Milne-Edwards, Henri. 1841–42. *Cours élémentaire d'histoire naturelle (zoologie)*. Paris: Fortin, Masson.

———. 1856. *A Manual of Zoology*. Translated by R. Knox. London: Henry Renshaw.

———. 1863. *A Manual of Zoology*. Translated by R. Knox. Edited by C. Carter Blake. 2nd ed. London: Henry Renshaw.

"Mineral Kingdom. Organic Remains." 1833. *Penny Magazine* 2:347–49.

"Miscellanea." 1844. *Athenæum*, no. 871, 629–30.

"Miscellanies." 1836. *Medico-Chirurgical Review*, n.s., 24:289–90.

"Miscellanies." 1839. *American Journal of Science* 37:369–71.

[Mitford, C.]. 1891. *Letters and Reminiscences of the Rev. John Mitford*. London: Sampson Low.

[Morley, Henry, and William Henry Wills]. 1852. "A Flight with the Birds." *Household Words* 5:381–84.

[Mozley, James Bowling]. 1869. "The Argument of Design." *Quarterly Review* 127:134–76.

"Mr. Huxley on the Negro-Question." 1864. *Reader* 3:287–88.

"Mr. Warburton's Bill." 1829. *Lancet*: 785–89.

Murchison, Charles. 1868. "Biographical Sketch." In *Palæontological Memoirs and Notes of the Late Hugh Falconer*, edited by Charles Murchison, 1:xxiii–liii. 2 vols. London: Robert Hardwicke.

Murchison, Roderick I. 1846. "Anniversary Address of the President" [1843]. *Proceedings of the Geological Society* 4:65–151.

———. 1865. "Notice." In Thomas H. Huxley and Robert Etheridge, *A Catalogue of the Collection of Fossils in the Museum of Practical Geology*, i. London: Her Majesty's Stationery Office.

"The Mysterious Tanks." 1916. *Times*, 19 September, 10.

"Names of Novels." 1867. *Pall Mall Gazette*, 11 September, 10.

"The National Cyclopædia of Useful Knowledge." 1847. *Hogg's Weekly Instructor* 5:315–17.

"Natural History.—Baron Cuvier." 1835. *Poor Man's Guardian* 3:454–55.

"Nature's Hints to Inventors." 1868. *Boy's Own Magazine* 10:33–39.

"A Newcastle Paper in 1765–6." 1852. *Chambers's Journal*, n.s., 17:105–7.

Newman, John Henry. 1837. *Lectures on the Prophetical Office of the Church*. London: J. G. & F. Rivington.

"Noctes Ambrosianæ." 1829. *Blackwood's Edinburgh Magazine* 25:371–400.

"Notice Concerning the Skeleton of a Very Large Species of Quadruped." 1796. *Monthly Magazine* 2:637–38.

Oliphant, Margaret. 1897. *Annals of a Publishing House: William Blackwood and His Sons*. 3 vols. Edinburgh: William Blackwood.

"One Hundred and First Anniversary Dinner." 1855. *Journal of the Society of Arts* 3:584–89.

"On Systems and Methods in Natural History." 1829. *Quarterly Review* 41:302–27.

"The Opening of the Crystal Palace." 1854. *Punch* 26:245.

Osborn, Henry Fairfield. 1893. *The Rise of the Mammalia in North America.* Boston: Ginn.

———. 1911. "Palaeontology." In *Encyclopædia Britannica,* 20:579–91. 11th ed. 29 vols. Cambridge: Cambridge University Press.

———. 1917. *The Origin and Evolution of Life.* New York: Charles Scribner's Sons.

"Our Weekly Gossip." 1838. *Athenæum,* no. 578, 841.

"Our Weekly Gossip." 1862. *Athenæum,* no. 1831, 699–700.

Owen, Richard. 1839. "Fossil Mammalia." In *Zoology of the Voyage of H.M.S. Beagle,* edited by Charles Darwin, vol. 1. 5 vols. London: Smith, Elder.

———. 1840a. "Exhibition of a Bone of an Unknown Struthious Bird from New Zealand." *Proceedings of the Zoological Society* 7:169–71.

———. 1840b. "On the Bone of an Unknown Struthian Bird of Large Size from New Zealand." *Annals of Natural History* 5:166–68.

———. 1842a. *Description of the Skeleton of an Extinct Gigantic Sloth.* London: John Van Voorst.

———. 1842b. "Notice of a Fragment of the Femur of a Gigantic Bird of New Zealand." *Transactions of the Zoological Society* 3:29–32.

———. 1842c. "Observations on the Fossils Representing the *Thylacotherium prevostii,* Valenciennes." *Transactions of the Geological Society,* n.s., 6:47–57.

———. 1843a. "On *Dinornis,* an Extinct Genus of Tridactyle Struthious Birds (I)." *Transactions of the Zoological Society* 3:235–71.

———. 1843b. "On *Dinornis novæ-zealandiæ.*" *Proceedings of the Zoological Society* 11:8–10 and 11:144–46.

———. 1846a. *History of British Fossil Mammals, and Birds.* London: John Van Voorst.

———. 1846b. *Lectures on the Comparative Anatomy and Physiology of the Vertebrate Animals,* part 1, "Fishes." London: Longman.

———. 1846c. "On *Dinornis,* an Extinct Genus of Tridactyle Struthious Birds (II)." *Transactions of the Zoological Society* 3:307–30.

———. 1848a. "Description of the Teeth and Portions of Jaws of Two Extinct Anthracotheroid Quadrupeds." *Quarterly Journal of the Geological Society* 4:103–41.

———. 1848b. "On *Dinornis,* an Extinct Genus of Tridactyle Struthious Birds (III)." *Transactions of the Zoological Society* 3:345–77.

———. 1848c. *On the Archetype and Homologies of the Vertebrate Skeleton.* London: John Van Voorst.

———. 1849–84. *History of British Fossil Reptiles.* 4 vols. London: Cassel.

———. 1851. "Description of the Impression on the Potsdam Sandstone." *Quarterly Journal of the Geological Society* 7:250–52.

———. 1852a. "Description of the Impressions and Foot-prints of the Protichnites from the Potsdam Sandstone of Canada." *Quarterly Journal of the Geological Society* 8:214–25.

[———]. 1852b. "Dr. Mantell." *Literary Gazette,* no. 1869, 842.

———. 1852c. "On the Raw Materials from the Animal Kingdom." In *Lectures on the Results of the Great Exhibition of 1851,* 77–131. London: David Bogue.

[———]. 1853. "Justice to the Hyæna." *Household Words* 6:373–77.

———. 1854. *Geology and Inhabitants of the Ancient World.* London: Bradbury and Evans.

———. 1857. "On the Affinities of the *Stereognathus ooliticus* (Charlesworth)." *Quarterly Journal of the Geological Society of London* 13:1–11.

————. 1858. "Palæontology." *Encyclopædia Britannica*, 17:91–176. 8th ed. 21 vols. Edinburgh: Adam and Charles Black.

————. 1860. *Palæontology*. Edinburgh: Adam and Charles Black.

————. 1861. *Memoir on the Megatherium*. London: Williams and Norgate.

————. 1863a. *Monograph on the Aye-aye*. London: Taylor and Francis.

————. 1863b. "On the *Archeopteryx* of Von Meyer." *Philosophical Transactions of the Royal Society* 153:33–46.

————. 1863c. "On the Character of the Aye-aye as a Test of the Lamarckian and Darwinian Hypothesis of the Transmutation and Origin of Species." In *Report of the Thirty-Second Meeting of the British Association for the Advancement of Science*, 114–16. London: John Murray.

————. 1866–68. *On the Anatomy of Vertebrates*. 3 vols. London: Longmans, Green.

————. 1870. "On the Fossil Mammals of Australia (I)." *Philosophical Transactions of the Royal Society* 160:519–78.

————. 1871a. *Monograph on the Fossil Mammalia of the Mesozoic Formations*. London: Palæontographical Society.

————. 1871b. "On the Fossil Mammals of Australia (IV)." *Philosophical Transactions of the Royal Society* 161:213–66.

————. 1879a. *Memoirs on the Extinct Wingless Birds of New Zealand*. 2 vols. London: John Van Voorst.

————. 1879b. "On the Extinct Animals of the Colonies of Great Britain." *Popular Science Review* 3:253–73.

Owen, Richard S. 1894. *The Life of Richard Owen*. 2 vols. London: John Murray.

Packard, Alpheus S. 1901. *Lamarck: The Founder of Evolution*. New York: Longmans, Green.

Paley, William. 1802. *Natural Theology*. London: R. Faulder.

Peck, Frederick B. 1904. "The Atlantosaur and Titanotherium Beds of Wyoming." *Proceedings of the Wyoming Historical and Geological Society* 8:25–41.

"Pencillings of Eminent Medical Men. Dr. Knox, of Edinburgh." 1844. *Medical Times* 10:245–46.

"The Penny Cyclopædia." 1834. *Mechanics' Magazine* 22:123–25.

"The People's Palace." 1854. *Leisure Hour* 3:231–35.

"The People's Palace and the Press." 1853. *Crystal Palace Herald* 1:49–50.

Phillips, Samuel. 1854. *Guide to the Crystal Palace and Park*. London: Bradbury and Evans.

————. 1855. *Guide to the Crystal Palace and Park*. 5th ed. London: Bradbury and Evans.

Pike, Luke Owen. 1866. *The English and Their Origins*. London: Longmans, Green.

————. 1868. "What Is a Teuton?" *Anthropological Review* 6:246–57.

[Playfair, John?]. 1814. "Cuvier on the Theory of the Earth." *Edinburgh Review* 22:454–75.

Poe, Edgar A. 1841. "The Murders in the Rue Morgue." *Graham's Magazine* 18:166–79.

Pope, Alexander. 1734. *An Essay on Man*. London: Lawton Gilliver.

Pound, Ezra. 1918. "Henry James." *Little Review* 3:5–41.

Powell, J. H. 1870. *William Denton, the Geologist and Radical*. Boston: J. H. Powell.

Prestwich, Grace. 1901. *Essays: Descriptive and Biographical*. Edinburgh: William Blackwood.

"Proceedings of Geological Societies." 1863. *Geologist* 6:31–35.

"Proceedings of Societies." 1832. *New Monthly Magazine* 35:350–51.

Proctor, Richard. 1870. *Other Worlds Than Ours*. London: Longmans, Green.

"Professor Huxley on Classification." 1869. *Popular Science Review* 8:284–85.

"Professor Owen's Lectures." 1857. *Illustrated Times* 4:251–52.

"Professor Owen's Lectures at the Museum of Practical Geology." 1857. *Saturday Review* 3:220–21.

"Pterodactyles." 1870. *Popular Science Review* 9:293–94.

Pycroft, James. 1844. *A Course of English Reading.* London: Longman.

Railway Readings. 1847. Oxford: J. Vincent.

Rees, Arthur J. 1919. *The Shrieking Pit.* London: Lane.

[Reeve, Henry]. 1855. "The Results of the Campaign." *Edinburgh Review* 102:572–92.

Report from the Select Committee on Scientific Instruction. 1868. London: Her Majesty's Stationery Office.

Report of the First and Second Meetings of the British Association for the Advancement of Science. 1833. London: John Murray.

"Retrospect of Domestic Literature—Medicine, Surgery, &c." 1803. *Monthly Magazine* 14: 588–90.

Roberts, Mary. 1837. *The Progress of Creation.* London: Smith, Elder.

Roscoe, Henry Enfield. 1871a. Preface [1867] to *Science Lectures for the People: First and Second Series,* [iii]. Manchester, UK: John Heywood.

———. 1871b. Preface to second series of *Science Lectures for the People: First and Second Series,* [v]. Manchester, UK: John Heywood.

———. 1906. *The Life and Experiences of Sir Henry Enfield Roscoe.* London: Macmillan.

Rossetti, William Michael. 1867. "The Epochs of Art as Represented in the Crystal Palace" [1854]. In *Fine Art, Chiefly Contemporary,* 51–92. London: Macmillan.

Routledge's Guide to the Crystal Palace and Park at Sydenham. 1854. London: George Routledge.

"Royal Institution Lectures." 1875. *Illustrated London News* 66:375, 66:402, 66:447.

Rule, J. 1843. "New Zealand." *Polytechnic Journal* 9:1–13.

Ruskin, John. 1854. *The Opening of the Crystal Palace.* London: Smith, Elder.

———. 1857. *Notes on the Turner Gallery at Marlborough House.* London: Smith, Elder.

———. 1865. *Sesame and Lilies.* London: Smith, Elder.

[Russell, William Howard]. 1855. "The Fall of Sebastopol." *Times,* 11 September, 6.

"Science and Religion." 1859. *Builder* 17:35–36.

"Scientific." 1854. *Athenæum,* no. 1386, 625–26.

"Scientific." 1865. *Punch* 49:252.

"Scientific Meeting at Oxford." 1832. *Fraser's Magazine* 5:750–53.

Sedgwick, Adam. 1850. *On the Studies of the University of Cambridge.* 5th ed. London: John W. Parker.

Seeley, Harry Govier. 1870a. *The Ornithosauria.* Cambridge: Deighton, Bell.

———. 1870b. "Remarks on Prof. Owen's Monograph on Dimorphodon." *Annals and Magazine of Natural History,* n.s., 6:129–52.

———. 1872. "The Origin of the Vertebrate Skeleton." *Annals and Magazine of Natural History,* n.s., 10:21–45.

Serres, É. R. A. 1842. *Précis d'anatomie transcendante appliquée à la physiologie.* Paris: Charles Gosselin.

"Sherlock Holmes's Predecessor." 1894. *Derby Mercury,* 10 January, 6.

Silliman, Benjamin. 1843. "Ornithichnites of the Connecticut River and the Dinornis of New Zealand." *American Journal of Science* 45:185–87.

[Simms, William Gilmore]. 1844. "Writings of Cornelius Mathews." *Southern Quarterly Review* 6:307–42.

"Sir Richard Owen." 1895. *Church Quarterly Review* 40:345–71.

Smiles, Samuel. 1859. *Self-Help.* London: John Murray.

————. 1876. *Life of a Scotch Naturalist: Thomas Edward.* London: John Murray.

Sotheby, Samuel Leigh. 1855a. *A Few Words by Way of a Letter Addressed to the Directors of the Crystal Palace Company.* London: John Russell Smith.

————. 1855b. *A Few Words by Way of a Letter Addressed to the Shareholders of the Crystal Palace Company.* London: Effingham Wilson.

[Spencer, Herbert]. 1857. "The Ultimate Laws of Physiology." *National Review* 10:332–55.

————. 1904. *An Autobiography.* 2 vols. London: Williams and Norgate.

Spittal, Robert, and Robert Stevenson. 1829. "Report on the Impression Made on the Ground by the Foot of the Sow." *Edinburgh New Philosophical Journal,* n.s., 7:285–87.

Sternberg, Charles H. 1905. "*Protostega gigas* and Other Reptiles and Fishes from the Kansas Chalk." *Transactions of the Kansas Academy of Science* 19:123–28.

Strickland, Hugh. 1845. "Report on the Recent Progress and Present State of Ornithology." In *Report of the Fourteenth Meeting of the British Association for the Advancement of Science,* 170–221. London: John Murray.

Swainson, William. 1834. *A Preliminary Discourse on the Study of Natural History.* London: Longman.

Tennyson, Alfred. 1842. *Poems.* 2 vols. London: Edward Moxon.

Tennyson, Hallam. 1897. *Alfred Lord Tennyson: A Memoir.* 2 vols. London: Macmillan.

[Thackeray, William Makepeace]. 1846. *Mrs. Perkins's Ball.* London: Chapman and Hall.

[————]. 1847a. "Brighton in 1847." *Punch* 12:153.

[————]. 1847b. "Punch's Prize Novelists." *Punch* 12:166.

[————]. 1848. "Travels in London." *Punch* 14:95.

————. 1853–55. *The Newcomes.* 2 vols. London: Bradbury and Evans.

"Thackeray's Pendennis." 1850. *Spectator* 23:1213–15.

Thiers, Adolphe. 1851. *Discours sur le régime commercial de la France.* Paris: Paulin, L'Heureux.

Third Report of the Department of Science and Art. 1856. London: Her Majesty's Stationery Office.

Thomson, Thomas. 1870. "John Fleming." In *A Biographical Dictionary of Eminent Scotsmen,* edited by Robert Chambers and Thomas Thomson, 2:26–30. 3 vols. London: Blackie.

Timbs, John. 1858. *Curiosities of Science.* London: Kent.

Todd, Robert B., ed. 1835–59. *Cyclopædia of Anatomy and Physiology.* 5 vols. London: Longman.

"To the Editor." 1828. *Lion* 2:29–31.

"To the Public." 1853. *Illustrated Crystal Palace Gazette* 1:1.

Townshend, Chauncey Hare. 1840. *Facts in Mesmerism.* London: Longman.

Turner, G. Grey. 1946. *The Hunterian Museum: Yesterday and To-morrow.* London: Cassell.

"Twenty-Second Ordinary Meeting." 1854. *Journal of the Society of Arts* 2:447–49.

Van Evrie, J. H. 1861. *Negroes and Negro "Slavery."* New York: Van Evrie, Horton.

[Verplanck, Guilian Crommelin]. 1819. *The State Triumvirate, a Political Tale.* New York: J. Seymour.

Voltaire, 1748. *Zadig, ou, la Destinée. Histoire orientale.* [Paris]: n.p.

————. 1964. *"Zadig" and "L'Ingénu."* Translated by John Butt. Harmondsworth, UK: Penguin.

Wagner, Andreas. 1862. "On a New Fossil Reptile Supposed to be Furnished with Feathers." *Annals and Magazine of Natural History,* n.s., 9:261–67.

"The Waiting Room." 1857. *Train* 4:60.

Wallace, William. 1832. *The History of the Life and Reign of George the Fourth.* 3 vols. London: Longman.

Ward, Mary. 1859. *Telescopic Teachings.* London: Groombridge.

Ward, Thomas Humphry. 1884. *Humphry Sandwith: A Memoir.* London: Cassell.

Warren, Samuel. 1851. *The Lily and the Bee*. Edinburgh: William Blackwood.

———. 1853. *The Intellectual and Moral Development of the Present Age*. Edinburgh: William Blackwood.

———. 1854. *The Lily and the Bee*. 2nd ed. Edinburgh: William Blackwood.

Whewell, William. 1837. *History of the Inductive Sciences*. 3 vols. London: John W. Parker.

———. 1842. "Address to the Geological Society" [1839]. *Proceedings of the Geological Society* 3:61–98.

———. 1858. *History of Scientific Ideas*. 2 vols. London: John W. Parker.

"William John Broderip." 1859. *Athenæum*, no. 1637, 357.

Williams-Ellis, Clough, and A. Williams-Ellis. 1919. *The Tank Corps*. London: George Newnes.

[Williamson, William Crawford]. 1855. "Animal Organization." *London Quarterly Review* 4:351–77.

[Wills, William Henry, and George Augustus Sala]. 1854. "Fairyland in 'Fifty-Four." *Household Words* 8:313–17.

"A Wonderful Bone." 1853. *Chambers's Journal*, n.s., 19:176.

Wood, J. G. 1859–63. *The Illustrated Natural History*. 3 vols. London: Routledge.

Woodward, Henry. 1862. "On a Feathered Fossil from the Lithographic Limestone of Solenhofen." *Intellectual Observer* 2:313–19.

———. 1875. "Birds with Teeth." *Popular Science Review* 14:337–50.

[———]. 1890. *A Guide to the Exhibition Galleries of the Department of Geology and Palæontology in the British Museum (Natural History)*, part 1, "Fossil Mammals and Birds." London: Trustees of the Museum.

[———]. 1893. "Sir Richard Owen." *Geological Magazine*, n.s., 10:49–54.

"Working Men's College." 1880. *Lloyd's Weekly Newspaper*, 14 March, 1.

[Wynter, Andrew]. 1853. "The New Crystal Palace at Sydenham." *Fraser's Magazine* 48:607–22.

"Your Company Is Requested in the Iguanodon." 1854. *Illustrated Crystal Palace Gazette* 1:52–53.

Secondary Sources

Adrian, Arthur A. 1966. *Mark Lemon: First Editor of "Punch."* London: Oxford University Press.

Alberti, Samuel J. M. M. 2007. "The Museum Affect: Visiting Collections of Anatomy and Natural History." In *Science in the Marketplace: Nineteenth-Century Sites and Experiences*, edited by Aileen Fyfe and Bernard Lightman, 371–403. Chicago: University of Chicago Press.

Altick, Richard D. 1978. *The Shows of London*. Cambridge, MA: Belknap Press.

———. 1980. "Varieties of Reader Response: The Case of *Dombey and Son*." *Yearbook of English Studies* 10:70–94.

———. 1991. *The Presence of the Present: Topics of the Day in the Victorian Novel*. Columbus: Ohio State University Press.

Anderson, Katharine. 2005. *Predicting the Weather: Victorians and the Science of Meteorology*. Chicago: University of Chicago Press.

Appel, Toby A. 1987. *The Cuvier-Geoffroy Debate: French Biology in the Decades before Darwin*. New York: Oxford University Press.

Atmore, Henry. 2004. "Utopia Limited: The Crystal Palace Company and Joint-Stock Politics, 1854–1856." *Journal of Victorian Culture* 9:189–215.

Auerbach, Jeffrey A. 1999. *The Great Exhibition of 1851: A Nation on Display*. New Haven, CT: Yale University Press.

Bailey, Isabel. 1996. *Herbert Ingram Esq. MP, of Boston*. Boston, UK: Richard Kay.

Baldwin, Melinda. 2012. "The Shifting Ground of *Nature:* Establishing an Organ of Scientific Communication in Britain, 1869–1900." *History of Science* 50:125–54.

Barrett, Paul H., Peter J. Gautry, Sandra Herbert, David Kohn, and Sidney Smith, eds. 1987. *Charles Darwin's Notebooks, 1836–1844.* Ithaca, NY: Cornell University Press.

Barrow, Logie. 1986. *Independent Spirits: Spiritualism and English Plebeians, 1850–1910.* London: Routledge.

Barton, Ruth. 1983. "Evolution: The Whitworth Gun in Huxley's War for the Liberation of Science from Theology." In *The Wider Domain of Evolutionary Thought,* edited by David Oldroyd and Ian Langham, 261–87. Dordrecht, Netherlands: Kluwer.

———. 1998a. "'Huxley, Lubbock, and Half a Dozen Others': Professionals and Gentlemen in the Formation of the X Club, 1851–1864." *Isis* 89:410–44.

———. 1998b. "Just before *Nature:* The Purposes of Science and the Purposes of Popularization in Some English Popular Science Journals of the 1860s." *Annals of Science* 55:1–33.

———. 2000. "Haast and the Moa: Reversing the Tyranny of Distance." *Pacific Science* 54:251–63.

Bartrip, Peter William. 2009. "*Lancet.*" In *Dictionary of Nineteenth-Century Journalism,* edited by Laurel Brake and Marysa Demoor, 343–44. London: British Library.

Bates, A. W. 2010. *The Anatomy of Robert Knox: Murder, Mad Science and Medical Regulation in Nineteenth-Century Edinburgh.* Brighton, UK: Sussex Academic Press.

Baym, Nina. 2002. *American Women of Letters and the Nineteenth-Century Sciences.* New Brunswick, NJ: Rutgers University Press.

Beer, Gillian. 1983. *Darwin's Plots: Evolutionary Narrative in Darwin, George Eliot and Nineteenth-Century Fiction.* London: Routledge and Kegan Paul.

Bellon, Richard. 2007. "Science at the Crystal Focus of the World." In *Science in the Marketplace: Nineteenth-Century Sites and Experiences,* edited by Aileen Fyfe and Bernard Lightman, 301–35. Chicago: University of Chicago Press.

Bennett, Scott. 1982. "Revolutions in Thought: Serial Publication and the Mass Market for Reading." In *The Victorian Periodical Press: Samplings and Soundings,* edited by Joanne Shattock and Michael Wolff, 225–57. Leicester, UK: Leicester University Press.

Bibby, Cyril. 1959. *T. H. Huxley: Scientist, Humanist and Educator.* London: Watts.

Black, Joel. 1991. "The Hermeneutics of Extinction: Denial and Discovery in Scientific Literature." *Comparative Criticism* 13:147–69.

Bonython, Elizabeth, and Anthony Burton. 2003. *The Great Exhibitor: The Life and Work of Henry Cole.* London: V&A Publications.

Bowler, Peter J. 2003. *Evolution: The History of an Idea.* 3rd ed. Berkeley: University of California Press.

———. 2009. *Science for All: The Popularization of Science in Early Twentieth-Century Britain.* Chicago: University of Chicago Press.

Boylan, Patrick J. 1977. *The Falconer Papers, Forres.* Leicester, UK: Leicestershire Records Service.

———. 1984. "William Buckland, 1784–1856: Scientific Institutions, Vertebrate Palaeontology and Quaternary Geology." PhD diss., University of Leicester.

———. 2008. "Evans, Geology and Palaeontology." In *Sir John Evans, 1823–1908: Antiquity, Commerce and Natural Science in the Age of Darwin,* edited by Arthur MacGregor, 52–67. Oxford: Ashmolean Museum.

Bramwell, Valerie, and Robert M. Peck. 2008. *All in the Bones: A Biography of Benjamin Waterhouse Hawkins.* Philadelphia: Academy of Natural Sciences.

Braude, Ann. 1989. *Radical Spirits: Spiritualism and Women's Rights in Nineteenth-Century America*. Boston: Beacon Press.

Brinkman, Paul D. 2010. *The Second Jurassic Dinosaur Rush: Museums and Paleontology in America at the Turn of the Twentieth Century*. Chicago: University of Chicago Press.

Brock, W. H., and A. J. Meadows. 1998. *The Lamp of Learning: Two Centuries of Publishing at Taylor and Francis*. 2nd ed. London: Taylor and Francis.

Broks, Peter. 1996. *Media Science before the Great War*. London: Macmillan.

Broman, Thomas H. 1991. "J. C. Reil and the 'Journalization' of Science." In *The Literary Structure of Scientific Argument: Historical Studies*, edited by Peter Dear, 13–42. Philadelphia: University of Pennsylvania Press.

Brooke, John Hedley. 1989. "Scientific Thought and Its Meaning for Religion: The Impact of French Science on British Natural Theology, 1827–1859." *Revue de synthèse* 110:33–59.

———. 1991a. "Indications of a Creator: Whewell as Apologist and Priest." In *William Whewell: A Composite Portrait*, edited by Menachem Fisch and Simon Schaffer, 149–73. Oxford: Clarendon Press.

———. 1991b. *Science and Religion: Some Historical Perspectives*. Cambridge: Cambridge University Press.

Brown, Michael. 2014. "'Bats, Rats and Barristers': *The Lancet*, Libel and the Radical Stylistics of Early Nineteenth-Century Medicine." *Social History* 39:182–209.

Buckland, Adelene. 2013. *Novel Science: Fiction and the Invention of Nineteenth-Century Geology*. Chicago: University of Chicago Press.

Buick, T. Lindsay. 1936. *The Discovery of Dinornis: The Story of a Man, a Bone, and a Bird*. New Plymouth, NZ: Thomas Avery.

Burkhardt, Frederick H., et al., eds. 1985–. *The Correspondence of Charles Darwin*. 22 vols. Cambridge: Cambridge University Press.

Burkhardt, Richard W., Jr., 1977. *The Spirit of System: Lamarck and Evolutionary Biology*. Cambridge, MA: Harvard University Press.

Burnett, D. Graham. 2007. *Trying Leviathan: The Nineteenth-Century New York Court Case That Put the Whale on Trial and Challenged the Order of Nature*. Princeton, NJ: Princeton University Press.

Burns, James. 2007. "John Fleming and the Geological Deluge." *British Journal for the History of Science* 40:205–25.

Butcher, Barry W. 1988. "Darwin's Australian Correspondents: Deference and Collaboration in Colonial Science." In *Nature in Its Greatest Extent: Western Science in the Pacific*, edited by Roy MacLeod and Philip F. Rehbock, 139–57. Honolulu: University of Hawaii Press.

Camardi, Giovanni. 2001. "Richard Owen, Morphology and Evolution." *Journal of the History of Biology* 34:481–515.

Cantor, Geoffrey. 2011. *Religion and the Great Exhibition*. Oxford: Oxford University Press.

Carey, John, ed. 1995. *The Faber Book of Science*. London: Faber and Faber.

Chase, Malcolm. 2007. *Chartism: A New History*. Manchester, UK: Manchester University Press.

Cohen, Claudine. 2002. *The Fate of the Mammoth: Fossils, Myth, and History*. Translated by William Rodarmor. Chicago: University of Chicago Press.

———. 2011. *La méthode de Zadig: La trace, le fossile, la preuve*. Paris: Éditions du Seuil.

Coleman, William. 1964. *Georges Cuvier, Zoologist: A Study in the History of Evolution Theory*. Cambridge, MA: Harvard University Press.

Collie, Michael. 1991. *Huxley at Work: With the Scientific Correspondence of T. H. Huxley and the Rev. Dr. George Gordon of Birnie, Near Elgin*. London: Macmillan.

Collini, Stefan. 1991. *Public Moralists: Political Thought and Intellectual Life in Britain, 1850–1930*. Oxford: Clarendon Press.

Cooter, Roger, and Stephen Pumphrey. 1994. "Separate Spheres and Public Places: Reflections on the History of Science Popularization and Science in Popular Culture." *History of Science* 32:237–67.

Corsi, Pietro. 1988a. *The Age of Lamarck: Evolutionary Theories in France, 1790–1830*. Translated by Jonathan Mandelbaum. Berkeley: University of California Press.

———. 1988b. *Science and Religion: Baden Powell and the Anglican Debate, 1800–1860*. Cambridge: Cambridge University Press.

Cowan, C. F. 1969. "Notes on Griffith's *Animal Kingdom of Cuvier* (1824–1835)." *Journal of the Society for the Bibliography of Natural History* 5:137–40.

Cox, Robert S. 2003. *Body and Soul: A Sympathetic History of American Spiritualism*. Charlottesville: University Press of Virginia.

Csiszar, Alex Attila. 2010. "Broken Pieces of Fact: The Scientific Periodical and the Politics of Search in Nineteenth-Century France and Britain." PhD diss., Harvard University.

Curwen, E. Cecil, ed. 1940. *The Journal of Gideon Mantell, Surgeon and Geologist*. London: Oxford University Press.

Daintith, John, and Derek Gjertsen, eds. 1999. *A Dictionary of Scientists*. Oxford: Oxford University Press.

Dawson, Gowan. 2004. "The *Review of Reviews* and the New Journalism in Late Victorian Britain." In Geoffrey Cantor, Gowan Dawson, Graeme Gooday, Richard Noakes, Sally Shuttleworth, and Jonathan R. Topham, *Science in the Nineteenth-Century Periodical: Reading the Magazine of Nature*, 172–95. Cambridge: Cambridge University Press.

———. 2007. *Darwin, Literature and Victorian Respectability*. Cambridge: Cambridge University Press.

———. 2009. "*Annals and Magazine of Natural History*." In *Dictionary of Nineteenth-Century Journalism*, edited by Laurel Brake and Marysa Demoor, 16. London: British Library.

———. 2011. "Literary Megatheriums and Loose Baggy Monsters: Paleontology and the Victorian Novel." *Victorian Studies* 53:203–30.

Dawson, Gowan, and Bernard Lightman. 2011–12. General introduction to *Victorian Science and Literature*, edited by Gowan Dawson and Bernard Lightman, 1:vii–xix. 8 vols. London: Pickering and Chatto.

———, eds. 2014. *Victorian Scientific Naturalism: Community, Identity, Continuity*. Chicago: University of Chicago Press.

Dean, Dennis R. 1969. "Hitchcock's Dinosaur Tracks." *American Quarterly* 21:639–44.

———. 1999. *Gideon Mantell and the Discovery of Dinosaurs*. Cambridge: Cambridge University Press.

de Beer, Gavin. 1960. *The Sciences Were Never at War*. London: Thomas Nelson.

Dell, Sharon, ed. 1983. "Gideon Algernon Mantell's Unpublished Journal, June–November 1852." *Turnbull Library Record* 16:77–94.

Desmond, Adrian J. 1975. *The Hot-Blooded Dinosaurs: A Revolution in Palaeontology*. London: Blond and Briggs.

———. 1979. "Designing the Dinosaur: Richard Owen's Response to Robert Edmond Grant." *Isis* 70:224–34.

————. 1982. *Archetypes and Ancestors: Palaeontology in Victorian London, 1850–1875.* Chicago: University of Chicago Press.

————. 1989. *The Politics of Evolution: Morphology, Medicine, and Reform in Radical London.* Chicago: University of Chicago Press.

————. 1994. *Huxley: The Devil's Disciple.* London: Allen Lane.

————. 1997. *Huxley: Evolution's High Priest.* London: Allen Lane.

————. 2001. "Redefining the X Axis: 'Professionals,' 'Amateurs' and the Making of Mid-Victorian Biology." *Journal of the History of Biology* 34:3–50.

Desmond, Adrian J., and James Moore. 2009. *Darwin's Sacred Cause: Race, Slavery and the Quest for Human Origins.* London: Allen Lane.

Di Gregorio, Mario A., ed. 1990. *Charles Darwin's Marginalia.* New York: Garland.

Douglas-Fairhurst, Robert. 2004. *Victorian Afterlives: The Shaping of Influence in Nineteenth-Century Literature.* Oxford: Oxford University Press.

Eddy, Matthew D., and David Knight. 2006. Introduction to William Paley, *Natural Theology*, edited by Matthew D. Eddy and David Knight, ix–xxix. Oxford: Oxford University Press.

Endersby, Jim. 2003. "Escaping Darwin's Shadow." *Journal of the History of Biology* 36:385–403.

————. 2008. *Imperial Nature: Joseph Hooker and the Practices of Victorian Science.* Chicago: University of Chicago Press.

Evans, Joan. 1943. *Time and Chance: The Story of Arthur Evans and His Forebears.* London: Longmans, Green.

Feely, Catherine. 2009. "Scissors and Paste Journalism." In *Dictionary of Nineteenth-Century Journalism*, edited by Laurel Brake and Marysa Demoor, 561. London: British Library.

Finney, Colin. 1993. *Paradise Revealed: Natural History in Nineteenth-Century Australia.* Melbourne: Museum of Victoria.

Forgan, Sophie, and Graeme Gooday. 1996. "Constructing South Kensington: The Buildings and Politics of T. H. Huxley's Working Environments." *British Journal for the History of Science* 29:435–68.

Fox, Robert. 2012. *The Savant and the State: Science and Cultural Politics in Nineteenth-Century France.* Baltimore: Johns Hopkins University Press.

Frank, Lawrence. 2003. *Victorian Detective Fiction and the Nature of Evidence: The Scientific Investigations of Poe, Dickens, and Doyle.* Houndmills, UK: Palgrave Macmillan.

Fyfe, Aileen, 1997. "The Reception of William Paley's *Natural Theology* in the University of Cambridge." *British Journal for the History of Science* 30:321–35.

————. 2004. *Science and Salvation: Evangelical Popular Science Publishing in Victorian Britain.* Chicago: University of Chicago Press.

————. 2007. "Reading Natural History at the British Museum and the *Pictorial Museum*." In Fyfe and Lightman 2007, *Science in the Marketplace*, 196–230.

————. 2012. *Steam-Powered Knowledge: William Chambers and the Business of Publishing, 1820–1860.* Chicago: University of Chicago Press.

Fyfe, Aileen, and Bernard Lightman, eds. 2007. *Science in the Marketplace: Nineteenth-Century Sites and Experiences.* Chicago: University of Chicago Press.

Gerstner, Patsy A. 1970. "Vertebrate Paleontology, an Early Nineteenth-Century Transatlantic Science." *Journal of the History of Biology* 3:137–48.

Ghiselin, Michael T. 1969. *The Triumph of the Darwinian Method.* Berkeley: University of California Press.

Gillispie, Charles Coulston. 2004. *Science and Polity in France: The Revolutionary and Napoleonic Years.* Princeton, NJ: Princeton University Press.

Gould, Stephen Jay. 1983. "The Stinkstones of Oeningen." In *Hen's Teeth and Horse's Toes,* 94–106. New York: W. W. Norton.

———. 1995. "Poe's Greatest Hit." In *Dinosaur in a Haystack: Reflections in Natural History,* 173–86. New York: Harmony.

———. 2002. *The Structure of Evolutionary Theory.* Cambridge, MA: Belknap Press.

Goulet, Andrea. 2006. *Optiques: The Science of the Eye and the Birth of Modern French Fiction.* Philadelphia: University of Pennsylvania Press.

Gray, Valerie. 2006. *Charles Knight: Educator, Publisher, Writer.* Farnham, UK: Ashgate.

Green, Margaret. 2004. "Brodie, Peter Bellinger (1815–1897)." In *Oxford Dictionary of National Biography,* edited by H. C. G. Matthew and Brian Harrison, 7:780–81. 60 vols. Oxford: Oxford University Press.

Green, Michelle. 1986. "Stephen Jay Gould: Driven by a Hunger to Learn and to Write." *People* 25:109–14.

Gruber, Jacob W. 1987. "From Myth to Reality: The Case of the Moa." *Archives of Natural History* 14:339–52.

Gruber, Jacob W., and John C. Thackray. 1992. *Richard Owen Commemoration.* London: Natural History Museum.

Hales, S. J. 2006. "Recasting Antiquity: Pompeii and the Crystal Palace." *Arion* 14:99–133.

Harden, Edgar F. 1979. *The Emergence of Thackeray's Serial Fiction.* London: George Prior.

———, ed. 1994. *The Letters and Private Papers of William Makepeace Thackeray: A Supplement.* 2 vols. New York: Garland.

Hayward, Jennifer. 1997. *Consuming Pleasures: Active Audiences and Serial Fictions from Dickens to Soap Opera.* Lexington: University Press of Kentucky.

Hilton, Boyd. 2000. "The Politics of Anatomy and an Anatomy of Politics, c. 1825–1850." In *History, Religion and Culture: British Intellectual History, 1750–1950,* edited by Stefan Collini, Richard Whatmore, and Brian Young, 179–97. Cambridge: Cambridge University Press.

———. 2006. *A Mad, Bad, and Dangerous People? England, 1783–1846.* Oxford: Oxford University Press.

Holland, Julian. 1996. "Diminishing Circles: W. S. Macleay in Sydney, 1839–1865." *Historical Records of Australian Science* 11:119–47.

Hollis, Patricia. 1969. Introduction to *The "Poor Man's Guardian," 1831–1835,* 1:vii–xxxix. 4 vols. London: Merlin.

Holmes, Frederick L. 2003. "The *Long Durée* in the History of Science." *History and Philosophy of the Life Sciences* 25:463–70.

Hopwood, Nick, and Soraya De Chadarevian. 2004. "Dimensions of Modelling." In *Models: The Third Dimension in Science,* edited by Soraya De Chadarevian and Nick Hopwood, 1–15. Stanford, CA: Stanford University Press.

Hopwood, Nick, Simon Schaffer, and James Secord. 2010. "Seriality and Scientific Objects in the Nineteenth Century." *History of Science* 48:251–85.

House, Madeline, Graham Storey, and Kathleen Tillotson, eds. 1965–2002. *The Letters of Charles Dickens.* 12 vols. Oxford: Clarendon Press.

Howsam, Leslie. 2006. "Literary Replication: Jim Secord's *Victorian Sensation* and Models of Book History." *SHARP News* 15:3–4.

Hughes, Linda K., and Michael Lund. 1991. *The Victorian Serial*. Charlottesville: University Press of Virginia.

Irmscher, Christoph. 2013. *Louis Agassiz: Creator of American Science*. Boston: Houghton Mifflin Harcourt.

Jacyna, L. S. 1983. "Immanence or Transcendence: Theories of Life and Organization in Britain, 1790–1835." *Isis* 74:311–29.

———. 1994. *Philosophic Whigs: Medicine, Science and Citizenship in Edinburgh, 1789–1848*. London: Routledge.

Jaffe, Mark. 2000. *The Gilded Dinosaur: The Fossil War between E. D. Cope and O. C. Marsh and the Rise of American Science*. New York: Three Rivers Press.

Jenkins, Alice. 2007. *Space and the "March of Mind": Literature and the Physical Sciences in Britain, 1815–1850*. Oxford: Oxford University Press.

Johns, Adrian. 1998. *The Nature of the Book: Print and Knowledge in the Making*. Chicago: University of Chicago Press.

Johnson, Thomas H., ed. 1955. *The Poems of Emily Dickinson*. 3 vols. Cambridge, MA: Belknap Press.

Jordanova, Ludmilla. 1984. *Lamarck*. Oxford: Oxford University Press.

Keene, Melanie. 2008. "Object Lessons: Sensory Science Education, 1830–1870." PhD diss., University of Cambridge.

Kete, Kathleen. 2012. *Making Way for Genius: The Aspiring Self in France from the Old Regime to the New*. New Haven, CT: Yale University Press.

Kinraide, Rebecca. 2006. "The Society for the Diffusion of Useful Knowledge and the Democratization of Learning in Early Nineteenth-Century Britain." PhD diss., University of Wisconsin–Madison.

Klancher, Jon P. 1987. *The Making of English Reading Audiences, 1790–1832*. Madison: University of Wisconsin Press.

Law, Graham, and Robert L. Patten. 2009. "The Serial Revolution." In *The Cambridge History of the Book in Britain*, vol. 6, *1830–1914*, edited by David McKitterick, 144–71. Cambridge: Cambridge University Press.

Le Guyader, Hervé. 2004. *Geoffroy Saint-Hilaire: A Visionary Naturalist*. Translated by Marjorie Greene. Chicago: University of Chicago Press.

Lenoir, Timothy. 1989. *The Strategy of Life: Teleology and Mechanics in Nineteenth-Century German Biology*. Chicago: University of Chicago Press.

Letteney, Michael James. 1999. "Georges Cuvier, Transcendental Naturalist: A Study of Teleological Explanation in Biology." PhD diss., University of Notre Dame.

Levine, George. 1981. *The Realistic Imagination: English Fiction from "Frankenstein" to "Lady Chatterley."* Chicago: University of Chicago Press.

———. 2002. *Dying to Know: Scientific Epistemology and Narrative in Victorian England*. Chicago: University of Chicago Press.

Lightman, Bernard. 1987. *The Origins of Agnosticism: Victorian Unbelief and the Limits of Knowledge*. Baltimore: Johns Hopkins University Press.

———. 2007a. "Lecturing in the Spatial Economy of Science." In *Science in the Marketplace: Nineteenth-Century Sites and Experiences*, edited by Aileen Fyfe and Bernard Lightman, 97–132. Chicago: University of Chicago Press.

———. 2007b. *Victorian Popularizers of Science: Designing Nature for New Audiences*. Chicago: University of Chicago Press.

————. 2009. *Evolutionary Naturalism in Victorian Britain: The "Darwinians" and Their Critics.* Farnham, UK: Ashgate.

Livingstone, David N. 2003. *Putting Science in Its Place: Geographies of Scientific Knowledge.* Chicago: University of Chicago Press.

Livingstone, David N., and Charles W. J. Withers, eds. 2011. *Geographies of Nineteenth-Century Science.* Chicago: University of Chicago Press.

Lloyd, Rosemary. 2011. "Georges Cuvier and the Power of Rhetoric." *Nineteenth-Century Prose* 38:13–34.

Lohrli, Anne. 1973. *"Household Words": A Weekly Journal 1850–1859 Conducted by Charles Dickens.* Toronto: University of Toronto Press.

Lubbock, Percy, ed. 1920. *The Letters of Henry James.* 2 vols. London: Macmillan.

Lurie, Edward. 1988. *Louis Agassiz: A Life in Science.* Baltimore: Johns Hopkins University Press.

Markovits, Stefanie. 2009. *The Crimean War in the British Imagination.* Cambridge: Cambridge University Press.

Marr, Andrew. 2009. *The Making of Modern Britain.* London: Macmillan.

Marshall, Nancy Rose. 2007. "'A Dim World, Where Monsters Dwell': The Spatial Time of the Sydenham Crystal Palace Dinosaur Park." *Victorian Studies* 49:286–301.

McCarthy, Steve, and Mick Gilbert. 1994. *The Crystal Palace Dinosaurs.* London: Crystal Palace Foundation.

McCracken-Flesher, Caroline. 2012. *The Doctor Dissected: A Cultural Autopsy of the Burke and Hare Murders.* Oxford: Oxford University Press.

McDonald, Peter D. 1997. *British Literary Culture and Publishing Practice, 1880–1914.* Cambridge: Cambridge University Press.

McGill, Meredith L. 2003. *American Literature and the Culture of Reprinting, 1834–1853.* Philadelphia: University of Pennsylvania Press.

McMaster, R. D. 1978. "'An Honorable Emulation of the Author of *The Newcomes*': James and Thackeray." *Nineteenth-Century Fiction* 32:399–419.

————. 1991. *Thackeray's Cultural Frame of Reference: Allusion in "The Newcomes."* London: Macmillan.

Moore, James R. 1979. *The Post-Darwinian Controversies: A Study of the Protestant Struggle to Come to Terms with Darwin in Great Britain and America, 1870–1900.* Cambridge: Cambridge University Press.

Morgan, Peter F., ed. 1973. *The Letters of Thomas Hood.* Edinburgh: Oliver and Boyd.

Morrell, Jack, and Arnold Thackray. 1981. *Gentlemen of Science: Early Years of the British Association for the Advancement of Science.* Oxford: Clarendon Press.

————, eds. 1984. *Gentlemen of Science: Early Correspondence of the British Association for the Advancement of Science.* London: Royal Historical Society.

Morus, Iwan Rhys. 1998. *Frankenstein's Children: Electricity, Exhibition, and Experiment in Early-Nineteenth-Century London.* Princeton, NJ: Princeton University Press.

————. 2007. "'More the Aspect of Magic Than Anything Natural': The Philosophy of Demonstration." In *Science in the Marketplace: Nineteenth-Century Sites and Experiences,* edited by Aileen Fyfe and Bernard Lightman, 336–70. Chicago: University of Chicago Press.

————. 2010. "Placing Performance." *Isis* 101:775–78.

Nieuwland, Ilja. 2010. "The Colossal Stranger: Andrew Carnegie and *Diplodocus* Intrude European Culture, 1904–12." *Endeavour* 34:61–68.

Norton, Charles Eliot, ed. 1894. *Letters of James Russell Lowell.* 2 vols. London: Osgood, McIlvaine.

O'Connor, Ralph. 1999. "Mammoths and Maggots: Byron and the Geology of Cuvier." *Romanticism* 5:26–42.

———. 2007a. *The Earth on Show: Fossils and the Poetics of Popular Science, 1802–1856*. Chicago: University of Chicago Press.

———. 2007b. "Young-Earth Creationists in Early Nineteenth-Century Britain? Towards a Reassessment of 'Scriptural Geology.'" *History of Science* 45:357–403.

———. 2009a. "Facts and Fancies: The Geological Society of London and the Wider Public, 1807–1837." In *The Making of the Geological Society of London*, edited by C. L. E. Lewis and S. J. Knell, 331–40. London: Geological Society of London.

———. 2009b. "Reflections on Popular Science in Britain: Genres, Categories, and Historians." *Isis* 100:333–45.

———. 2012. "Victorian Saurians: The Linguistic Prehistory of the Modern Dinosaur." *Journal of Victorian Culture* 17:492–504.

O'Gorman, Francis. 2008. "Victorian 'Afterlives.'" *Journal of Victorian Culture* 13:277–78.

Orr, M. 2007. "Keeping It in the Family: The Extraordinary Case of Cuvier's Daughters." In *The Role of Women in the History of Geology*, edited by C. V. Burek and B. Higgs, 277–86. London: Geological Society of London.

Ospovat, Dov. 1981. *The Development of Darwin's Theory: Natural History, Natural Theology, and Natural Selection, 1838–1859*. Cambridge: Cambridge University Press.

Outram, Dorinda. 1978. "The Language of Natural Power: The 'Eloges' of Georges Cuvier and the Public Language of Nineteenth-Century Science." *History of Science* 16:153–78.

———. 1980. *The Letters of Georges Cuvier: A Summary Calendar of Manuscript and Printed Materials Presented in Europe, the United States of America, and Australasia*. Chalfont St Giles: British Society for the History of Science.

———. 1984. *Georges Cuvier: Vocation, Science and Authority in Post-revolutionary France*. Manchester, UK: Manchester University Press.

———. 1986. "Uncertain Legislator: Georges Cuvier's Laws of Nature in Their Intellectual Context." *Journal of the History of Biology* 19:323–68.

———. 1996. "New Spaces in Natural History." In *Cultures of Natural History*, edited by N. Jardine, J. A. Secord, and E. C. Spary, 249–65. Cambridge: Cambridge University Press.

Owen, Alex. 1989. *The Darkened Room: Women, Power and Spiritualism in Late Victorian England*. London: Virago.

Padian, Kevin. 1997. "The Rehabilitation of Sir Richard Owen." *BioScience* 47:446–53.

Pantin, C. F. A. 1963. *Science and Education*. Cardiff: University of Wales Press.

Patten, Robert L. 1978. *Charles Dickens and His Publishers*. Oxford: Clarendon Press.

Piggott, J. R. 2004. *Palace of the People: The Crystal Palace at Sydenham, 1854–1936*. London: Hurst.

Potter, Jane. 2007. "For Country, Conscience and Commerce: Publishers and Publishing, 1914–18." In *Publishing in the First World War*, edited by Mary Hammond and Shafquat Towheed, 11–26. Houndmills, UK: Palgrave Macmillan.

Qureshi, Sadiah. 2011. *Peoples on Parade: Exhibitions, Empire, and Anthropology in Nineteenth-Century Britain*. Chicago: University of Chicago Press.

Rachootin, Stan P. 1985. "Owen and Darwin Reading a Fossil: *Macrauchenia* in a Boney Light." In *The Darwinian Heritage*, edited by David Kohn, 155–83. Princeton, NJ: Princeton University Press.

Ray, Gordon N., ed. 1946. *The Letters and Private Papers of William Makepeace Thackeray*. 4 vols. London: Oxford University Press.

Rea, Tom. 2001. *Bone Wars: The Excavation and Celebrity of Andrew Carnegie's Dinosaur.* Pittsburgh: University of Pittsburgh Press.

Rehbock, Philip F. 1983. *The Philosophical Naturalists: Themes in Early Nineteenth-Century Biology.* Madison: University of Wisconsin Press.

Reid, James J. 2000. *Crisis of the Ottoman Empire: Prelude to Collapse, 1839–1878.* Stuttgart: Steiner.

Reiss, John O. 2009. *Not by Design: Retiring Darwin's Watchmaker.* Berkeley: University of California Press.

Richards, Evelleen. 1987. "A Question of Property Rights: Richard Owen's Evolutionism Reassessed." *British Journal for the History of Science* 20:129–71.

———. 1989. "The 'Moral Anatomy' of Robert Knox: The Interplay between Biological and Social Thought in Victorian Scientific Naturalism." *Journal of the History of Biology* 22:373–436.

Richards, Robert J. 2002. *The Romantic Conception of Life: Science and Philosophy in the Age of Goethe.* Chicago: University of Chicago Press.

Richardson, Ruth. 1987. *Death, Dissection and the Destitute.* London: Routledge.

Rieppel, Lukas. 2012. "Bringing Dinosaurs Back to Life: Exhibiting Prehistory at the American Museum of Natural History." *Isis* 103:460–90.

Riley, David. 2003. "The Manchester Science Lectures for the People, *c.* 1866–79." *Bulletin of the John Rylands University Library of Manchester* 85:127–45.

Rosner, Lisa. 1991. *Medical Education in the Age of Improvement: Edinburgh Students and Apprentices, 1760–1826.* Edinburgh: Edinburgh University Press.

Rudwick, Martin J. S. 1985. *The Meaning of Fossils: Episodes in the History of Palaeontology.* 2nd ed. Chicago: University of Chicago Press.

———. 1992. *Scenes from Deep Time: Early Pictorial Representations of the Prehistoric World.* Chicago: University of Chicago Press.

———. 1997. *Georges Cuvier, Fossil Bones, and Geographical Catastrophes: New Translations and Interpretations of the Primary Texts.* Chicago: University of Chicago Press.

———. 2000. "Georges Cuvier's Paper Museum of Fossil Bones." *Archives of Natural History* 27:51–68.

———. 2005. *Bursting the Limits of Time: The Reconstruction of Geohistory in the Age of Revolution.* Chicago: University of Chicago Press.

———. 2008. *Worlds before Adam: The Reconstruction of Geohistory in the Age of Reform.* Chicago: University of Chicago Press.

———. 2013. "Deciphering the Prehistoric Past." *Metascience* 22:363–65.

Rupke, Nicolaas A. 1983. *The Great Chain of History: William Buckland and the English School of Geology, 1814–1849.* Oxford: Clarendon Press.

———. 1994. *Richard Owen: Victorian Naturalist.* New Haven, CT: Yale University Press.

———. 1997. "Oxford's Scientific Awakening and the Role of Geology." In *The History of the University of Oxford,* vol. 6, *Nineteenth Century Oxford, Part 1,* edited by M. G. Brock and M. C. Curthoys, 543–62. Oxford: Clarendon Press.

———. 2000. "Translation Studies in the History of Science: The Example of *Vestiges.*" *British Journal for the History of Science* 33:209–22.

———. 2009. *Richard Owen: Biology without Darwin.* Chicago: University of Chicago Press.

Russell, E. S. 1916. *Form and Function: A Contribution to the History of Animal Morphology.* London: John Murray.

Ruston, Sharon. 2005. *Shelley and Vitality.* Houndmills, UK: Palgrave Macmillan.

Salmon, Richard. 2013. *The Formation of the Victorian Literary Profession.* Cambridge: Cambridge University Press.

Schaffer, Simon. 1989. "The Nebular Hypothesis and the Science of Progress." In *History, Humanity, and Evolution: Essays for John C. Greene,* edited by James R. Moore, 131–64. Cambridge: Cambridge University Press.

Secord, Anne. 1994. "Science in the Pub: Artisan Botanists in Early Nineteenth-Century Lancashire." *History of Science* 32:269–315.

———. 2003. "'Be What You Would Seem to Be': Samuel Smiles, Thomas Edward, and the Making of a Working-Class Scientific Hero." *Science in Context* 16:147–73.

Secord, James A. 1991. "Edinburgh Lamarckians: Robert Jameson and Robert E. Grant." *Journal of the History of Biology* 24:1–18.

———. 2000. *Victorian Sensation: The Extraordinary Publication, Reception, and Secret Authorship of "Vestiges of the Natural History of Creation."* Chicago: University of Chicago Press.

———. 2004a. "Knowledge in Transit." *Isis* 95:654–72.

———. 2004b. "Monsters at the Crystal Palace." In *Models: The Third Dimension in Science,* edited by Soraya De Chadarevian and Nick Hopwood, 138–69. Stanford, CA: Stanford University Press.

———. 2005. "The Electronic Harvest." *British Journal for the History of Science* 38:463–67.

———. 2007. "How Scientific Conversation Became Shop Talk." In *Science in the Marketplace: Nineteenth-Century Sites and Experiences,* edited by Aileen Fyfe and Bernard Lightman, 23–59. Chicago: University of Chicago Press.

———. 2014. *Visions of Science: Books and Readers at the Dawn of the Victorian Age.* Oxford: Oxford University Press.

Semmel, Stuart. 2004. *Napoleon and the British.* New Haven, CT: Yale University Press.

Sera-Shriar, Efram. 2013. *The Making of British Anthropology, 1813–1871.* London: Pickering and Chatto.

Shanahan, Timothy. 2004. *The Evolution of Darwinism: Selection, Adaptation, and Progress in Evolutionary Biology.* Cambridge: Cambridge University Press.

Sheets-Pyenson, Susan. 1985. "Popular Science Periodicals in Paris and London: The Emergence of a Low Scientific Culture, 1820–1875." *Annals of Science* 42:549–72.

Shpayer-Makov, Haia. 2011. *The Ascent of the Detective: Police Sleuths in Victorian and Edwardian England.* Oxford: Oxford University Press.

Shteir, Ann B. 1996. *Cultivating Women, Cultivating Science: Flora's Daughters and Botany in England, 1760 to 1860.* Baltimore: Johns Hopkins University Press.

Simon, Bart. 2002. *Undead Science: Science Studies and the Afterlife of Cold Fusion.* New Brunswick, NJ: Rutgers University Press.

Sinnema, Peter W. 1998. *Dynamics of the Pictured Page: Representing the Nation in the "Illustrated London News."* Farnham, UK: Ashgate.

Slater, Michael. 1972. "The Composition and Monthly Publication of *Nicholas Nickleby.*" In Charles Dickens, *"The Life and Adventures of Nicholas Nickleby": Reproduced in Facsimile from the Original Monthly Parts, 1838–39,* 1:vii–lxxii. 2 vols. London: Scholar Press.

Smith, Jean Chandler. 1993. *Georges Cuvier: An Annotated Bibliography of His Published Works.* Washington, DC: Smithsonian Institution Press.

Smith, Jonathan. 1994. *Fact and Feeling: Baconian Science and the Nineteenth-Century Literary Imagination.* Madison: University of Wisconsin Press.

Snyder, Laura J. 2004. "Sherlock Holmes: Scientific Detective." *Endeavour* 28:104–8.

Socarides, Alexandra Anne. 2007. "Emily Dickinson and the Problem of Genre." PhD diss., Rutgers University.

Sommer, Marianne. 2007. *Bones and Ochre: The Curious Afterlife of the Red Lady of Paviland.* Cambridge, MA: Harvard University Press.

Spokes, Sidney. 1927. *Gideon Algernon Mantell, Surgeon and Geologist.* London: John Bale.

Stafford, Robert A. 1990. *Scientist of Empire: Sir Roderick Murchison, Scientific Exploration and Victorian Imperialism.* Cambridge: Cambridge University Press.

Stauffer, R. C., ed. 1975. *Charles Darwin's "Natural Selection": Being the Second Part of His Big Species Book Written from 1856 to 1858.* Cambridge: Cambridge University Press.

Stearn, William T. 1981. *The Natural History Museum at South Kensington: A History of the British Museum (Natural History), 1753–1980.* London: William Heinemann.

Stevens, Peter F. 1994. *The Development of Biological Systematics: Antoine-Laurent de Jussieu, Nature, and the Natural System.* New York: Columbia University Press.

Taquet, Philippe. 2006. *Georges Cuvier: Naissance d'un genie.* Paris: Odile Jacob.

———. 2009. "Cuvier's Attitude toward Creation and the Biblical Flood." In *Geology and Religion: A History of Harmony and Hostility,* edited by M. Kölbl-Ebert, 127–34. London: Geological Society of London.

Tate, Trudi. 1997. "The Culture of the Tank, 1916–1918." *Modernism/Modernity* 4:69–87.

Thackray, John C. 1999. *To See the Fellows Fight: Eyewitness Accounts of the Meetings of the Geological Society of London and Its Club, 1822–1868.* Stanford in the Vale: British Society for the History of Science.

Topham, Jonathan R. 1992. "Science and Popular Education in the 1830s: The Role of the *Bridgewater Treatises.*" *British Journal for the History of Science* 25:397–430.

———. 1993. "'An Infinite Variety of Arguments': The *Bridgewater Treatises* and British Natural Theology in the 1830s." PhD diss., University of Lancaster.

———. 1998. "'Beyond the 'Common Context': The Production and Reading of the Bridgewater Treatises." *Isis* 89:233–62.

———. 1999. "Evangelicals, Science, and Natural Theology in Early Nineteenth-Century Britain: Thomas Chalmers and the *Evidence* Controversy." In *Evangelicals and Science in Historical Perspective,* edited by David N. Livingstone, D. G. Hart, and Mark A. Noll, 142–74. New York: Oxford University Press.

———. 2004. "The *Mirror of Literature, Amusement and Instruction* and Cheap Miscellanies in Early Nineteenth-Century Britain." In Geoffrey Cantor, Gowan Dawson, Graeme Gooday, Richard Noakes, Sally Shuttleworth, and Jonathan R. Topham, *Science in the Nineteenth-Century Periodical: Reading the Magazine of Nature,* 37–66. Cambridge: Cambridge University Press.

———. 2007. "Publishing 'Popular Science' in Early Nineteenth-Century Britain." In *Science in the Marketplace: Nineteenth-Century Sites and Experiences,* edited by Aileen Fyfe and Bernard Lightman, 135–68. Chicago: University of Chicago Press.

———. 2009. "Rethinking the History of Science Popularization / Popular Science." In *Popularizing Science and Technology in the European Periphery, 1800–2000,* edited by Faidra Papanelopoulou, Agustí Nieto-Galan, and Enrique Perdiguero, 1–20. Farnham, UK: Ashgate.

———. 2011. "Science, Print, and Crossing Borders: Importing French Science Books into Britain, 1789–1815." In *Geographies of Nineteenth-Century Science,* edited by David N. Livingstone and Charles W. J. Withers, 311–44. Chicago: University of Chicago Press.

———. 2013. "Anthologizing the Book of Nature: The Origins of the Scientific Journal and Circulation of Knowledge in Late Georgian Britain." In *The Circulation of Knowledge between Britain, India and China: The Early-Modern World to the Twentieth Century*, edited by Bernard Lightman, Gordon McOuat, and Larry Stewart, 119–52. Leiden, Netherlands: Brill.

Torrens, Hugh. 2012. "Politics and Paleontology: Richard Owen and the Invention of Dinosaurs." In *The Complete Dinosaur*, edited by Michael K. Brett-Surman, Thomas R. Holtz Jr., and James O. Farlow, 25–44. 2nd ed. Bloomington: Indiana University Press.

Turner, Frank Miller. 1974. *Between Science and Religion: The Reaction to Scientific Naturalism in Late Victorian England*. New Haven, CT: Yale University Press.

———. 1980. "Public Science in Britain, 1880–1919." *Isis* 71:589–608.

———. 1993. *Contesting Cultural Authority: Essays in Victorian Intellectual Life*. Cambridge: Cambridge University Press.

van der Klaauw, C. J. 1932. "The Scientific Correspondence between Professor Jan van der Hoeven and Professor Richard Owen." *Janus* 36:327–51.

Viénot, John. 1932. *Le Napoléon de l'intelligence: Georges Cuvier, 1769–1832*. Paris: Fischbacher.

Vincent, David. 1989. *Literacy and Popular Culture: England, 1750–1914*. Cambridge: Cambridge University Press.

Wallen, John. 2013. "The Cannibal Club and the Origins of Nineteenth-Century Racism and Pornography." *Victorian* 1:1–13.

Watts, Iain P. 2014. "'We Want No Authors': William Nicholson and the Contested Role of the Scientific Journal in Britain, 1797–1813." *British Journal for the History of Science* 47:397–419.

White, Paul. 2003. *Thomas Huxley: Making the "Man of Science."* Cambridge: Cambridge University Press.

Wilson, Leonard G., ed. 1970. *Sir Charles Lyell's Scientific Journals on the Species Question*. New Haven, CT: Yale University Press.

Winter, Alison. 1998. *Mesmerized: Powers of Mind in Victorian Britain*. Chicago: University of Chicago Press.

Wolfe, Richard. 2003. *Moa: The Dramatic Story behind the Discovery of a Giant Bird*. Auckland: Penguin.

Wright, Willard Huntington. 1927. Introduction to *The Great Detective Stories*, edited by Willard Huntington Wright, 3–37. New York: Charles Scribner's Sons.

Yanni, Carla. 1999. *Nature's Museums: Victorian Science and the Architecture of Display*. New York: Princeton Architectural Press.

Yeo, Richard. 1993. *Defining Science: William Whewell, Natural Knowledge and Public Debate in Early Victorian Britain*. Cambridge: Cambridge University Press.

———. 2001. *Encyclopaedic Visions: Scientific Dictionaries and Enlightenment Culture*. Cambridge: Cambridge University Press.

Young, Paul. 2009. *Globalization and the Great Exhibition: The Victorian New World Order*. Houndmills, UK: Palgrave Macmillan.

Index

Page numbers in italics refer to figures.